D1708918

Spatial Patterns in Catchment Hydrology

Observations and Modelling

For many years now, modelling tools have been available to simulate spatially distributed hydrological processes. These tools have been used for testing hypotheses about the behaviour of natural systems, for practical applications such as erosion and transport modelling, and for simulation of the effect of land use and climate change. However, so far the quality of the simulations and spatial process representations has been difficult to assess because of a lack of appropriate field data.

Spatial Patterns in Catchment Hydrology: Observations and Modelling brings together a number of recent field exercises in research catchments, that illustrate how the understanding and modelling capability of spatial processes can be improved by the use of observed patterns of hydrological response. In addition the introductory chapters review the nature of the hydrological variability, and introduce basic concepts related to measuring and modelling spatial hydrologic processes. This introductory material provides the conceptual and theoretical background needed to move into this exciting area of research for a general earth sciences/water engineering audience. The book demonstrates that there is rich information in patterns that provide much more stringent tests of the models and much greater insight into hydrological behaviour than traditional methods.

Written in an intuitive and coherent manner, the book is ideal for researchers, graduate students and advanced undergraduates in hydrology, and a range of water related disciplines such as physical geography, earth sciences, and environmental and civil engineering as related to water resources and hydrology.

Rodger Grayson is an Associate Professor and Senior Research Fellow at the Center for Environmental Applied Hydrology and the Cooperative Research Center for Catchment Hydrology, both of which are in the Department of Civil and Environmental Engineering at the University of Melbourne. His professional interests include research, teaching and consulting related to environmental hydrology, the modelling and monitoring of water quality and quantity from research catchment to continental scales, and integrated catchment management. He has published over 100 papers and reports in international and national journals and conferences, as well as an edited book, several book chapters and this current book. He is an associate editor of *Water Resources Research* and the *Journal of Hydrology*.

Günter Blöschl is an Associate Professor at the Institute of Hydraulics, Hydrology and Water Resources Management of the Technical University of Vienna. His professional interests include measuring and modelling spatial hydrologic processes at a range of scales as well as engineering hydrology and water resources management. He is an author of over 100 scientific papers and has received the Schrödinger and Lise Meitner awards from the Austrian Science Foundation. He is an associate editor of *Water Resources Research*, the *Journal of Hydrology* and an editorial board member of *Environmental Modelling and Software*. He is a Vice President of sections of both the European Geophysical Society and the International Association of Hydrological Sciences.

Spatial Patterns in Catchment Hydrology

Observations and Modelling

Edited by

RODGER GRAYSON
University of Melbourne

GÜNTER BLÖSCHL
Technische Universität Wien

PUBLISHED BY THE PRESS SYNDICATE OF THE UNIVERSITY OF CAMBRIDGE
The Pitt Building, Trumpington Street, Cambridge, United Kingdom

CAMBRIDGE UNIVERSITY PRESS
The Edinburgh Building, Cambridge, United Kingdom
40 West 20th Street, New York, NY 10011–4211, USA
10 Stamford Road, Oakleigh, Melbourne 3166, Australia
Ruiz de Alarcón 13, 28014 Madrid, Spain
Rock House, The Waterfront, Cape Town 8001, South Africa

http://www.cup.org

First published 2001

Printed in the United Kingdom at the University Press, Cambridge.

Typefaces Times New Roman PS 10.5/13 and Helvetica Neue Condensed   *System* 3B2 [KW]

A catalogue record for this book is available from the British Library

Library of Congress Cataloguing-in-Publication Data

Spatial patterns in catchment hydrology : observations and modelling / edited by Rodger Grayson, Günter Blöschl.
    p. cm.
  ISBN 0-521-63316-8
    1. Watersheds–Mathematical models. 2. Hydrology–Mathematical models. 3. Spatial analysis (Statistics) I. Grayson, Rodger, 1963–II. Blöschl, Günter, 1961–

GB980.S66 2001
551.48′01′5118–dc21   00-040357

Contents

Preface

Our everyday lives are dominated by patterns. The all too persistent temporal pattern of sleep, work, relax; the nightly weather maps and satellite images that might mould the shape of the forthcoming weekend's activities; or the intricate pattern represented by the arrangement of features on a human's face that lets us recognise a friend in a crowd. Some patterns contain simple information like the isobars on a map of surface pressure, while others are breathtakingly rich. Some of this information we can understand and interpret, while some is well beyond us. This book is about patterns. It is about how we measure, interpret and model aspects of spatial hydrological response. It is motivated by a belief that to advance the knowledge base of scientific hydrology, and to answer many of the questions of environmental management that are being asked by the broader community, we have to better exploit the information that resides in the myriad of patterns observable in nature.

For many years now, modelling tools have been available to simulate spatially distributed hydrological processes. The quality of the simulations and spatial process representations has been difficult to assess because of a lack of appropriate field data. In recent years there have been several major field exercises in research catchments, aimed specifically at improving our understanding and modelling capability of spatial processes. This book seeks to bring some of those studies together within the context of reviewing our understanding of spatial hydrological processes and presenting research work aimed at improving that understanding. In addition, we hope it provides a reference and source of motivation for others interested in undertaking detailed spatial data collection in combination with distributed modelling to improve our understanding and prediction of hydrological processes.

Specifically this book seeks to:

(i) Provide readers with an introduction to the nature and representation of spatial patterns in hydrological processes;

(ii) Show, through example, how the comparison of measured and simulated spatial patterns of hydrological response can be used both to improve our understanding of processes and to inform model development; and

(iii) Provide an avenue for expanding upon the experiences of those who have undertaken major collection and collation exercises of spatial field data, for the purpose of spatial modelling and gaining insight into spatial processes.

This book is aimed at two types of readers. The first will have a general knowledge of catchment hydrology and be keen to develop their understanding of the nature of hydrological variability, and be introduced to some methods and models that can assist in quantifying that variability. We have deliberately kept the introductory chapters free of detailed mathematics, preferring to concentrate on an intuitive understanding of the underlying concepts, many of which are quite rich and complex. We do not intend to provide a complete description of all available techniques or models, rather we seek to equip the reader with the knowledge needed to assess the *types* of tools and models that may be appropriate for their particular application, and to understand the basic approaches to modelling and analysis used in the case-study chapters. The second type of readers will be hydrologists who already have a sound knowledge of methods for spatial data analysis and of distributed modelling, but are thinking of undertaking studies similar to those presented in the book. For these readers, the case studies provide a wide range of measurement techniques, analysis methods, model types used, and approaches to the comparison of observed and simulated patterns, which should help them decide on the best approaches for their own work.

The book is presented in three parts. The first part (Chapters 1–5) starts with three introductory chapters (Chapters 1–3) on fundamentals that are key to putting later chapters in context. Chapters 4 and 5 deal with spatial patterns in precipitation and evaporation, respectively. These two processes were singled out because they are so critical to spatial hydrological response, yet are relatively poorly represented in most models. The two chapters include discussions about the state of the art in measurement and analysis of spatial information and the synthesis of point data. In the second part, case studies in research catchments are presented (Chapters 6–12). These chapters cover a range of environments from the tropics to Alpine regions; a range of dominant processes from Hortonian runoff to surface–groundwater interaction; a range of spatial data including remote sensing and multiple-point measurements; and a range of modelling structures including fully distributed grid and contour-based models of different complexities. An important feature of all the case studies, and something that makes them relatively rare in the hydrological literature, is that they directly compare *observed* with *simulated* spatial patterns. We asked the authors to focus on the collection and interpretation of patterns and their implications for model testing, while providing only a brief description of the models themselves. For full descriptions of the models, references are given in each of the chapters. The final part (Chapters 13 and 14) focuses on implications of the material presented in the earlier chapters. Chapter 13 addresses the implications when one moves away from the small research catchments to larger scales where practical predictions from distributed models are needed, focusing on issues of calibration and validation of these models. The final chapter (Chapter 14) is a summary of the case studies and a discussion of broader implications from the work, highlighting what we have learnt from pattern comparisons and the challenges that remain.

In preparing this book we are greatly indebted to a large number of people. First and foremost, the contributors, for their dedication in addressing the central theme of the book and for providing their insights for us all to share. Thanks to the reviewers who provided timely feedback, and several colleagues who reviewed sections and were able to see the big picture when we were lost in detail. We are particularly grateful to Erich Plate for his thoughtful comments and willingness to look over the entire manuscript. Andrew Western and Ralf Merz deserve special mention for their help with a range of tasks from stimulating discussions on technical matters, to figure preparation. Dieter Gutknecht and Tom McMahon provided continued support for this project and both the Technical University of Vienna and the University of Melbourne assisted in a number of ways, not least being to help us work in the same office for extended periods of time. It was during these periods that the book really came together, over hours of discussion and friendly argument. Matt Lloyd from Cambridge University Press provided much needed assistance in all matters related to the production of the book.

Finally, love and thanks, to our long suffering partners and families, Ali Dedman and Elisabeth, Roman, Agnes and Margit Blöschl for their patience and encouragement.

Rodger Grayson & Günter Blöschl
August 2000

Contributors

YVES BESSARD École Polytechnique Fédérale de Lausanne, Switzerland

KEITH BEVEN Environmental Sciences, Lancaster University, United Kingdom

GÜNTER BLÖSCHL Institut für Hydraulik, Gewässerkunde und Wasserwirtschaft, Technische Universität Wien, Austria

KEITH COOLEY USDA – Agricultural Research Service, Northwest Watershed Research Center, Boise, Idaho, USA

HELMUT ELSENBEER Department of Civil and Environmental Engineering, University of Cincinnati, Ohio, USA

EFI FOUFOULA-GEORGIOU St. Anthony Falls Laboratory, Department of Civil Engineering, University of Minnesota, Minneapolis, Minnesota, USA

PHILIPPE GINESTE L'Unité Mixte de Recherche 3S "Systèmes et Structures Spatiaux", Montpellier, France

DAVID GOODRICH USDA – Agricultural Research Service, Southwest Watershed Research Center, Tucson, Arizona, USA

RODGER GRAYSON, Centre for Environmental Applied Hydrology, CRC for Catchment Hydrology, Department of Civil and Environmental Engineering, University of Melbourne, Australia

LAWRENCE HIPPS Department of Plants, Soils and Biometeorology, Utah State University, Logan, Utah, USA

PAUL HOUSER NASA Goddard Space Flight Center, Greenbelt, Maryland, USA

ROBERT KIRNBAUER Institut für Hydraulik, Gewässerkunde und Wasserwirtschaft, Technische Universität Wien, Austria

WILLIAM KUSTAS USDA – Agricultural Research Service, Hydrology Laboratory, Beltsville, Maryland, USA

ANDREAS LACK Swiss Landscape Fund, Bern, Switzerland

ROBERT LAMB Centre for Ecology and Hydrology (formerly Institute of Hydrology), Wallingford, United Kingdom

JOHN LEVINE Department of Geography, Boston University, Massachusetts, USA

CHARLES LUCE US Forest Service, Rocky Mountain Research Station, Boise, Idaho, USA

PHILIPPE MÉROT Unité mixte de recherche Sol – agronomie, Rennes, France

STEINAR MYRABØ NSB Gardermobanen AS, Oslo, Norway

CLAUDIO PANICONI CRS4 (Center for Advanced Studies, Research and Development in Sardinia), Cagliari, Italy

JENS CHRISTIAN REFSGAARD Department of Hydrology, Geological Survey of Denmark and Greenland, Copenhagen, Denmark

GUIDO SALVUCCI Departments of Earth Sciences and Geography, Boston University, Massachusetts, USA

KAMRAN SYED Department of Meteorology, University of Maryland, College Park, Maryland, USA

DAVID TARBOTON Utah Water Research Laboratory, Department of Civil and Environmental Engineering, Utah State University, Logan, Utah, USA

PETER TROCH Environmental Sciences, Wageningen University, The Netherlands

NIKO VERHOEST Laboratory of Hydrology and Water Management, Ghent University, Belgium

ROBERT VERTESSY CSIRO Land and Water, CRC for Catchment Hydrology, Canberra, Australia

VENUGOPAL VURUPUTUR Center for Ocean-Land-Atmosphere Studies, Calverton, Maryland, USA

ANDREW WESTERN Centre for Environmental Applied Hydrology, CRC for Catchment Hydrology, Department of Civil and Environmental Engineering, University of Melbourne, Australia

PART ONE

FUNDAMENTALS

1

Spatial Processes, Organisation and Patterns

Rodger Grayson and Günter Blöschl

1.1 INTRODUCTION

Observation and interpretation of spatial patterns are fundamental to many areas of the earth sciences such as geology and geomorphology, yet in catchment hydrology, our historic interest has been more related to temporal patterns and in particular, that of streamflow. The fact that patterns are everywhere in hydrology hardly needs explanation. From the rich RADAR images of precipitation, to the photographs from dye studies illustrating preferential flow (e.g. Flury et al., 1994), there is a wide range of spatial arrangements present in hydrologic systems. But because of an interest in streamflow, that wonderful integrator of variability, we have until recently managed to avoid confronting the challenges of spatial heterogeneity. It is worth noting that a similar history is apparent in groundwater hydrology where pumping tests have long provided a measure of integrated aquifer response and distracted researchers from the quantification of aquifer heterogeneity (Anderson, 1997). The last few decades, however, have heralded an explosion of interest in spatial patterns in hydrology, from the pioneering work on spatial heterogeneity in runoff producing processes during the sixties and early seventies (e.g. Betson, 1964; Dunne and Black, 1970a, b), through the development of spatially distributed hydrological models that provide a way to interpret spatial response, to the ever increasing capabilities of remote sensing methods which provide information on state variables of fundamental importance to catchment hydrology.

Two important areas of work are arguably the catalyst that brought the issues of patterns to the forefront of hydrologists' minds.

The first is the ready availability of digital elevation models (DEMs) and the attendant analysis that is possible with these data (e.g. Beven and Moore, 1992), made all the easier by the ever decreasing cost of computing power and availability of Geographic Information System (GIS) software. DEMs and powerful computers have made it possible for every hydrologist with an appropriate com-

Rodger Grayson and Günter Blöschl, eds. *Spatial Patterns in Catchment Hydrology: Observations and Modelling* © 2000 Cambridge University Press. All rights reserved. Printed in the United Kingdom.

puter package to generate pages of impressive looking patterns that are intuitively meaningful, using, for example, topographic wetness indices. These indices can be computed from a DEM alone and are designed to represent the spatial patterns of soil moisture deficit in a catchment (e.g. Beven and Kirkby, 1979). The potential of distributed parameter models can now be exploited via automating the element representations that are central to these models and assisting in the management and manipulation of often enormous spatial data sets. Also, off-the-shelf software for spatially distributed catchment models is now available at low cost.

The second is the rise in environmental awareness of the broader community and its subsequent impact on research into, and the management of, natural resources. We now want to know not only what is the quantity and quality of water in a stream, but also from *where* any contaminants came and *where* best to invest scarce financial resources to help rectify the problem. We now need predictions of the hydrological (and ecological) impacts of land use and climate change – predictions that must account for the *spatial variability* we see in nature if they are to be of any practical use. Natural resource agencies are amassing large amounts of spatial data to complement the temporal data traditionally measured, and are eagerly looking to use this for predictive spatial modelling of environmental response. In principle, we have the tools available to undertake this work and already, the spatial models and impressive colour graphics that our GISs generate can seduce even the most sceptical of politicians and administrators (Grayson et al., 1993). But in many cases, the scientific credibility of these predictions is questionable. We need better ways to develop and assess spatial predictions, as well as to exploit the information that is becoming available from new measurement methods, which often provide us with very different information to that we are using today.

However, while our ability to *generate* patterns might be impressive, it is not of itself useful. It is the extent to which these patterns represent reality and to which they provide us with new insights into hydrological behaviour that is important. Where we can actually observe patterns in nature, they provide us with a powerful test of our distributed modelling capabilities and so should significantly improve the confidence we have in subsequent predictions. But observed spatial patterns of hydrologically important variables (other than land use, terrain and in some cases soils) are not very common. To progress, we will need to make quite different measurements from those used in the past, perhaps requiring the development of new measurement methods. We will also need to develop more sophisticated approaches to the testing of spatial predictions against spatial measurements.

The number of papers that have presented comparisons of *observed and simulated* spatial patterns of catchment hydrological processes is relatively small. In 1991, Blöschl et al. (1991b) used photographs of snow cover to assess the performance of a spatially distributed energy and water balance model of the snowpack. Along similar lines, Wigmosta et al. (1994) and Davis et al. (1995) have used snow patterns in analyses of alpine hydrology models. But other than snow cover and comparisons between shallow piezometers and models such as Topmodel (e.g. Quinn and Beven, 1993), there have so far been only a few attempts to compare measured and simulated patterns. For example, Moore and Grayson (1991) compared observed saturated source areas to simulations from a distributed parameter model in a small laboratory sand bed. Whelan and Anderson (1996) simulated the spatial variability of throughfall and compared it to measurements from an array of ground collectors. Bronstert and Plate (1997) compared observed and simulated soil moisture patterns in a small German research catchment. These were all simple visual comparisons of observations versus simulations, but the insights gained into model performance were new and could never have been achieved through comparison with traditional measures such as runoff. We predict that testing models by comparing simulated and observed patterns will eventually become commonplace and will provide a quantum advance in the confidence we are able to place on predictions from distributed models.

With rapid developments in measurement methods and tools for analyses, we should have all the ingredients to give us new insights into how nature behaves. But just how do we go about it? Are the models we develop able to use the information we have available? Is the understanding of fundamental processes that stood us in good stead in the laboratory, suitable once we move into the realm of three-dimensional landscapes? Even when we have both simulated patterns and observations for comparison, how do we quantitatively assess model performance? How do we determine how well the processes are understood and represented?

While we cannot hope to answer all of these issues, we hope that through the following chapters, you will see some specific examples of how data, modelling and our basic understanding of processes can be combined to develop new insights into hydrological behaviour.

1.2 PROCESS AND PATTERN

What are the links between process and pattern? We recognise patterns because of some form of organisation. This may be highly ordered such as in a map of elevation of a large river basin, or totally disordered as might be apparent in an elevation map of micro-topography of a rough surface. Throughout this book we will use the term "organisation" to denote a non-random spatial pattern that becomes apparent when examined visually. But can we explain the organisation (or lack thereof) in a pattern, via an understanding of the processes that underlie its creation? Can we infer process behaviour from observed patterns? Some of these links between process and pattern are obvious while others are not.

Figure 1.1 shows a map of the drainage network of a section of southern Germany. It is clear that there are areas of distinctly different drainage density, in this case, caused by a region of limestone in a landscape that is otherwise sandstone and mudstone. The limestone geology creates much higher infiltration rates and hence less surface runoff, which translates into a lower drainage density.

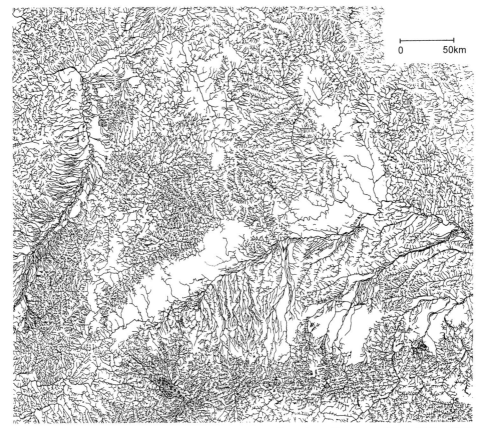

Figure 1.1. Map of the drainage network of a section of southern Germany. (From Keller, 1978; reproduced with permission.)

Were we to try to model drainage in this area without such a map, one would hope that knowledge of the geological structure of the region and an understanding of the type of drainage that occurs in different rock types, would enable us to predict the observed difference in drainage density and treat the regions differently in our model. In the absence of such process understanding we would probably treat the region as homogeneous and have substantial difficulty in reproducing observed flows.

In other cases, different processes produce different patterns on the same landscape. Figure 1.2 is an example of soil moisture measurements in a 10 ha catchment in S.E. Australia (see Chapter 9). It shows the measured soil moisture patterns during a period in early winter when surface runoff was occurring (top), and a pattern in mid-summer (bottom). In winter (when it is wet), surface and subsurface lateral flow occurs, particularly in the gullies, which produces a topographically organised pattern. In summer, however, (when it is dry) there is a minimum of lateral redistribution and fluxes are essentially vertical, which produces a random pattern that is not related to topography (Grayson et al., 1997; Western et al., 1999a). Here we either know the underlying process (organised wet winter patterns dominated by topographic effects on lateral flow) and can

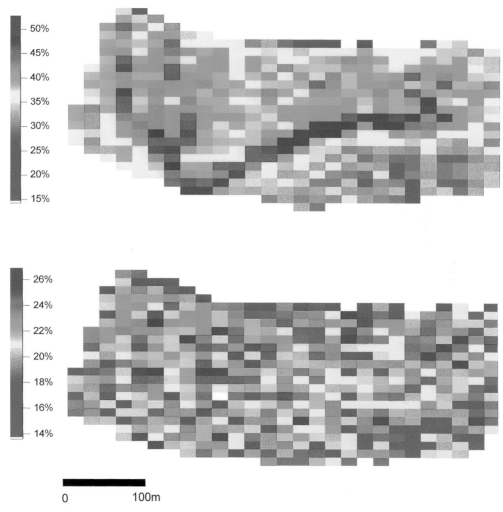

Figure 1.2. Soil moisture measurements from the top 300 mm in a 10 ha catchment in S.E. Australia, collected during wet conditions in winter (upper) and dry conditions in summer (lower).

represent it (Western et al., 1999a) or know that the pattern can be considered random (dry summer case), with soil moisture varying over a narrow range. It is therefore possible to confidently incorporate the spatial variability into any modelling or further analysis, either by deterministically representing the effects of topography or making an assumption of randomness.

In some cases we may be able to observe patterns and have some knowledge of the controlling processes but our ability to represent them is severely limited. An important example in catchment hydrology is preferential flow through soil. Figure 1.3 shows horizontal slices and a vertical slice of dye patterns observed in a block of soil in the field (Flury et al., 1994). Water containing dye was applied to the surface and infiltrated for some time, after which the soil block was excavated revealing the patterns of water flow. The silty loam soil contained many cracks and earthworm channels; the infiltrating water bypassed the soil

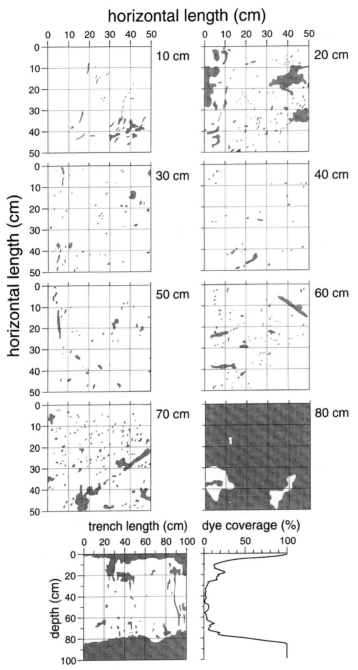

Figure 1.3. Horizontal slices and a vertical slice of dye patterns observed in a block of soil in the field, showing preferential flow paths. (From Flury et al., 1994; reproduced with permission.)

matrix almost completely and was channelled into the subsoil. The dye patterns in Figure 1.3 are extremely complex and their prediction poses a major challenge (Flury et al., 1994). There are still other cases where we might be aware of processes that should lead to particular patterns in the landscape, but have difficulty measuring the real patterns to test the hypothesis. A common example is the subsurface lateral redistribution of soil moisture leading to patterns of saturated areas (see Chapters 9 and 11), but these are rarely measured over large areas (see Chapter 8).

The examples of patterns and processes we have discussed in the previous paragraphs (e.g. Figures 1.1–1.3) span a wide range of space and time scales, and this is typical of the processes we need to deal with in catchment hydrology. Often different types of patterns are encountered at different time and space scales and these are associated with different processes. Figure 1.4 shows a schematic representation of a number of processes at various space–time scales. At the lower left of the figure are processes with short characteristic time and space scales, such as infiltration excess runoff, that will lead to patterns that are very "patchy". These compare to the slower, larger scale processes such as groundwater flow (top right of figure) where we would expect patterns of, for example piezometric head, to be spatially more coherent and slowly varying. Given the relationship between process and pattern, it is worth briefly considering some key processes in catchment hydrology that lead to patterns in hydrological behaviour. Precipitation dominates hydrological response and its patterns are highly dependent on the types of storms (Austin and Houze, 1972). Convective thunderstorms display patterns with localised, high intensities and short durations. Figure 1.4 indicates typical space scales of 1–10 km and typical time scales of 1 minute to 1 hour. Maps of rain depth tend to be "patchy" for convective storms with subsequently great variability in spatial patterns of soil moisture and runoff (see Chapter 6). On the other hand, frontal weather systems tend to produce long bands of relatively uniform rainfall. Figure 1.4 indicates typical space scales of 100–1000 km and typical time scales of 1 day. These result in patterns of runoff that are more spatially uniform, at least at the scale of the weather system. While Figure 1.4 is a schematic, it is possible to quantitatively derive similar space-time diagrams for some processes. One example is given later in this book in Figures 4.5 and 4.13 for the case of precipitation. It is interesting that the lines in Figures 4.5 and 4.13 plot directly on the band for precipitation shown in Figure 1.4.

The processes of runoff generation also lead to very different patterns. In humid catchments with relatively low rainfall intensities compared to infiltration rates of the soil, surface runoff is usually generated from saturated areas (called saturated source area or saturation excess runoff). These are formed due to the concentration of subsurface flows, so their patterns in the landscape depend on the bedrock and surface topographies, differences in soil properties and, to a lesser extent, vegetation characteristics. Runoff is focused in and around the drainage lines appearing as patterns of connected linear features that expand

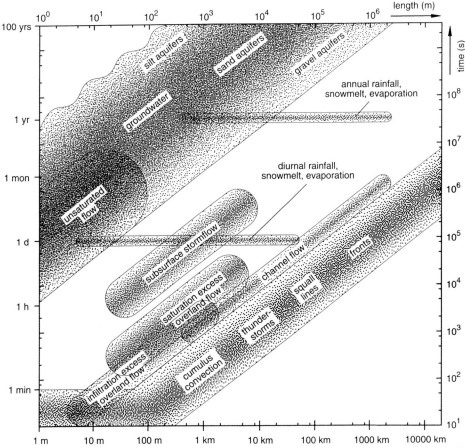

Figure 1.4. Schematic relationship between spatial and temporal process scales for a number of hydrological processes. (From Blöschl and Sivapalan, 1995; reproduced with permission.)

and contract seasonally and within storms (e.g. Dunne et al., 1975). Different patterns result from runoff generated by the infiltration excess mechanism (i.e. where rainfall intensity exceeds the infiltration rate of the surface, sometimes called Hortonian runoff). Instead of being focused on drainage lines, runoff can occur from anywhere on the surface, dependent only on the pattern of infiltration characteristics. These in turn are related to the patterns of soil, vegetation, microtopographic features and the patterns of rainfall, all of which may be highly organised or apparently random. Runoff may never reach a drainage line, perhaps re-infiltrating in a patch of porous soil resulting in highly disconnected patterns of runoff. Eventually, with enough high intensity rain, gravity will ensure that runoff reaches drainage lines, producing more linear features similar to the saturated source process.

In snow dominated environments, spatial variations in energy inputs and wind exposure tend to dominate patterns of melt and accumulation (see Chapter 7). Exposure to direct and indirect solar radiation is affected by latitude, time of year, terrain slope and aspect, with large differences between north and south facing slopes, as well as shading and emission/reflection from surrounding terrain. The interaction between terrain and prevailing wind conditions leads to depositional and erosional patterns with major impacts on the distribution of snow water equivalent (the amount of water stored in the snowpack). Some of these controls are predictable (e.g. from geometry) while others are not, such as emission and reflection from surrounding terrain, due to large temporal and spatial variability in specific properties of the surface that are difficult to define quantitatively.

Perhaps the most complex interrelationship between processes and patterns occurs for evaporation. We have a general understanding of the quantities that influence evaporation such as soil moisture, vegetation characteristics, radiative inputs, air humidity and speed of the wind, and for many of these we can determine spatial patterns. But just how these factors combine to produce patterns of evaporation is complicated by the fact that each depends on the other and the atmosphere itself tends to smooth out differences in a way that cannot be easily described (see Chapter 5). What is more, unlike for example precipitation or snow cover, we have no means yet of accurately measuring the patterns of actual evaporation.

Thus there are degrees to which we can observe and explain patterns, due to limitations in our knowledge of processes and/or our measurement and modelling methods. It is important to realise that the scale at which we measure phenomena will also affect the extent to which we are able to observe and describe patterns. If, in Figure 1.2, we had only a few data points rather than over 500, we would be unlikely to identify a meaningful pattern. With just a few measurements, we might be tempted to treat the distribution of soil moisture as a random field – an assumption that might be acceptable in summer when it was dry, but definitely not in winter when it was wet. We must be confident that the measurements we are interpreting are capturing the nature of the underlying variability of the system we seek to represent, and are not simply a function of our sampling density (see Chapter 2). Because we can rarely sample densely enough to fully define the underlying variability, we must exploit our understanding of dominant processes and their manifestations at different scales. We generally formulate our understanding of processes in the form of models, which in turn need measurements for proper testing, and so we have observations, understanding and modelling linked in an iterative loop. This theme is central to the chapters that follow.

1.3 MODELLING AND PATTERNS

There are many distributed parameter hydrological models available today and they should provide us with the tools to undertake the detailed spatial analyses that we are arguing should occur. The large modelling development exercises of the 1980s such as SHE (Abbott et al., 1986) have turned Freeze and Harlan's blueprint of 1969 for a comprehensive spatial model into a reality (Freeze and

Harlan, 1969). Algorithms that were developed by discipline specialists for the various processes to convert precipitation to runoff, infiltration and evaporation, now have a framework within which they can be linked. We have a variety of methods for representing terrain (see Chapter 3), we can choose from an array of sub-process representations for evapotranspiration, infiltration and surface ponding, vertical and lateral flow through porous media, overland and channelised flow and so on (e.g. Singh, 1995).

But how well do the process descriptions, built up in this reductionist approach, represent the spatial reality? As mentioned earlier, there are few examples of explicit comparisons of spatial reality with spatial simulated response. There have, of course, been innumerable applications of these models, using other methods of testing, but just how well have we really exploited the spatial capabilities of distributed hydrological models?

Every time we use a model of hydrological response, we are forced to accept (and make) a series of assumptions about spatial heterogeneity. It is most common to assume that parameters are uniform within the elementary spatial unit of the models we use. For the lumped and semi-lumped conceptual models (see Chapter 3) that still prevail in engineering practice, these elementary units might be large subcatchments, while in detailed distributed models, they might be 100s of m^2, nevertheless, the "uniform" assumption is generally made at some point. Furthermore, when we use distributed models it is often necessary (due to lack of data) to make the uniform assumption over large parts of the area being modelled – in doing so we "impose" some spatial organisation which may restrict the interpretations we can make about the simulation results. For example, it is quite common for the only source of variability represented in a spatial model to be terrain, and to assume that soil hydraulic properties, rainfall etc. are uniform (perhaps because we have limited information to say otherwise). It would therefore not be surprising to conclude from the simulations that terrain is a dominant source of variability – indeed it was the only one represented! While this sounds obvious, it has often been done with (mis)applications of topographically based models where a good hydrograph fit is provided as evidence of the importance of topography despite the fact that all the other spatial variables were constant. Additionally, we might represent soil characteristics as a random field or as a function of soil type. Again, this is not a "value free" decision. If the soil properties are indeed randomly distributed or highly correlated with soil type, the simulation results may be meaningful but if they are not, if in reality there is some different organisation in the landscape, the simulation results will be highly distorted (e.g. Chapter 6). Grayson et al. (1995) show how important spatial organisation can be for runoff simulations. Two patterns of soil moisture deficit, each with the same properties of mean, variance and correlation length (see Chapter 2), but one spatially random and the other "organised" by a wetness index, produce very different responses to given rainfall input (Figures 1.5 and 1.6). The organised pattern gives higher runoff peaks than the random case for small precipitation events, while the reverse is true for larger rainfall events. This highlights the importance of properly defining spatial organisation where it exists

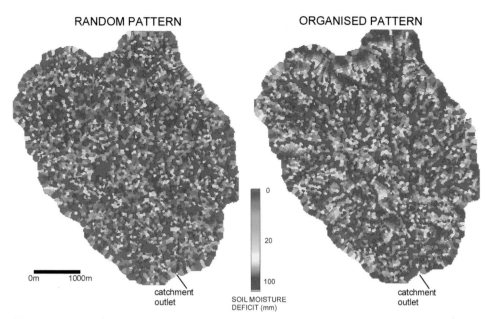

RANDOM PATTERN   ORGANISED PATTERN

0

20

100

SOIL MOISTURE
DEFICIT (mm)

0m   1000m

catchment
outlet

catchment
outlet

Figure 1.5. Two simulated patterns of soil moisture deficit, one spatially random and the other 'organised' by a wetness index. Red values correspond to dry areas (high deficits) and blue areas correspond to wet areas (low deficits).

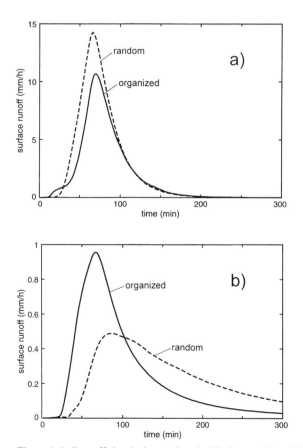

Figure 1.6. Runoff simulations using the Thales model and the saturation deficit scenarios in Figure 1.5, for a rainfall event of (a) 30 mm and (b) 5 mm over 1 hour.

13

and underlines the fact that, while the advent of distributed hydrological models has opened up enormous potential for spatial analysis, it has brought with it a requirement for careful interpretation and thoughtful representation of spatial characteristics.

1.4   REPRESENTATION OF PATTERNS

Even if we are able to observe a pattern, how do we represent it numerically? In some limited cases, we can directly use an observed data set to produce a single, "deterministic" pattern. Other deterministic patterns could be produced if we assume, for example, that a wetness index is a true representation of distributed soil moisture deficit. Alternatively, we may know very little about the underlying pattern, or believe that it is random, and so wish to represent the variability in a statistical manner. This is done by either the generation of a random field (where just mean and variance are preserved) or perhaps one in which some higher level of spatial correlation is also preserved (see Chapter 2). In these cases, we can generate any number of patterns, each with the same statistical properties (see Chapter 2). The deterministic and statistical approaches can also be combined to account for the fact that while we might expect a certain level of "deterministic pattern" based on process understanding, there will be a significant amount of uncertainty (e.g. Chapter 10). The influence of these different representations of patterns on the resulting hydrological simulations will depend on the extent to which the deterministic and statistical measures capture features of hydrological significance.

Not only must we choose the basic approach to pattern representation (deterministic or statistical) but must also decide on the scale at which heterogeneity is to be represented. A central question in representing spatial heterogeneity is whether the processes that dominate the hydrological response of interest change as we change scale. Would runoff from a 1 m^2 plot be dominated by the same sort of heterogeneity that dominates continental streamflow? Our quest is not one for universal laws, but rather for approaches to identify and represent the dominant processes at different scales. This is vital for the representation of patterns in models where we must be regularly making decisions about what variability to explicitly represent, what to ignore, and what to incorporate in some other manner. For example, if our interest is in a general representation of surface flow across a landscape we may use a readily available DEM as a sufficient descriptor of surface flow paths. But if our interest is in, for example, explicitly determining the erosive power of the surface flow, we would need to represent far more detail of the surface micro-topography. This could be done explicitly, using detailed DEMs to define the micro-flowpaths, or we might be able to represent the *effect* of the micro-channels and surface roughness of a real surface by some other, non-explicit, means, e.g. by defining a distribution of flow depths (Abrahams et al., 1989) or conceptualising the surface as a series of rills (Moore and Burch, 1986; Chapter 3). This is the notion of "sub-grid variability" wherein we represent the effects of variability without explicitly representing the variability itself. These ideas are explored further in Chapter 3.

In many applications, the spatial capability of distributed models is used to change spatial parameters and assess the impact of the change on an output variable such as streamflow. This is particularly common in studies of the impact of land use change. In other applications, spatially distributed predictions are required. The different levels to which the spatial capabilities of models are exploited must also be considered when deciding how best to use information on patterns. This is a question of "horses for courses" – of choosing the appropriate tool for the job. Land surface schemes (i.e. models of the water and energy balances at the land surface) as used in atmospheric General Circulation Models (GCMs) are perhaps an extreme example of this issue. These models seek to describe the effects of spatial heterogeneity at very large scales. They have complex vertical process representations including multi-layer soils, variable stomatal resistance and aerodynamic functions (for evaporation estimation), but the parameters are spatially lumped at the order of 10000s of km^2. For the purpose of representing the general circulation of the atmosphere, these models do a reasonable job, because general circulation is dominated by surface processes at these large spatial scales, but in terms of describing surface hydrology for terrestrial purposes, these models are poor. We are interested in outputs for which these models were not designed (e.g. local runoff) and which are dominated by heterogeneity that they generally ignore (local terrain and soils). These are the wrong tools for catchment hydrology. On the other hand, hydrologists working on land surface schemes recognise that the heterogeneity we take for granted may need to be somehow incorporated at these larger scales. But this cannot be done simply by applying catchment hydrology's distributed models and deterministic representation of patterns because such fine scale detail would render the schemes too unwieldy. Methods of pattern representation must be tailored to the model scale and the types of outputs required, based on an understanding of the dominant controls.

It is therefore clear that investigations which utilise the information available in spatial patterns must have four key features:

1. A model that has the structure to represent spatial variability at a scale appropriate for the dominant processes and required output.
2. Methods for the realistic representation of spatial variability; be they deterministic or statistical; be they explicit or in the form of "sub-grid" representations.
3. Measurements that enable the parameters in the representations from (2) to be defined.
4. Methods for the comparisons between observations and predictions of spatial response.

1.5   DATA AND PATTERNS

As the sophistication of the modelling has progressed, so too has our need for appropriate data to assess the quality of simulations. In this regard we are less well advanced. Remote sensing is a tool for rapid mapping of variables like vegetation and to some extent precipitation, but it is not routinely used for other key variables. Surprisingly, the forthcoming chapters in this book make relatively modest use of remote sensing (RS) information. This is largely because the type of information that these instruments provide is quite different from what we are used to using as input or state variables in our models. We cannot directly obtain a map of root zone porosity or hydraulic conductivity, yet it is parameters of this type that the models need. As recently demonstrated by Mattikalli et al. (1998), some of these instruments can provide information on characteristics related to these variables, but not the variables themselves. This presents a major challenge for hydrologists of the 21st century – to build models that are able to exploit the information that is (and will be) coming from RS platforms. We predict that these will not simply be extensions of the models presently in use, but rather be tailor-made to utilise what is often more subjective information. Chapter 12 is an example of model outputs compared with subjective data (in this case combined field observations rather than remote sensing) and illustrates the power of this type of information. Chapters 6, 7 and 8 utilise RS data in both a "traditional" manner (i.e., as directly replacing measurements normally used) and in a more subjective sense.

But there is still a great deal that can be done using the (essentially point) measurement techniques on which we have traditionally relied. We need to choose data that give us the best insight into spatial behaviour – e.g. if shallow groundwater tables are well linked across a catchment, point information from piezometers can provide information to reduce the uncertainty in model response (see Chapter 11). In other cases we need to apply interpolation methods that produce realistic spatial patterns from point data, to provide both spatially varied input information (such as soil hydraulic properties) and spatial patterns of hydrological response for comparison with simulated patterns. The case studies presented in later chapters are examples of combined field and modelling programmes that were specifically designed to use comparisons of observed and simulated patterns for gaining insight into hydrological behaviour and to inform spatial model development. The next two chapters in this introductory section expand on these broad ideas of spatial data and modelling. They provide a more detailed discussion of the concepts, and some examples of the tools needed to better utilise spatial patterns of catchment hydrological response.

2

Spatial Observations and Interpolation

Günter Blöschl and Rodger Grayson

2.1 INTRODUCTION

Spatial patterns of hydrological processes are a rich source of variability which in some instances is quite obvious to the observer, as in the case of spatial patterns of a seasonal snow cover; and in other instances is hidden from the eye and very difficult to identify by even the most sophisticated measurement techniques, as is the case with patterns of subsurface preferential flow paths. Part of the richness comes from the diversity in the spatial arrangement of hydrologically relevant variables. It is important to understand this arrangement to design measurement strategies adequately, to interpret the data correctly, to build and/or apply a model of catchment dynamics, and ultimately to use these data in predictions of the hydrological behaviour of catchments. There is a wide spectrum of "measurement techniques" (in a general sense) available for exploring these complex patterns, ranging from traditional stream gauging to remote sensing. Ideally, a measurement technique should be designed to take into account the type of natural variability one would expect to encounter. Depending on the nature of the hydrological variability, certain measurement techniques will be more suitable than others.

Measurement techniques that are potentially capable of capturing hydrological patterns differ in terms of their scale and their accuracy. The scale relates to the area and the time that the measurements represent. Point measurements are representative over a small area or volume, such as measurements using Time Domain Reflectometry (TDR), raingauge or infiltrometer measurements. Other measurements average over a larger area or volume, such as runoff data (which averages over a catchment), or remote sensing images where each pixel is representative of a certain area. Many point measurements make up a measured pattern. In fact, most observations of patterns in this book are essentially multiple point measurements. Similarly, in the time domain, measurements can be representative over a very short interval in time (snap shots) such as a single set of soil moisture measurements by TDR (see Chapter 9) or they can give an integrated measure over time such as vegetation patterns which represent the integrated effects of soil moisture, nutrient availability, climate and other controls over a number of years (see Chapter 12).

The accuracy of the measurements can vary greatly depending on the type of measurement technique. Measurements can be a direct measure of a hydrological variable (such as the stage of a stream, rainfall depth, or snow water equivalent measured by weighing a snow core), and they can be indirect measures where some feature that is closely related to the variable of interest is recorded. Strictly speaking most measurement methods are indirect methods, where electrical resistance (e.g. for temperature measurements), signal travel time (e.g. for TDR soil moisture measurements), electromagnetic emission (for sensors used in remote sensing) are used, which in turn are converted by a rating function to the variable of interest (e.g. the stage–discharge curve for streamflow, dielectric constant–volumetric moisture content curve for TDR). The conversions can introduce additional measurement errors. The use of indirect measures can be taken further by using "surrogate" or auxiliary variables (also termed "proxy data") that may perhaps show only a limited degree of correlation to the variable of interest but are available in much spatial detail. The classical example in hydrology is the use of topography as a surrogate for rainfall variations.

Section 2.2 reviews a number of fundamental sampling issues related to scale and accuracy of measuring the spatial patterns of hydrologic variables. The scale at which data are collected is often not equivalent to the scale of the model used to describe the process of interest, so some sort of interpolation is needed before the observed patterns can be used either as model input, for estimating parameter values, or for testing of the model. Interpolation issues will be dealt with in Section 2.3 where the focus is on how the methods work, on their advantages and disadvantages and what information is needed for them. More detailed reviews of measurement techniques in hydrometry can be found in Herschy (1999) and in Sorooshian et al. (1997). Practical methods for interpolations in a GIS context are given in Meijerink et al. (1994) as well as in the extensive literature on geostatistics (e.g. Journel and Huijbregts, 1978; Cressie, 1991; Armstrong, 1998).

2.2 SAMPLING ISSUES

2.2.1 Scale and Patterns

Observed patterns are usually obtained by multiple measurements at discrete locations (and discrete points in time). This implies that their spatial (and temporal) dimensions can be characterised by three scales as depicted in Figure 2.1. These scales are the spacing, the extent, and the support, and have been termed the "scale triplet" by Blöschl and Sivapalan (1995). The spacing refers to the distance (or time) between samples, the extent refers to the overall coverage of the

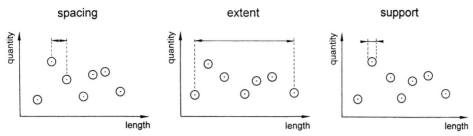

Figure 2.1. Definition of the scale triplet (spacing, extent and support). This scale triplet can apply to samples (i.e. measurement scale) or to a model (i.e. modelling scale). (Redrawn from Blöschl and Sivapalan, 1995; reproduced with permission.)

data (in time or space), and the support refers to the averaging volume or area (or time) of the samples. All three components of the scale triplet are needed to uniquely specify the space and the time dimensions of the measurements of a pattern. For example, for TDR soil moisture samples in a research catchment, the scale triplet in space may have typical values of, say, 10 m spacing (between the samples), 200 m extent (i.e. the length of the plot sampled), and 10 cm support (the diameter of the region of influence of a single TDR measurement). Similarly, for a remotely sensed image, the scale triplet in space may have typical values of, say, 30 m spacing (i.e. the pixel size), 10 km extent (i.e. the overall size of the image), and 20 m support (i.e. the "footprint" of the sensor). The footprint of the sensor is the area over which it integrates the information to record one pixel value. It is usually on the order of the pixel size but not necessarily identical to it. There are more complex cases such as measurements of evapotranspiration where the support is difficult to define and may vary in time (see Chapter 5). While the terms: spacing, extent and support are commonly used in spatial analysis, the analogous terms in time series analyses are: sampling interval, length of record and smoothing or averaging interval (e.g. Blackman and Tukey, 1958).

Ideally, the measurements should be taken at a scale that is able to resolve all the variability that influences the hydrological features in which we are interested. In general, due to logistic constraints, this will rarely be the case so the measurements will not reflect the full natural variability. For example, if the *spacing* of the data is too large, the small-scale variability will not be captured. If the *extent* of the data is too small, the large-scale variability will not be captured and will translate into a trend in the data. If the *support* is too large, most of the variability will be smoothed out. These examples are depicted schematically in Figure 2.2 where the sine wave relates to the natural variability of some hydrological variable and the wavelength is related to the scale of the true hydrological features. The points in Figure 2.2 relate to the scale triplet of the measurements. Similar concepts apply to the time domain. For example, if for an air temperature sensor the time constant (the time the sensor averages over, i.e. the support) is too large (say of the order of 1 second) it will not be possible to measure the short term fluctuations of air temperature due to turbulent eddies with that sensor and the measured response will be much smoother than the actual fluctuations. It is clear that some sort of filtering is involved, i.e. the true patterns are filtered by the

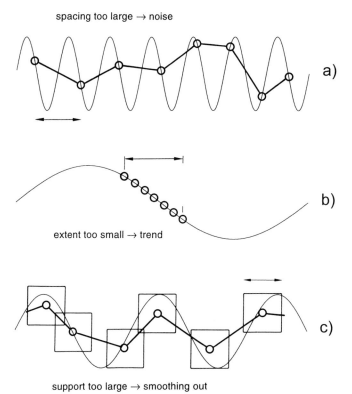

Figure 2.2. The effect of measurement scale on capturing the 'true' hydrologic pattern. The circles are the measurements and the thin line is the 'true' hydrologic pattern with the characteristic length scale equal to the wavelength.

properties of the measurement which are then reflected in the data. The effect of this filtering is to smooth out variability if the scale of the measurement does not match the scale of the process. There is a substantial body of literature that deals with methods for defining and predicting the way in which variability is captured (or not captured) by the measurement characteristics (e.g. Wiener, 1949; Krige, 1951; Matheron, 1965, 1973; Blackman and Tukey, 1958; Federico and Neuman, 1997). An important practical outcome of that work is the development of methods to: (i) assess how many measurements are needed to capture (to a certain accuracy and under particular assumptions) a natural pattern; and (ii) to quantify the variability that is lost due to filtering. This second method, known as regularisation (Journel and Huijbregts, 1978; Vanmarcke, 1983), provides a tool for quantifying the variability expected for different measurement supports and spacings, under particular assumptions about the underlying pattern being measured. Western and Blöschl (1999) show some examples of regularisation methods applied to spatial measurements of soil moisture and indicate that they work well, provided that the statistics of the underlying patterns are well known.

While a quantitative treatment will often not be needed, a qualitative consideration of the scale of the natural variability and that of the measurements is important to assess at least the magnitude of information on variability that is not resolved by the sampling. In Figure 2.3 the spatial spacing and extent of a range of measurement methods are plotted versus their temporal spacing and extent. The shaded area refers to the domain between spacing and extent of the measurements. Taking the example of daily raingauges from a typical hydrometric network, the domain covers ranges, in time, from 1 day to, say, 100 yr, and in space, from 10 km (average spacing of the gauges) to 2000 km (size of the region). Figure 2.3 also shows the typical scales of TDR measurements of soil moisture in research catchments as well as a number of space-borne sensors relevant to hydrology. When comparing Figure 2.3 with Figure 1.4, areas in the space–time domain that overlap are those where we have measurement techniques that are appropriate for describing the process of interest, whereas areas that do not overlap are not described well. In other words, from a particular measurement one can only "see" processes within a limited window (determined by the scale triplet), and processes at larger and smaller scales will not be reflected in the data. For example, daily raingauges cannot capture atmospheric dynamics at the 10 km scale as the temporal spacing is too large, but on the other hand the

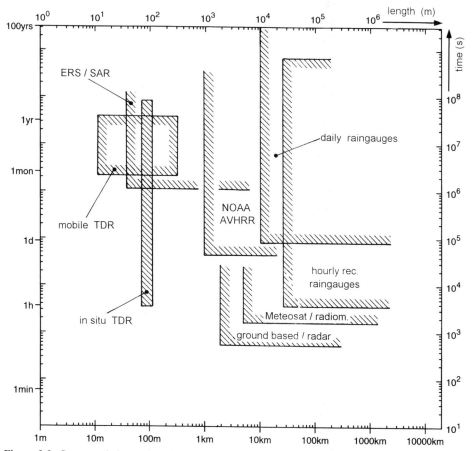

Figure 2.3. Space and timescales of rainfall and soil moisture variability that can be captured by different instruments represented as the domain between spacing and extent of the measurements.

Meteosat satellite sensor is commensurate with atmospheric processes from thunderstorms to fronts and one would expect it to capture these processes with little bias due to scale incompatibility. The comparison also indicates that the TDR measurements can potentially capture runoff generation processes in a research catchment setting. Clearly, experimental research catchments and operational hydrometric networks provide samples at vastly different scales and hence provide information on different processes.

The type of comparison illustrated by Figure 2.3 and Figure 1.4 allows an assessment of how representative are multiple-point measurements of the underlying spatial pattern of hydrologic processes. In general, this depends on the scale of the measurements and on whether the pattern varies smoothly in space which implies large-scale variability (e.g. groundwater heads) or whether there is a lot of erratic (small-scale) variability (e.g. soil hydraulic conductivity). In the case of groundwater which varies smoothly, a few samples at a large spacing will be quite representative of the pattern, while many more samples at shorter spacings will be needed for erratically varying quantities such as soil hydraulic conductivity. If a large number of samples in space are available (implying a relatively large extent and a relatively small spacing) it is much more likely that we can capture the spatial processes of interest. The key to a successful representation of spatial patterns in catchment hydrology, therefore, is to maximise the number of sampling points in space that cover an extent sufficient to capture the processes of interest.

The issue of data being commensurate with the scale of the underlying patterns can be generalised to the *spatial arrangement* of the patterns, i.e. the issue of identifying the level of "organisation" of the underlying patterns. Here we use the term "organisation" to describe the complexity of the pattern. If the spatial pattern is purely random it is not organised, while if the pattern does show features such as elongated bands of high soil moisture values in gullies, it is organised (see Figure 1.2, and Journel and Deutsch, 1993). Most spatial measurements are essentially point measurements and the number of measurements available in most practical cases is often small. Because of this, the spatial complexity of natural patterns cannot be identified very well. Often, the apparent variability of the data is then interpreted as an evidence of spatially random processes, but this tends to be a consequence of poor sampling density rather than a reflection of the underlying hydrologic variability. Williams (1988) commented that in the case of subsurface hydrology, the apparent disorder is largely a consequence of studying rocks through point measurements such as boreholes, while a visual examination of the rocks in mines or through outcrops almost always shows clearly discernible organisation. This statement is also valid in catchment hydrology, implying again that, in order to properly define patterns, *a large number* of such point measurements will be needed if we are to avoid the trap of trying "to squeeze the nonexistent information out of the few poor anaemic point measurements" (Klemeš, 1986a, p. 187S).

2.2.2 Accuracy and Patterns

In the measurement of hydrologic patterns there is often not only a scale problem but also a problem with the accuracy of the measurements, i.e. there often exists a considerable measurement error. If the measurement error is large, the patterns apparent in the data will be a poor representation of the true underlying pattern. The presence of measurement errors confounds the identification of patterns. See the soil moisture data described in Chapter 9 for an example. During winter conditions, the measurement error is relatively small as compared to the true underlying variability (3 $(\%V/V)^2$ as compared to 20 $(\%V/V)^2$), and the data give a good appreciation of the real soil moisture patterns. However, during summer conditions, the measurement error is relatively large as compared to the true underlying variability (3 $(\%V/V)^2$ as compared to 5 $(\%V/V)^2$), and the data do not allow us to infer any underlying soil moisture pattern very well.

There are two types of measurement errors, systematic and random. A systematic measurement error may be introduced either by an improper measurement setup (such as the catch deficit of raingauges caused by wind exposure), or by improper rating functions (e.g. the TDR calibration curves). In many cases it will be possible to correct for such systematic errors, provided additional (more accurate) data are available for comparison. A random measurement error may be introduced, for example, by air gaps around the probes of a TDR and by inaccurate readings of an observer who reads off the stage of a stream gauge. While it is not possible to remove random errors by applying a correction, these errors can be significantly reduced by taking multiple measurements of the same variable. For example, if there is a measurement error variance of 3 $(\%V/V)^2$ attached to a single TDR measurement, ten such measurements at the same location and time pooled together only have a measurement error of 0.3 $(\%V/V)^2$, provided the errors of these ten measurements are statistically independent. More generally speaking, the measurement error variance decreases with the inverse of the number of samples that are aggregated (see any basic statistics text, e.g. Kottegoda and Rosso, 1996). Geostatistical methods use this property to "optimally" estimate true values from samples that are "contaminated" by measurement errors. However, this error reduction is contingent on the measurement errors being truly random, i.e. uncorrelated and symmetrically distributed. If they are correlated or possess some organised structure, the actual error reduction may be much smaller than is implied by these methods. From a practical point of view, the presence of random measurement errors can be countered, to some degree, by increasing the number of independent samples; i.e. many samples using a method with a particular random error can give a similar accuracy to few samples which have less error in an individual measurement.

This idea can be extended to the use of surrogates (or proxy data, or auxiliary data). Surrogates are variables that show some (often limited) degree of correlation to the pattern of interest but are much easier to collect in a spatially distributed fashion. Examples of surrogates include soil texture to infer hydraulic properties (e.g. Rawls et al., 1983) and are discussed in more detail later in the chapter. The conceptual point of importance is that the lack of correlation between the original variable and the surrogate can be interpreted as a sort of measurement error. Ideally we would have surrogates that are both easy to measure and are highly correlated (low error) to the original variable. Unfortunately, it is often the case that the easier the method of data collection (and hence the larger the number of points it is feasible to collect) the poorer is the correlation (or equivalently the larger the measurement error).

There is therefore usually a trade-off between a few points of great accuracy (and hence a poor resolution/coverage), and many points of poorer accuracy. An important example in hydrology is the use of remote sensing data. For example, weather radar (see Chapter 4) does not, strictly speaking, measure rainfall, but radar reflectivity which is correlated with rainfall intensity but also depends on other factors (such as drop size distribution), only some of which are known. As a consequence, there is often a substantial error introduced when converting reflectivity to rainfall. Other examples in remote sensing include soil moisture as estimated from SAR sensors (see Chapter 8) where a huge number of points (pixels) in space are available, but correlations between the SAR backscatter and soil moisture tend to be poor. The same is true with some ground data. For example, in an Alpine environment, it typically takes on the order of 3 minutes to measure snow depth, but it may take 30 minutes or more to collect a sample of snow water equivalent. Similarly, it is much faster (and hence cheaper) to make a TDR measurement of soil moisture than a gravimetric measurement. Hence we can collect many snow depth samples (or TDR samples) and have a chance of seeing patterns, yet have to accept a larger error in an individual measurement than is possible with a more accurate technique that takes more time to use.

As mentioned above, averaging (or aggregation) can help improve the accuracy of surrogates, and this is part of the trade-off. In the example of inferring patterns of rainfall intensity from radar reflectivity, often, multiple images (for many points in time) are aggregated. The aggregated (average) image is then more reliable than the individual images. Similarly, passive microwave data have been aggregated over time to improve the accuracy of rainfall estimation (Negri et al., 2000). SAR images are sometimes aggregated in space for more reliably estimating soil moisture. Aggregated pixels that are on the order of $20\,km \times 20\,km$ rather than the original $20\,m \times 20\,m$ have been shown to be much better related to soil moisture than the individual images (e.g. Wagner, 1998). In the snow depth sampling example mentioned above, it is common practice to measure depth at a minimum of 10 locations in close vicinity to a sample site and to average these values to increase their accuracy. Clearly, this is at the cost of reducing the number of sites given fixed time/resources.

An important question for sampling design, therefore, is whether there is an optimum in the trade-off between accuracy and number of points. In general, such an optimum will depend on the relationship between the accuracy of a sample and its cost of collection, but it has been illustrated in Kupfersberger and Blöschl (1995) that the value of surrogates also depends on the level of "organisation" (or complexity) of the underlying pattern. For their case study of aquifer variability, they found that 190 samples of auxiliary data (in this case subsurface electrical resistance) with a correlation of $r^2 = 0.36$ (between hydraulic conductivity and electrical resistance), outweighed the information from 11 error-free measurements of hydraulic conductivity when random spatial variability was present, but outweighed 25 error-free measurements when the underlying variability was not random, but exhibited preferential flow paths.

The influence of the underlying pattern complexity on optimum sampling strategies is a "chicken and egg" dilemma. We need to know a lot about an underlying pattern to design an optimal sampling scheme, but we need data to know the underlying pattern. This implies an iterative approach where sampling can be refined as more is known about the pattern being measured – this can be assisted by an understanding of the processes or features leading to the patterns. For example, if the aquifer in the example above was largely clay but with some highly permeable sand lenses present, knowing the location of the lenses (say via a geophysical method) would enable a much more efficient sampling approach than if it was assumed that the lenses in the area were randomly distributed.

The notion of surrogates can be taken even further to incorporate process understanding to improve pattern estimation. The classical example in catchment hydrology involves the use of terrain parameters computed from digital elevation models (which generally represent a very high "sampling density" compared to many field measurements). There are a range of terrain parameters that have direct causal links to the driving processes in catchment hydrology and some parameters are deemed to be useful because of feedbacks between different processes. Examples include the terrain aspect for representing snowmelt and evapotranspiration processes (since aspect is related to solar radiation exposure), and combinations of slope and upslope contributing area to represent soil saturation or erosion processes. A comprehensive review of terrain parameters that can be used as surrogates (or indices) in catchment hydrology is provided in Moore et al. (1991) and the use of indices is discussed further in Section 2.3.3. The important point is that if an index is able to capture some key features of a pattern, the amount of sampling needed is greatly reduced. If, for example, we know that aspect is perfectly correlated to the pattern of snowmelt, we may need only a few field measurements to calibrate the relationship.

In summary, the number of sampling points that we need to adequately represent a spatial hydrologic pattern depends on:

- the scale of the processes compared with the extent we are interested in (small-scale processes require a larger number of points, while for large-scale (smoothly varying) processes a single point is representative of a large area),
- accuracy of the data/correlation of surrogates with the values of interest (accurate data require a smaller number of points), and

- complexity of patterns/process (rich pattern requires a larger number of points unless we have some process understanding that can define certain features of the pattern).

In the case studies of this book (Chapters 6–12) the number of sampling points in space varies greatly with the type of measurement method used. A number of case studies use remotely sensed data, both for testing hydrologic models and for being tested by ground data. For remotely sensed images the number of sampling points in space (i.e. the number of pixels) is very large but this is at the cost of a single pixel not providing much hydrologically relevant information. In Walnut Gulch (Chapter 6) airborne sensors (ESTAR and PBMR) are used for estimating soil moisture and rainfall; in Kühtai (Chapter 7) aerial photographs of snow cover patterns (snow/no snow) are used; and in Zwalmbeek and Coët-Dan (Chapter 8) satellite data (SAR) are used for estimating saturated source areas. The majority of case studies use multiple point values measured in the field. In Walnut Gulch (Chapter 6) more than 90 recording raingauges are used; in Reynolds Creek (Chapter 7) snow water equivalent was sampled at about 300 points in space; at Tarrawarra (Chapter 9) TDR soil moisture was measured at about 500 points in space; in La Cuenca (Chapter 10) runoff occurrence (for a single event, runoff occurred/did not occur) was measured by runoff detectors at 72 locations in the catchment; at Minifelt (Chapter 11), the shallow groundwater table was observed for 108 piezometers; and at Trochu (Chapter 12) recharge/ discharge observations as derived from chemical/vegetation indicators are used from 48 locations. Finally, two case studies in Chapter 8 use qualitative data mapped in the field on a continuous basis. These are Coët-Dan where patterns of saturated source areas from a field survey are used and Zwalmbeek where soil drainage classes based on field mapping are used.

In practice, there are no hard and fast answers to the problems of sampling design – of how many points and where, when and how to make measurements. Any design will include compromise. The best we can do is to ensure that we consider the implications of that compromise for the use to which we put the data. At the end of this chapter (Section 2.4), we present a detailed description and practical example of the steps that can be taken to try to optimise sampling within given constraints, as a guide to the type of thinking that should go into sampling design in catchment hydrology.

2.3 FROM POINTS TO PATTERNS

2.3.1 The Interpolation Challenge

We are often able to obtain multiple point measurements but what we would like to have is some estimate of the variable of interest *everywhere* in the catchment or region of interest. This really amounts to generating a pattern from point values and involves interpolation and often extrapolation in space, and sometimes also in time. It is rare to measure an input or model parameter at the same scale as it is to be used in a model, so again some sort of interpolation (requiring a variety of assumptions) must virtually always be undertaken. This section summarises the concepts that underpin interpolation and common techniques for the interpolation of spatial data, discussing some of the key issues to be considered in choosing and using the methods.

Consider two scenarios. In the first scenario, snow depths are sampled at a regular spacing of 1000 m and are subsequently interpolated on to a 10 m grid. In the second scenario, snow depths are sampled on a 10 m grid. It is easy to envisage that the spatial pattern of (interpolated) snow depths at the 10 m spacing of the first scenario will be much smoother than the patterns in the second scenario. This example illustrates that interpolation involves a change of scale (from 1000 m spacing to 10 m spacing in the above example) and that interpolations tend to smooth measured patterns, i.e. interpolations can be thought of as a kind of filter. There is some similarity between *sampling* which is a filter on the process as discussed in 2.2.1 and *interpolation* which is a filter on the data. As with the sampling case, it is possible to define a scale triplet for interpolations consisting of spacing, extent and support. This triplet has been denoted the *model scale triplet* by Blöschl and Sivapalan (1995) (as opposed to a *measurement scale triplet* in the case of sampling).

In interpolating samples of snow depths (which involves a model in a general sense) the scale triplet may have typical values of, say, 10 m spacing (i.e. the resolution to which the map is produced), 10 km extent (i.e. the overall size of the map), and 10 cm support (i.e. the area over which each point in the map is representative). Clearly, it is possible to draw many maps of snow depth of the same area, ranging from ones showing point values (very small support) to ones with various levels of averaging (i.e. increases in interpolation model support). The larger the interpolation model support, the smoother the map will be. This notion of a model scale triplet can also be used for dynamic models of catchment response (see Chapter 3). For example, for a spatially distributed runoff model, the scale triplet may have typical values of, say, 25 m spacing (i.e. the model element size), 1 km extent (i.e. the size of the catchment to be modelled), and 25 m support (the element size). The support is the spatial dimension over which the variables in each model element are representative. This is equal to the element size in most distributed models but in some models the variables are defined at points which means that the support is very small.

The concept that interpolation is a filter having a scale triplet associated with it is critical to interpreting the results of interpolation and how they represent the underlying data. Let us again consider Figure 2.2. In many applications in catchment hydrology, the situation resembles the case where the spacing of the data is much coarser than the scale of the underlying hydrological variability of interest and consequently the choice of interpolation method becomes exceedingly important. Most of the interpolation methods for obtaining patterns from observed point values are "dumb" or black box approaches that do not take into account the type of pattern to be expected, i.e. for different hydrologic variables (that obviously possess very different spatial patterns) one would often use the same interpolation method. However, if we have some understanding of the underlying process we can introduce some prior knowledge, for example by introducing auxiliary data (or equivalently surrogates) and using them in a clever way. Auxiliary data are useful in spatial interpolations if an underlying relationship to the variable of interest is present and can be defined. There are a wide range of interpolation methods that are, in principle, capable of doing this. In fact, the availability and accessibility of mathematically complex interpolation techniques in GIS and specialist software has made the application of interpolation methods relatively simple. The challenge for the user is to have a sufficient understanding of the concepts on which the methods are based and to apply them in a way that best exploits the information in the data. Below we will briefly summarise the concepts that underlie the more widely used spatial interpolation techniques in hydrology. It is important to realise that the quality of an interpolated pattern depends on both the accuracy of the original point data and on how well the method of interpolation/extrapolation reflects the underlying spatial structure of the measurement – something that depends on our understanding of the phenomena being measured.

2.3.2 A Brief Summary of Concepts that Underlie Spatial Interpolation Methods

A hierarchy of interpolation techniques can be thought of in terms of the extent to which information beyond that inherent in an individual data point is used. The very simplest methods use only the nearest data points for the interpolation. Methods that exploit information on the total data set in estimating values at any particular location are a further step up in complexity. Even more complex are methods that utilise auxiliary data or some theory of system behaviour to assist in the interpolation. These are all interpolation techniques that attempt to estimate the *most likely* true pattern from the points. Sometimes one is interested in a pattern that perhaps is *not* the most likely one, but possesses a *variability* that is close to the real one. For this purpose stochastic simulation methods can be used which produce numerous realisations of equally likely patterns, i.e. a suite of patterns that are all deemed representative of important features of the true pattern. Finally, one sometimes needs to generate spatial patterns from an areal average value (rather than from multiple point values) for which disaggregation methods can be used. We will review these methods below.

(a) Interpolation Methods That Do Not Use Auxiliary Data

The simplest spatial interpolation methods are usually based on deterministic concepts. These assume that one true but unknown spatial pattern exists which we attempt to estimate by either some ad hoc assumptions or some optimality criterion (e.g. Meijerink et al., 1994). The methods are illustrated schematically in Figure 2.4 for the one-dimensional case where the dots are observed values and the lines are interpolations by various methods. Deterministic interpolation methods include the following.

- The Thiessen (1911) method, or equivalently, nearest neighbour method where each point to be interpolated is assigned the value of the data point that is closest in space. This results in patterns that have polygon shaped patches of uniform values with sharp boundaries between the patches (Figure 2.4a). These are clearly very unrealistic for hydrological variables but the method is simple and robust. The method assumes that the data are error free (i.e. the value of the interpolated surface at the measurement point matches the measurement itself).
- The inverse distance squared method where the interpolated value is estimated by a weighted mean of the data and the weights are inversely proportional to the squared distance between the interpolated value and each data point. This results in a pattern that is smooth between the data points but can have a discontinuity in slope at the data points. For typical hydrologic data that are unevenly spaced, this method produces artefacts as is exemplified in Figure 2.4b. It is clear that hydrological variables are very unlikely to vary in the way this method assumes. Even though this method is very easy to use, it cannot be recommended.

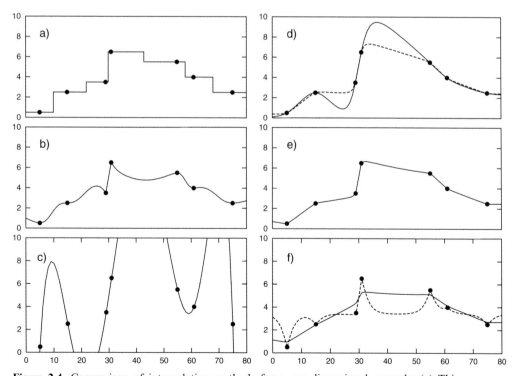

Figure 2.4. Comparison of interpolation methods for a one-dimensional example: (a) Thiessen method; (b) inverse distance squared; (c) example of overfitting using a sixth-order polynomial; (d) thin plate splines with different tension parameters; (e) kriging (zero nugget, large range); (f) kriging (solid line: large nugget, large range; dashed line: zero nugget, very short range).

- Moving polynomials where a trend surface is locally (i.e. for each interpolated value) fitted to the data within a small local (i.e. 'moving') window
  about each point to be interpolated. The fitting is usually done by least
  squares and the trend surface is represented as a polynomial (Tabios and
  Salas, 1985). The higher the order of the polynomial, the closer the fit to the
  data but the more irregular the trend surface and the more likely the
  occurrence of 'overfitting' will be. Overfitting occurs when the particular
  data points are very well represented but representation of the underlying
  (true) pattern is poor, i.e. more credibility is given to the data than is
  merited due to measurement error and the fact that the true surface may
  not vary like a polynomial. An example of overfitting is given in Figure 2.4c
  where a sixth-order polynomial has been fitted to the seven data points.
  This polynomial is clearly an unrealistic representation of hydrologic variability. Overfitting can be avoided by selecting the order of the polynomial
  to be much smaller than the number of local data points which will result in
  an interpolated (trend) surface that is smoother and usually does not
  exactly match the data points. This is consistent with an implicit assumption that the data may be in error and need not be fitted exactly. As the
  method only uses data within a local neighbourhood, it is computationally
  efficient and can therefore be used for large data sets. There are numerous
  variants of this method, some of them having discontinuities in the interpolated surface when the boundary of the window moves across a data
  point that is very different from the rest of the data points within the
  window. Some of the methods of this type are discussed in Meijerink et
  al. (1994).
- Thin plate (or Laplacian) splines (Wahba and Wendelberger, 1980;
  Hutchinson, 1991, 1993) where a continuously differentiable surface is
  fitted to *all* the data, i.e. this is a global rather than a local method. The
  name of 'thin plate' derives from the minimisation function used which has
  a physical analogy in the average curvature or bending energy of a thin
  elastic sheet. A low-order polynomial is usually used and a minimum curvature criterion implies finding the smoothest possible function. This
  method therefore does not suffer from the overfitting or oscillation problems of the moving polynomial method. There are two main variants of
  this method. The simpler one assumes that the data are error free and hence
  the interpolated surface goes through the data points. It has one 'smoothing' or 'tension' parameter which can be used to control the smoothness of
  the interpolated surface (Figure 2.4d). The other variant allows for a measurement error by introducing an additional parameter representing the
  error term (e.g. Hutchinson, 1991). This parameter can be used to control
  the balance between smoothness of the surface versus how close the interpolated surface is to the data. There are automated methods available for
  determining these parameters such as minimum generalised cross-validation (GCV, Hutchinson, 1991) which is based on withholding each data
  point in turn from the fitting procedure and calculating the error between

the surface and the data point. Thin plate splines work very well in most hydrologic applications, unless the data are too widely spaced as compared to the scale of the underlying phenomenon. They are robust and operationally straightforward to use. While early implementations were computationally burdensome, more recent implementations that subdivide the domain into subregions can be used efficiently for very large data sets (Hutchinson, 1991). Of the deterministic interpolation methods, this is the one that can be generally recommended. The only drawback, as compared to geostatistical methods (see below), is that it is not as straightforward to explicitly consider measurement error and estimation error with spline interpolations (e.g. Cressie, 1991).

The notion of filtering or a change of scale triplet can be used to highlight some capabilities and limitations of interpolation methods. Clearly, a change of extent occurs when extrapolation beyond the domain of observation is attempted. Thin-plate splines generally perform adequately, provided that the extrapolation is not too great. A change of support occurs implicitly as the interpolated function is always smoother than the original data, hence the support notionally increases to about the size of the data spacing. Methods such as splines allow for adjusting the smoothness and hence the support of the interpolated surface. A change of spacing occurs in all the methods, from the spacing of the data to the spacing of the numerical grid to which one interpolates. Often, in catchment hydrology, the spacing of the data is much larger than the scale of the underlying process and therefore the interpolated (small spacing) grid will not show all the small scale detail.

Stochastic interpolation methods are an alternative to deterministic approaches. The most widely used stochastic approaches are termed geostatistical techniques and are based on the notion that the interpolated pattern is a random variable, which can be described by the variogram. The variogram is the variance between pairs of points spaced at a certain distance (or lag), or equivalently a measure of spatial correlation. Pairs of points that are close to each other tend to be similar, hence the value of the variogram (termed gamma) at small lags tends to be small. As the lag increases so does gamma as the values become increasingly dissimilar at larger lags. Figure 2.5a shows a typical variogram of a stationary pattern (i.e. a pattern for which the mean does not change with space). There are three parameters that need to be specified for a variogram, two of which are the sill (which is the spatial variance of the overall pattern) and the range (which is the spatial correlation length and a measure of how continuous or smoothly varying the patterns are – large ranges relating to smoothly varying patterns, short ranges relating to erratically varying patterns). If the pattern is not stationary in the mean (i.e. exhibits a spatial trend) it neither possesses a sill nor a range (Figure 2.5b) but it may still be used for geostatistical analyses. The third parameter is the nugget (Figure 2.5a), which represents the variance of pairs of points that are at essentially the same location. Each of these parameters, the sill, range and nugget can be interpreted in terms of the physical

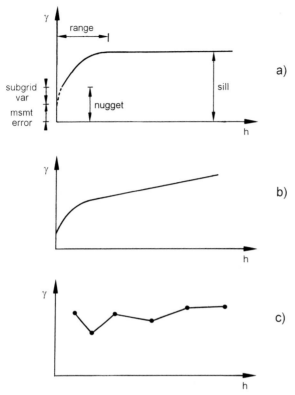

Figure 2.5. (a) Variogram of a stationary random function (i.e. no spatial trend) showing a nugget consisting of subgrid (small scale) variability and measurement error, sill, and range; (b) variogram for a random function that is non-stationary in the mean (i.e. possesses a spatial trend); (c) variogram as it is often derived from hydrologic data with zero range.

processes that lead to the pattern. The sill is a measure of the overall variability of the process. The range is a measure of the spatial scale of the process. The nugget can be interpreted as the sum of two variances related to two distinct phenomena (see e.g. de Marsily, 1986, p. 304). First, a non-zero nugget can be due to random measurement errors rather than a feature of the physical process. In this case, even if the samples are very closely spaced, there will be some variance between the measured data. This part of the nugget is equal to the measurement error variance. Second, a non-zero nugget can be due to the data not having been collected at sufficiently small spacings to reveal the continuous behaviour of the phenomenon, i.e. the measurement spacing is too coarse to represent the underlying process of interest. This part of the nugget is equal to variability at scales smaller than the sample spacing and is also termed sub-grid variability or "apparent nugget" (de Marsily, 1986). An apparent nugget will disappear if the data are collected at sufficiently small spacings.

Geostatistical approaches consist of two phases. In the first phase (termed structural analysis) a variogram is estimated from the observed data. This is done by plotting the variances of differences of the values of data pairs versus their lag (i.e. plotting the sample variogram) and fitting a smooth function, known as the *theoretical variogram,* which is assumed to be the variogram of the population. There are a number of functions that can be used for the theoretical variogram, but it is common to use an exponential function or a power law function. The parameters of the chosen function define the sill, range and nugget used for the interpolation. In the second phase, a spatial pattern is estimated from both the data and the characteristics of the variogram, based on Best Linear Unbiased Estimation (BLUE). Linearity implies that the estimated value at any point is a linear combination of all of the measurements, with a different weight for each measurement (these weights are calculated as part of the method). The estimates are required to be unbiased (i.e. the mean of the data and the mean of the interpolated pattern must be identical) and the estimates are required to be "best" or optimal in the sense that the variance of the estimation error is minimised. Combination of these three criteria (linear combinations of data points, unbiased and optimal estimates) results in a system of equations which is solved for the unknown weights. There are a wide range of geostatistical estimation methods which differ in the assumptions about the way the random function varies spatially and in the way they are constrained by other information, resulting in different levels of complexity of the interpolation method. One of the simpler and widely used methods is Ordinary Kriging (Journel and Huijbregts, 1978).

The main difference between kriging and splines is that spline interpolation, being a deterministic approach, assumes that the surface is *one* unknown function, while kriging assumes that the surface is a random variable and attempts to estimate the expected (most likely) value at any point. In practice, kriging and splines can give very similar results as exemplified in Figure 2.4d,e. The advantage of kriging is that measurement errors can be more directly introduced through the nugget (Figure 2.4f, solid line). However, kriging is less robust than splines as it heavily depends on the proper selection of the theoretical variogram which is sometimes not well defined. If, for example, too small a value for the range is specified, the interpolated pattern may look like Figure 2.4f (dashed line) where the interpolated surface approaches the mean at a certain distance from the data points.

Geostatistical methods are probably the most widely used interpolation methods in catchment hydrology, but in practice there are three main pitfalls that should be recognised:

1. Often the nugget of the estimated variogram is of similar size to the sill, i.e. closely spaced pairs of points are no better correlated than points that are far apart (Figure 2.5c). This undermines the central assumption of geostatistics that the spatial correlation (i.e. the variogram) is useful for spatial interpolations. If one does use this type of variogram for interpolation, the interpolated value will be equal to the mean of all data points everywhere (except near the data points if the nugget is zero), i.e. it is similar to the dashed line in Figure 2.4f. As a remedy, one can either use auxiliary data to improve the interpolation (see later) or one can resort to hand-drawn

contours from the data, thereby implicitly introducing expert knowledge on the expected appearance of the underlying pattern. An example where this is often necessary is the spatial interpolation of extreme storm rain-gauge readings where hand-drawn contours are used because more "objective" methods will usually give very unrealistic patterns. This is because the spatial scale of the phenomenon (extreme rainfall) is usually much smaller than the spacing of the raingauge data.

2. The theoretical variogram estimated from the data not only depends on the underlying process, but also on the scale of the sampling, and on how well the range, sill and nugget can be defined. Provided that the sample spacing is sufficiently small to capture the variability of the phenomena of interest, the range parameter provides a measure of the scale of the underlying process. But if the sample spacing is too large to define the variability of the phenomena, the range of the variogram tells us nothing about the scale of the underlying process. When comparing case studies at vastly different scales, it is common to find a significant increase of the estimated range with the spacing of the data (e.g. Gelhar, 1993, Figure 6.5). This may well be just an artefact of the sampling scale, resulting from a sample spacing that was too big to properly define the small-scale variability of the process and a sample extent that was too small to properly define the large-scale variability of the process (Blöschl, 1999). The key point is that a variogram can be derived for *any* data, but the range only has significance for interpretation of physical processes if the data are spaced closely enough and cover a large enough area to capture the process scale variability. This means that a variogram can not be simply transposed across scales and should be estimated from data at the same scale as the study of interest.

3. The third problem is when the spacing of the data is uneven in *x* and *y* and/ or if the underlying patterns show linear features. This is best explained through example. Consider a series of transects of soil moisture down a hillslope and across a wet gully (Figure 2.6). Along the transect, there is a rapid increase in soil moisture as we move from the hillside into the gully. We have closely spaced the sampling points to reflect this rate of change and any interpolation algorithms would work adequately along the transect. In the across-transect direction (i.e. along the axis of the gully) the rates of change might be slower, hence our choice of using transects. But the distance between adjacent transects is too great for the interpolation algorithm to "fill the gaps" between transects. We might be able to look at the data and intuitively draw contours (implicitly using our understanding of the phenomena) but the automated interpolation algorithm cannot do so, and the resulting patterns will be very poor (Figure 2.6, bottom). For the simple example in Figure 2.6 this problem can be overcome by accounting for anisotropy (i.e. where variograms computed in different directions differ in their range), but this is not possible for real world problems in catchment hydrology, which are always more complex, invol-

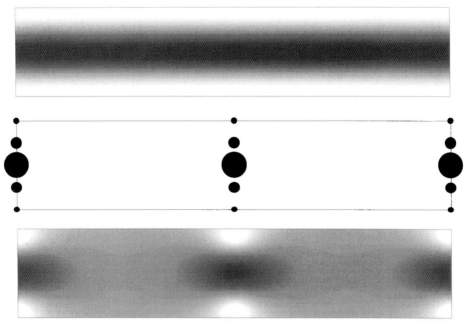

Figure 2.6. Schematic example illustrating effect of uneven spacing on interpolation. Top: Hypothetical spatial pattern of soil moisture in a valley (the valley is along the left–right direction in the figure) with larger soil moisture (dark) in the gully and lower soil moisture (light) on the hillslope. Centre: three transects of soil moisture samples across the valley where dot size represents the magnitude of the soil moisture values. Bottom: Interpolated pattern of soil moisture based on the samples in the centre of the figure.

ving a number of valleys with different directions. The most efficient alternative to hand drawing contours is to use auxiliary data to improve the interpolation.

(b) Interpolation Methods That Use Auxiliary Data

There are two ways in which spatial interpolation methods can use auxiliary data: dual-step and single-step methods. Dual-step methods treat the relationship to the auxiliary variable and the actual spatial interpolation separately. One widely used method, representative of this genre, is the external regression approach for spatially interpolating, say, mean annual rainfall in mountainous terrain. In the first step, a regression between terrain elevation and mean annual rainfall is calculated for those locations where both rainfall and terrain elevation are available, and the regression line is used to estimate rainfall everywhere. The regression can be either made over the entire domain or over a moving neighbourhood (a window). At the locations where data are available, the rainfall so estimated will be different from the measured rainfall as the regression line does not exactly fit the data. These differences (i.e. the residuals) can, in a second step, be spatially interpolated with any of the methods discussed above. The final interpolated surface is then made up of the sum of the regressed values for any point and the interpolated residuals. The advantage of this approach over simply interpolating the data without auxiliary information, is that it extracts the small-scale patterns from the auxiliary data in addition to using the large-scale, smoothly varying component from the original data, while a simple interpolation will neglect the small-scale variability. This type of approach is fairly robust and can give excellent results (see also the example in Chapter 9 where the method was used for smoothing out sample noise).

Single-step methods are more elegant and tend to give slightly better results as they are based on *joint* optimality criteria for the original data and the auxiliary data (Deutsch and Journel, 1997). Conceptually, single-step methods can be based on an extension of two-dimensional interpolation methods that do not use auxiliary data to a third dimension. Often the third dimension is elevation. In both spline interpolation (e.g. Hutchinson, 1993) and kriging (e.g. Jensen, 1989), this can be done by introducing a generalised lag (or distance) between two points, that is the square root of the sum of the weighted squared distances in x, y and z. The weights (or factors) account for anisotropy (i.e. the interpolated surface usually varies more quickly with elevation than horizontally) but otherwise the method is the same as for splines and kriging, described earlier. An alternative is to use some sort of submodel, i.e. a relationship between the auxiliary data and the original data that is built into the interpolation scheme. Again this can be done for splines (which are then termed partial splines, Hutchinson, 1991) and for various variants of kriging. The variants of kriging include external drift kriging (Ahmed and de Marsily, 1987) where the existence of a linear relationship between the additional information and the original data is postulated and the auxiliary variable is assumed to be error free. The interpolated patterns hence look very similar to the pattern of the auxiliary variable, i.e. a lot of spatial structure is imposed. The linear relationship is implicitly calculated by the method from the data and only the variogram of the original data needs to be specified by the user. The variants of kriging also include co-kriging where the covariance (or the cross-variogram) between the auxiliary variable and the original variable is exploited. Both the auxiliary and the original variables may be subject to measurement error. This method imposes less structure than external drift kriging and hence the interpolated patterns tend to be smoother. However, in co-kriging, the appropriate choice of the variograms (of the auxiliary data and the original data) and the cross-variogram is not straightforward and needs to meet certain criteria for the method to work (e.g. Journel and Huijbregts, 1978; Deutsch and Journel, 1997). Both external drift kriging and co-kriging require the additional information to be numerical, such as a wetness index, rather than categorical such as land use or soil type. Methods such as Bayes Markov Kriging (Zhu and Journel, 1993), or the simpler Bayes Markov Updating (e.g. Bárdossy and Lehmann, 1998) can be used to incorporate this type of categorical information and so enable a wide range of auxiliary information to be utilised.

Bárdossy and Lehmann (1998) also illustrate what the implications are for soil moisture patterns of a particular choice of an interpolation method. They show that, with sparse data, the interpolated patterns vary enormously depending on which method is used, highlighting the fact that the modeller must use a large amount of judgement in determining which interpolated pattern is the "best". It is clear that no matter which method of spatial interpolation using auxiliary data is used, the key question to be addressed is whether, for the particular application under consideration, the patterns imposed by the auxiliary information are indeed those likely to be present in the phenomena being represented – i.e. how valid are the assumptions underlying the relationship between the auxiliary data and the parameters of interest, and, even if the relationships are sound, how will the errors associated with them affect the modelling exercise.

(c) Stochastic Simulations

Geostatistical interpolation methods such as kriging are "best" estimators (i.e. they give the most likely value of the variable between observations) and hence they smooth out the small-scale variability between the observations. There is a class of methods that preserve the small-scale variability. These are generally referred to as conditional (stochastic) simulation and do not give the most likely pattern, but rather a suite of equally likely patterns that all exhibit realistic spatial variability (i.e. multiple realisations). Each realisation is one possible scenario, which represents both the individual observations and the variogram structure of the set of observations. The term 'conditional' refers to the patterns being 'conditioned' to the observations, i.e. they reproduce the observations exactly. Examples of conditional simulations are Sequential Gaussian Simulation (SGS) and Sequential Indicator Simulation (SIS) (Deutsch and Journel, 1997). SGS assumes that all values (both low and high values) are well represented by a single variogram. It is termed 'sequential' because the stochastic simulation procedure first assigns the observed values to the nearest interpolation grid nodes, and then determines the value at a randomly chosen grid node on the basis of the variogram and the grid values that have previously been assigned. SIS is similar, but different variograms are used for low and high values, based on the indicator approach (see e.g. Loague and Kyriakidis, 1997; Western et al., 1998b). There are also unconditional simulations which satisfy the variogram structure but do not match the observations. One example of an unconditional simulation technique is the Turning Bands Method (Mantoglou and Wilson, 1981). It is based on one-dimensional stochastic simulations along lines (or bands) in different directions which are then projected onto the two-dimensional grid. There exists a wide spectrum of both conditional and unconditional stochastic simulation techniques based on various assumptions on the type of variability to be represented. A detailed review with applications to hydrology is given in Koltermann and Gorelick (1996).

(d) Disaggregation and Aggregation

While the approaches discussed above are methods for estimating patterns from points, disaggregation methods estimate patterns from spatial average values. For example, if we know an estimate of catchment average soil moisture from water balance calculations, one may be interested in estimating the spatial pattern of soil moisture from this average. This is done by using disaggregation methods. While interpolation involves a change of scale in terms of the spacing, disaggregation involves a change of scale in terms of support, i.e. the support is decreased. The opposite transformation of disaggregation is aggregation (i.e. a number of point values in space are combined to form one average value) which corresponds to an increase in support scale. Aggregation reduces the spatial variance and this reduction can be estimated from the variogram by regularisation methods referred to earlier (e.g. Journel and Huijbregts, 1978; Vanmarcke, 1983). Conversely, disaggregation increases the spatial variance. The spatial pattern of say rainfall in a region will always have a larger variance than the catchment average rainfalls in the same region (Sivapalan and Blöschl, 1998). In order to disaggregate average values into spatially variable values, additional information is needed for which assumptions must be made. Process understanding and auxiliary data can also be used in disaggregation approaches. Disaggregation methods based on auxiliary data are very similar to interpolation methods and can involve, for example, relationships between soil moisture and the topographic wetness index, or between snow depth and terrain elevation. In these examples, a catchment average soil moisture (or snow depth) would be spatially disaggregated based on the spatial pattern of wetness index (or terrain elevation). Stochastic disaggregation methods are very similar to stochastic simulation methods but are conditioned on spatially averaged rather than on point values as discussed above, i.e. the patterns generated reproduce both the variogram and the spatial averages exactly. An example for the rainfall case is given in Chapter 4. For many cases, one can assume that the aggregated value is simply the arithmetic average of the individual values, in which case the variable is said to average linearly. In catchment hydrology there are many processes that do average linearly; in particular, those for which a conservation law (of mass or energy) holds. Examples include rainfall or snow water equivalent. However, other variables and, in particular, model parameters, do not average linearly (i.e. the aggregated average value is a more complicated function of the individual values). For example, if we aggregate snow albedo in a physically realistic manner, the aggregated value will not simply be the arithmetic average of the individual point values. In a similar vein, model parameters such as hydraulic conductivity do not average linearly, and neither do landscape surface parameters, used for estimating evapotranspiration (e.g. see Chapter 5). These parameters can therefore not be simply (linearly) disaggregated, but need more complicated procedures (see e.g. Wen and Gómez-Hernández, 1996; Michaud and Shuttleworth, 1997; Becker et al., 1999).

This question of linearity/nonlinearity in aggregation and disaggregation is central to the use in modelling of "effective" parameter values. An effective parameter refers to a single parameter value assigned to all points within a domain, such that the model based on the uniform parameter field will yield the same output as the model based on the heterogeneous parameter field (Blöschl and Sivapalan, 1995). This is an important issue in modelling spatial patterns in catchment hydrology and is revisited in Chapter 3.

In the following, we will discuss specific problems with the spatial interpolation of a number of variables that are important in catchment hydrology, and discuss types of terrain information and other auxiliary data that can be used to improve the spatial interpolation of these variables.

2.3.3 DEMs, Terrain Indices, and Other Surrogates

The most commonly used data in measuring, analysing and modelling spatial processes in catchment hydrology are probably Digital Elevation Models (DEMs). Digital Elevation Models are spatial fields of terrain elevation values that are usually arranged in a regular square grid or in other arrangements such as a Triangulated Irregular Network (TIN) (see Chapter 3). DEMs are the basis of catchment representations in most distributed dynamic models of catchment processes and they can be used for calculating terrain indices that may assist in the spatial interpolation of hydrological variables. It is important to realise that DEMs are always obtained by interpolation, and interpolation artefacts may affect the dynamic models and terrain indices in which they are used. There are a number of ways in which DEMs can be derived, including digitising contour lines from topographic maps, ground surveys using theodolites or levels, and stereo interpretation of pairs of aerial photographs, all of which will contain measurement errors in position and elevation. Common to all of these methods is that the elevation readings are point values (i.e. the spatial support is small) usually at irregular locations. In order to make them useful for applications they are almost always interpolated to a regular grid or to other DEM structures. This interpolation involves filtering which effectively increases the support of the spot height, i.e. each (interpolated) value in a DEM is then no longer representative of a single point, but of an area around it which may be on the order of the grid size (due to the interpolation). This may look like a theoretical issue of little practical relevance, but on closer inspection the support has very practical implications when it comes to any sort of further interpolation or modelling of dynamic catchment response. Here it becomes important to know just what the DEM represents. For example, does the pixel value represent the average, the lowest point within the pixel or some other measure of the variability within the pixel? (see e.g. Rieger, 1998). This affects how well features such as lines of steepest descent (i.e. flow paths) are described, which are important when calculating the upslope contributing area for a grid element as flow accumulation algorithms are very sensitive to the way in which the terrain surface is conceptualised (Costa-Cabral and Burges, 1994). This is a particular problem when the resolution of the DEM is coarse relative to the scale of variability of the real terrain such as in heavily incised landscapes. So in addition to the positional and elevation errors of the DEM (which could be random or systematic), there will be artefacts introduced by the manner in which the DEM was interpolated from the raw data. There is a range of software for deriving DEMs that takes into account most of these problems. These packages can be based on spline interpolation (e.g. ANUDEM, Hutchinson, 1989) or kriging interpolation (e.g. SCOP, Molnar, 1992). The quality of DEMs, for hydrological applications, can be improved by making use of the stream network (e.g. Hutchinson, 1989) and there are a range of algorithms to remove pits (artificial depressions) in DEMs.

Once a Digital Elevation Model has been established for a catchment, it can be used for deriving terrain indices. Terrain indices are variables that usually combine a number of terrain attributes (such as local slope) in a way that represents the most important spatial features of a hydrological process (Moore et al., 1991). Terrain indices have been suggested for numerous processes and there is some debate in the literature on how accurate terrain indices can be (e.g. Western et al., 1999a). While it is always useful to invoke process interpretations for giving guidance on selecting a particular index, the main reason for the popularity of terrain indices stems from the general availability of Digital Elevation Models and the ease with which terrain indices can be derived. In some instances the relevance of a particular surrogate is not obvious and may depend on the timescale considered and/or may change with time. For example, elevation is often used as the main surrogate for spatially interpolating rainfall, the rationale being that orographic barriers tend to enhance rainfall. From a physical perspective one might expect terrain slope to be the more significant parameter, but it is true that elevation is often very well correlated with *mean annual* rainfall. However, this correlation drastically decreases to next to zero as one moves down in timescale to *daily* rainfall or *hourly* rainfall, as in most climates the increase in mean annual rainfall with elevation is mainly due to more frequent rainfall events rather than higher rainfall intensities (e.g. Obled, 1990). This implies that for estimating spatial patterns of hourly rainfall, elevation will be a very poor surrogate.

Another example of the use of terrain indices is runoff generation, for which a widely used surrogate is the $\ln(a/\tan\beta)$ wetness index of Beven and Kirkby (1979) where a is the specific upslope contributing area (i.e. the area above a segment of a terrain contour divided by its length) and β is the local slope of the terrain. Both quantities can be derived from a DEM (e.g. Costa-Cabral and Burges, 1994; Rieger, 1998). The assumptions underlying this index are, among others, that the dominating runoff-generating mechanism is saturation excess, and the surface slope is an accurate measure of the gradient-driving subsurface lateral flow. However, in many climates the dominating mechanism (Dunne, 1978) not only depends on soil type and depth but also on rainfall intensity and duration which will clearly vary seasonally and from event to event, as well as spatially. The $\ln(a/\tan\beta)$ wetness index will therefore only be a useful surrogate for those situations where the underlying assumption of saturation excess is valid. If other runoff generation mechanisms (such as infiltration excess overland flow) prevail, other surrogates should be used to estimate spatial patterns of runoff generation.

There are a large number of other surrogate or qualitative measures used in catchment hydrology that do not use terrain. These include patterns of the vegetation type for inferring moisture availability patterns or recharge patterns (see Chapter 12) and geophysical information (e.g. ground-penetrating radar) for inferring patterns of subsurface flow (e.g. Copty et al., 1993). Another application for the use of auxiliary data is patterns of soil hydraulic properties which are needed for any spatial hydrological modelling. A commonly used approach to deriving patterns is based on relationships between soil type (often defined by % sand, silt, clay, organic matter, and perhaps bulk density) and soil hydraulic properties (e.g. saturated conductivity, porosity, soil water release characteristics). These are called "pedo-transfer functions" (e.g. Rawls et al., 1983; Puckett et al., 1985; Romano and Santini, 1997). A map of soil type provides the spatial patterns for estimates of soil hydraulic properties, resulting in "patchy" maps like those from Thiessen polygons. The rationale behind the use of pedo-transfer functions is that the grain size distribution (defined by the soil type) should also be relevant to the pore size distribution (which in turn is related to soil hydraulic properties). Unfortunately, this is not often the case because peds and cracks, rather than the grain size distribution, tend to dominate the hydraulic properties. It is therefore not uncommon for soil properties to vary as much between soil types as within a soil type (e.g. Chapter 10; Warrick et al., 1990) and for other influences such as terrain to be important to soil hydraulic properties (Gessler et al., 1995). Ignoring these considerations has a direct impact on the subsequent modelling. For example, simulations of soil moisture may have sharp boundaries at the interface between different soil types with different porosities (Chapters 6, 9, 10) and infiltration excess runoff will never occur if a single value of hydraulic conductivity is used that is greater than the input rainfall rates (see Chapter 10 where more realistic methods that incorporate differences both within and between soil types are discussed).

To illustrate some of the issues of interpolation related to sampling and the use of terrain indices as auxiliary data, we will use some very high-resolution soil moisture data. This data was collected during a field experiment during October 25–26, 1996 on a 102×68 m plot in the Tarrawarra catchment (see Chapter 9). The sampling grid was 2×2 m which gave a total of 1734 measurement points. TDR probes were inserted vertically to 30 cm depth at each location. The measurement error was estimated as $3\%(V/V)^2$. The soil moisture data are shown in Figure 2.7a. The plot includes a terrain convergence (centre left) where the soil was wettest but additional controls such as evaporation and soil properties have produced a soil moisture pattern with complex features.

The first example illustrates the effect of noise due both to measurement error and to small-scale variability and how it can be reduced by interpolation schemes. Figure 2.7b shows a pattern where the data have been filtered using kriging with a variogram nugget of $3\%(V/V)^2$ to represent the measurement error. This is simply an application of ordinary kriging where values are estimated at the same locations as the samples rather than in between as with kriging used for interpolation. This filtering can be thought of as an aggregation of neighbouring samples (see Section 2.2.2) to remove the measurement error. We would have obtained a similar pattern had we sampled multiple times at each grid location (to reduce measurement error), but this was not possible for logistical reasons. A comparison of patterns of original data and with the error removed also indicates that in this case the noise problem was not significant. To show the effect of a larger measurement error, a random error (variance equal to

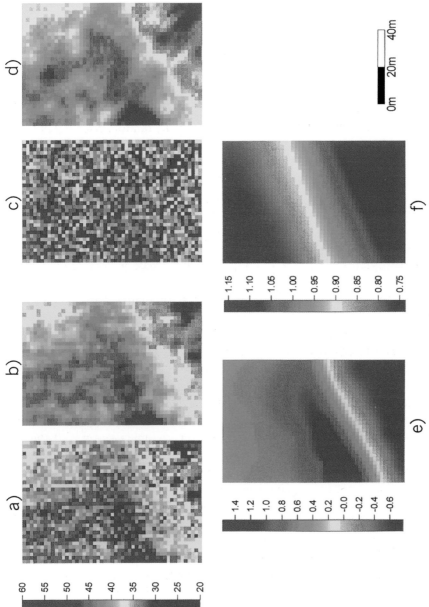

Figure 2.7. (a) Measured soil moisture data on a 2×2 m grid (Western et al., 1998a); (b) data from (a) filtered using ordinary kriging with a nugget of 3 $(\%V/V)^2$ to smooth out the measurement error; (c) measured soil moisture data with a random error of variance $= 40$ $(\%V/V)^2$ added; (d) data from (c) filtered using ordinary kriging with a nugget of 43 $(\%V/V)^2$ to smooth out the "contamination" and measurement error; (e) pattern of a radiation weighted wetness index for the measurement domain; (f) pattern of topographic aspect index for the measurement domain.

$40\,\%(V/V)^2$ was added to the original data (Figure 2.7c). This reduced the "signal to noise ratio". It is clear that the ability to resolve the pattern visually significantly decreases. This pattern was also filtered using kriging and a nugget of $43\,\%(V/V)^2$ (i.e. the estimated measurement error plus noise) which gave the pattern in Figure 2.7d. It is noteworthy that even for the large noise case, filtering can very efficiently remove the noise, as the pattern of the filtered image is remarkably similar to the pattern of the original data with the measurement error removed (Figure 2.7b). However, this is only possible because a truly random error ("white noise") has been added. If the error shows consistent spatial patterns, high-noise cases will be greatly in error and filtering will not improve the pattern.

Figure 2.8 shows the effects of different types of interpolation. For the scenarios, nine sampling points at a 30×40 m spacing were assumed to be known from the pattern and the other values were assumed to be unknown. 30×40 m might be a more typical spacing for soil moisture measurements in small research catchments. The patterns shown in Figure 2.8 are based on the nine samples from the original data (Figure 2.7a) and nine samples from the data with added noise (Figure 2.7c).

Consider first the samples from the original data. In Figure 2.8a, we have assumed that we know nothing extra about the data and applied the Thiessen (nearest-neighbour) method to produce 2×2 m interpolated patterns from the 30×40 m data. The interpolated pattern consists of rectangles because of the regular location of the samples. Although the main feature (higher soil moisture in the centre of the plot) is retained, one would clearly not consider this pattern to be a good representation of the true pattern (Figure 2.7a,b). A more judicious choice of interpolation method is ordinary kriging (Figure 2.8b) which produces much smoother and more "likely" patterns. However, the main features of the spatial arrangement remain unchanged, and the interpolated pattern is still significantly different from the true pattern. In Figure 2.8c, we have used auxiliary data consisting of a radiation weighted wetness index (Western et al., 1999a) shown in Figure 2.7e. External Drift kriging was used to interpolate the 30×40 m soil moisture samples, using the 2×2 m radiation weighted wetness index, onto a 2×2 m grid (Figure 2.8c). This pattern is quite similar to the real pattern with the wet band in the gully being obvious. Next we repeated the External Drift kriging but this time using a topographic aspect index (Figure 2.7f) (Western et al., 1999a) as the auxiliary data. The interpolated pattern (Figure 2.8d) does not improve over the case without auxiliary data (Figure 2.8b). Clearly, selection of the "right" auxiliary variable is important for improving the interpolated pattern, and when an auxiliary variable that does not represent the main features is used (in this case aspect index) the interpolated pattern will remain poor or can even deteriorate as compared with not using auxiliary data. The linear features in Figure 2.8d are an artefact of the method which uses a finite local neighbourhood (search radius) for obtaining a relationship between soil moisture and the auxiliary data. When this search radius is increased to a large value, greater than the plot size (Figure 2.8e), the artefacts vanish, but the interpolated pattern is no more accurate than the pattern without auxiliary data (Figure 2.8b).

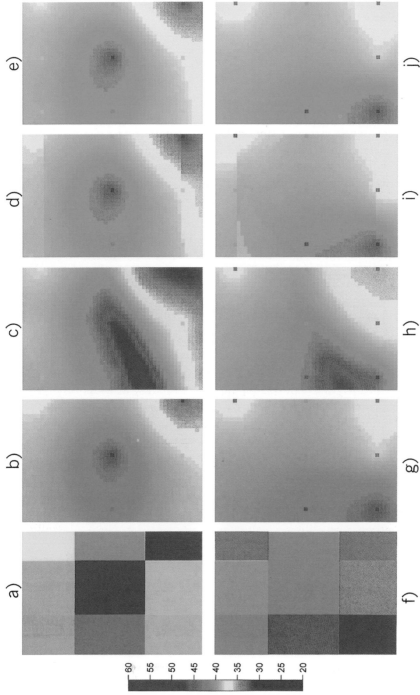

Figure 2.8. Interpolation methods applied to nine sample points on a 30 × 40 m spacing. (a) to (e) are based on nine samples from Figure 2.7(a); (a) Thiessen method; (b) ordinary kriging; (c) External Drift kriging with the radiation weighted wetness index; (d) External Drift kriging with topographic aspect index; (e) as for (d) but with a larger search radius; (f) to (j) are based on nine samples from Figure 2.7(c); (f) Thiessen method; (g) ordinary kriging; (h) External Drift kriging with the radiation weighted wetness index; (i) External Drift kriging with topographic aspect index; (j) as for (i) but with a larger search radius.

For the non-contaminated data we can resolve the broad patterns of soil moisture using interpolation of the widely spaced data if we choose an appropriate auxiliary variable (in this case a radiation weighted wetness index), but this ability is reduced with the contaminated data. We repeated the preceding scenarios using the "noisy" data of Figure 2.7c. The Thiessen method produced a poor representation (Figure 2.8f). Using ordinary kriging (2.8g) was also poor. Using the radiation-wetness index as an auxiliary variable in External Drift kriging only slightly improves the interpolated pattern (Figure 2.8h). Use of topographic aspect index as the auxiliary variable has no positive effect (Figure 2.8i,j) and the artefacts are worse than in the non-contaminated case. Hence, if the data are poor (low signal to noise ratio) using auxiliary data cannot significantly improve the interpolation. This is because the External Drift kriging procedure derives the relationship between auxiliary data and the variable of interest from the two data sets (i.e., the data sample and the auxiliary data) and with a lot of noise, the relationship is poor, hence the interpolated pattern is not very good.

Whatever interpolation technique and/or auxiliary information is used, the important point to realise is that by choosing a particular technique for developing a spatial pattern, we have implicitly made some assumptions about the spatial structure of the variable. These assumptions will carry through all subsequent simulations. If the structure is wrong, and the variable is important, the modelling exercise will be severely constrained from the outset. On the other hand, a prudent choice can significantly improve the results of a spatial modelling exercise. The decreasing cost of computer power has enabled the more widespread use of sophisticated interpolation methods. GISs have in-built analysis tools allowing a range of interpolation methods to be used with a minimum of effort. But *all* of these are based on some assumption about the distribution of the parameter in question. Simple approaches are inherently no more or less value-free than the complex approaches. The important question is whether the assumed spatial pattern that underlies the interpolation/extrapolation method best represents the nature of the phenomena controlling the pattern. This can be assessed by using as much process understanding as possible.

2.4 GUIDANCE ON SAMPLING AND INTERPOLATION IN PRACTICE

This chapter has been presented in two sections, Sampling and Interpolation. Our desire to measure patterns means that these two issues are intimately linked – we want to use measurements (usually at points) to derive patterns (usually via some sort of interpolation). Both of these issues depend on the depth of understanding we have about the underlying processes of which we are making measurements (note that strictly we do not measure the process itself but some feature of it that lends itself to measurement). This is the "chicken and egg" dilemma referred to earlier. We cannot define an "ideal" sampling scheme or choose the best interpolation method without knowing about the variability of the feature being measured, but we probably do not know this variability without measuring it! We are forced to draw on understanding of the processes involved and be prepared to modify sampling or interpolation methods as more information becomes available.

The next section presents a list of basic questions which help to gather the information needed to determine a sampling programme or choose an ideal interpolation method. We then work through an example, applying this list to a real sampling problem to illustrate the procedure and highlight the compromises that are required in practice.

A)   *What are the processes we are trying to capture with the measurement programme? What is the variability in time and space of the feature of the process that we will be measuring?*

- Which variables should we measure and how representative are they of the process?
- What is the typical length scale of the feature of interest?
- How quickly does the feature change and are there particularly important timescales (e.g. diurnal, seasonal etc.)?
- What are the minimum and maximum values that are expected to be measured?
- Do we have *predictive methods* for defining the variation of the feature and how accurate are these?

B)   *For sampling, the next step is to define in more detail the specific requirements of the exercise.*

- What measurement device (or devices) should be used?
- What is the accuracy of the measurement device?
- What is the sampling support (time and space) of the device?
- Over what extent (time and space) do we want to make measurements?
- What are the practical constraints related to time, cost and the logistics of the measurements to be made? (these will indicate the possible number of samples and so the spacing of measurements).
- Are there alternative variables to be measured that perhaps are less representative of the process but can be more easily collected?

C)   *We then need to try to match the needs of the sampling exercise with the variability in the feature being measured and the characteristics of the measurement device. In this step we need to recall that:*

- if the spacing is too big compared to the feature of interest, we will not characterise small-scale variability (it will become "noise");
- if the extent is too small, we will miss out on the large-scale pattern and so measure a trend;
- if the support is too large, small-scale variability is smoothed out.
- if the sampling error is large compared to the variance of the feature, we will not detect the pattern.

Compromises will always be needed and, because of lack of knowledge, there will be some guesswork (see example below).

We can then collect the data.

D) *For interpolation, the next step is to look at the data set more closely and ask how well did the sampling programme capture the underlying variability?*

- If the data very well define the patterns, we can use one of the simpler methods described above. The choice of method is unlikely to be too important in this case. Interpolation could be used to take account of the measurement error by smoothing the pattern.
- If not, do we have auxiliary data or understanding available that could be used to "add information" using one of the more sophisticated interpolation techniques?
- If not, we can still interpolate the data but we will have little idea of how well the resulting pattern represents reality. Regrettably this is often the case!

The following example relates to the sampling exercise that produced the soil moisture data used in Chapter 9 (see also Western and Grayson, 1998**).**

A) *We were interested in runoff processes, the water balance and the seasonal changes to patterns of soil moisture in the root zone (say 300 mm) of a small (10.5 ha) pasture covered catchment in a humid climate. We did not know the detail of variability in time and space but we expected that:*

- We should measure volumetric soil moisture content which would very well define spatial soil moisture patterns and should allow important insights into runoff processes in the catchment.
- Variability would be due to rainfall, topographic position, time of year and soil characteristics. Soils were quite uniform but initial tests indicated that there was short-scale variability of moisture content of the order of 1–2 %V/V. Slope lengths were of the order of 50–100 m and rainfall occurred throughout the year.
- Soil moisture would change immediately in response to rainfall but slowly in response to evaporation and drainage, e.g. an evaporation rate of 2 mm/d would change the water content of the top 300 mm by only 1.5 %V/V per day. Seasonal changes would be large, due to big differences in evaporation between summer and winter. We expected that topography would be important to the lateral redistribution of soil moisture.
- Soil moisture would vary between permanent wilting point and saturation (approximately 10 %–50 %V/V).
- We did have predictive methods for defining variability but we did not know their accuracy – we wanted the data to help test and develop these methods.

B)   *We identified the specific requirements of the sampling exercise by initial field and laboratory tests and by using the manufacturer's specifications:*

- We chose Time Domain Reflectometry equipment for sampling soil moisture. We were intending to use 300 mm probes to be inserted vertically into the soil.
- The accuracy of the TDR was tested by taking multiple field measurements and comparing them to gravimetric samples. The combined error of the measurement and small-scale soils variability was estimated to be 4–5 $(\%V/V)^2$. With a small-scale variability of 1–2 $(\%V/V)^2$, the measurement error is about 3 $(\%V/V)^2$.
- The support of the TDR is a volume defined approximately by a cylinder of length 300 mm and diameter 100 mm.
- The overall extent we wanted to cover was the whole catchment of 10 ha. We were planning to sample for at least one year to get a complete seasonal cycle.
- We wanted to be able to sample the whole area in a day to minimise errors due to evaporation or drainage between the start and the end of sampling. We could reliably sample at the rate of about 60 measurements per hour.
- In this case, soil moisture was the key variable as we were interested in the catchment water balance and it could be relatively easily measured with the equipment we had available. Often the choice of variables to be measured is not so simple, as the measurement methods can vary greatly in terms of resources needed. An example is hydraulic conductivity where methods range from undertaking soil particle size analysis combined with pedo-transfer functions for estimating hydraulic properties, through to using field infiltrometers, to taking soil cores for analysis in a laboratory, with each method having large differences in speed of application, measurement accuracy and measurement support.

C) We tried to match the needs of the sampling exercise with the hydrologic variability as follows: Given the sampling rate and desire to complete sampling in a day, it appeared that around 500–600 measurements could be taken. We wanted to return to the same places each time, so a grid sampling seemed appropriate. Because of the shape of the catchment, the direction of most of the hillslopes and our interest in topographic effects, it made sense to sample on a regular, rectangular grid. A simple calculation leads to a grid spacing of 10×20 m. This gave us around ten measurements on every hillslope and a few measurements in the gully, so this should capture topographic effects. The small support compared to the spacing meant that we had to accept a reasonable degree of noise. With an expected minimum soil moisture of 20 %V/V and a maximum of 40 %V/V for one survey we guessed that the spatial variance would be on the order of $10-20\,(\%V/V)^2$ which is still large as compared with the measurement error variance of 3 $(\%V/V)^2$ and implies a signal-to-noise ratio on the order of 5. We could have taken multiple measurements at each location (and subsequently a coarser grid) to reduce this noise. For example, five samples at each spot would have reduced the error to one-fifth, i.e. $0.6\,(\%\,V/V)^2$ which gives a much better signal-to-noise ratio of about 25 but only 100 locations sampled at a spacing of about $25 \times 50\,\mathrm{m}$. At this spacing only very few samples would have been located in the gully area. This was a classic compromise between accuracy and pattern detail. The noise problem was significant in the summer when the overall range of soil moisture was low and of a similar order to the noise. In these conditions we had to accept that we would not be able to resolve patterns well. Nevertheless, we judged, based on the generally expected variability and range of soil moisture over the whole sampling period, that it was better to keep the spacing small than improve the accuracy at the cost of pattern detail.

As for temporal sampling, we were interested in seasonal changes, could afford 10–12 sample runs per year (so could resolve seasonal effects), and needed to try to minimise the effects of particular rainfall events. We therefore did not sample during significant rain. At times when the catchment wetness was rapidly changing (spring and autumn) we sampled more regularly than during the summer and winter when overall changes were slower. We could not totally remove the effects of particular events, but had detailed meteorological and runoff measurements that were used to interpret how representative the measured patterns were.

D) As for interpolation of the data sets collected over the whole catchment, we decided that we would not interpolate but rather assume that for visual presentation of the data, the measurement support was actually $10 \times 20\,\mathrm{m}$ (see Figures in Chapter 9). We did undertake interpolation for one piece of analysis and this is explained in Chapter 9 – we undertook a two-step process where the data were regressed against a combination of terrain parameters and the residuals were smoothed on the basis of estimates of measurement error, then added back to the regression values. Its effect was to produce a pattern smoothed on the basis of measurement error which could then be directly compared to the model, where this error was not represented. In other cases we interpolated the modelled soil moisture to match the measurement locations because the variability in modelled output was smoothly varying and therefore less affected by the interpolation method.

We will finish this chapter by reiterating that no data transformation method for generating patterns from point measurements, no matter how sophisticated, can generate knowledge. The method must be chosen so as to maximise the use of available information (both in the data itself, and the users' knowledge of the processes operating). The ideal interpolation approach exploits both the data and an understanding of the processes that lead to patterns in the variable of interest, and so mimics the shape and spatial arrangement of the patterns expected from the process. A poor interpolation approach is one that simulates a spatial structure that is at odds with knowledge about the system. The numerical sophistication of the technique is no measure of its quality.

The important point is that there are interpolation schemes of different complexity (in the way they use information contained within the data, the extent to which they utilise additional information, and how faithfully they reproduce the data as opposed to the statistical characteristics of the data); by choosing any of them we make an assumption about the nature of the underlying pattern, and the quality of interpolation reflects the validity of that assumption. Similar considerations apply to sampling. The extent to which we are likely to capture the 'true' hydrological patterns by a sampling exercise depends on: the scale of the processes compared with the scales we are interested in; the accuracy of the data and quality of correlations between surrogates and the values of interest (accurate or more representative data requiring a smaller number of points); and the complexity of patterns/processes (rich patterns requiring a larger number of points, unless we have some process understanding that can define certain features of the pattern). While there are no hard and fast answers to the practical problems of sampling design – of how many points and where, when and how to make measurements – careful consideration of the issues raised in this chapter should assist in designing a sampling scheme that best meets the requirements and constraints of a particular study.

The emphasis in this chapter has been on using process understanding to guide sampling and interpolation as this is the key to successfully capturing patterns in catchment hydrology. This understanding can be further exploited (and challenged) when we move into dynamic catchment modelling.

3

Spatial Modelling of Catchment Dynamics

Rodger Grayson and Günter Blöschl

3.1 INTRODUCTION

In the previous chapter, we addressed the issues associated with data and sampling of physical phenomena as well as methods for the interpolation of spatial patterns. We believe that these spatial observations are vital to improving our understanding of catchment hydrological response. But measured patterns (or patterns interpolated from measurements) are of limited use without a framework within which they can be exploited. We need to use the patterns in a way that lets us test hypotheses. Mathematical modelling provides a powerful tool for this purpose. This modelling can take many forms, the most relevant to the topic of this book being spatially explicit models of catchment surface and subsurface response. These models are used for a range of purposes from tools for testing hypotheses about the behaviour of natural systems and putting data from research catchments into a consistent framework (the chapters in this book are examples), to practical applications such as erosion and transport modelling and simulation of the effect of land use change – in fact anything influenced by hydrological response.

This chapter is about the basic structure of, and approaches to process representation in, spatial models. We introduce the concepts of model calibration and testing, focussing on the way in which spatial patterns can be used to inform model development and reduce model uncertainty. Finally we look at approaches to the evaluation of spatial predictions from spatially explicit models.

3.2 SPATIAL MODELLING CONCEPTS

As scientists and engineers, our interest in the hydrological response of catchments ranges from basic understanding of processes to prediction under changed conditions. We use models for each purpose – to see how well we can simulate measured responses (i.e. to test how well our conceptual understanding and its manifestation via the model reproduces "reality"); and to predict what might happen in the future. These models combine our understanding of the natural system with observations of the system.

The observations we make are, and will always be, just peep-holes into the complexity of the natural system. Even if we were able to measure every aspect of the water and energy cycles that occur in a catchment, we would still have a picture only of what is happening *now* under *present* conditions and could not determine what would occur under changed conditions without some sort of extrapolation. In order to make sense of our "peep-hole" measurements, or to extrapolate beyond what we have actually observed, we need to make use of our basic understanding of physical systems and a convenient way to do that is with models.

Every model is a simplified representation of some part of reality. The *art* of the modeller is to determine, for the "part of reality" in which he or she is interested, what are the key processes that dominate the response at the scale of interest. These processes may be represented in detail, while others may either be ignored or represented simply. This step is usually called "conceptualisation" and represents the modeller's hypothesis about the way nature works in the context of the modelling problem. The resulting model is a mathematical expression of this hypothesis and is in a form that can be tested.

Figure 3.1 shows a photograph of a soil cross-section in which lateral subsurface flow is occurring, and a typical conceptualisation of this process. This conceptualisation can be turned into a mathematical model via the application of equations for flow through porous media, such as are described later in the chapter. The model does not show all the intricate detail of the real world. Indeed, the reality and the model are very different. The art of the modeller is to know just how these differences might affect the modelling results, and whether a different sort of model might be more appropriate.

In principle, *anything* can be modelled. The question is "how well does the model represent reality?", and we cannot answer that without knowing what is the "reality". In this regard, model building shares the "chicken and egg" problem referred to in Chapter 2; i.e. in order to build (or choose) an appropriate model, we need to understand a lot about the processes that are important, yet often these cannot be identified until we have tried some modelling. This situation is shown schematically in Figure 3.2 where we have also linked in sampling. In an ideal world, we begin with some process understanding, do some sampling to improve that understanding, then, when we have enough understanding to be able to attempt a conceptualisation, we build a model. Hopefully this model helps us understand the processes a little more and, perhaps with the help of some more sampling, we iteratively refine our modelling and understanding. In the "real world", this loop cannot go on forever and when it is broken, all of the remaining inconsistencies between reality and the model have to be accepted and somehow dealt with by the modeller. As with the sampling and interpolation issues described in Chapter 2, knowing when to "break the loop" and how to deal with the remaining inconsistencies between reality and the model can only be done effectively with a sound appreciation of the dominant physical processes. In

(a)

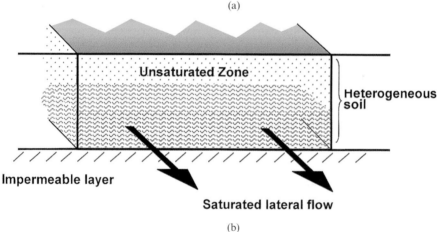

(b)

Figure 3.1. (a) Photograph of subsurface flow in the Löhnersbach catchment, Austria (courtesy of Robert Kirnbauer and Peter Haas) showing exfiltration of saturated flow over an impermeable layer; (b) a typical conceptualisation of the process shown in (a) used in distributed catchment models.

modelling catchment hydrological response, this implies that we need a sound understanding of all components of the hydrological cycle and have mathematical descriptions for each – a daunting task indeed. Models are usually built to focus on particular ranges of time and space scales and this helps us to make the task somewhat more tractable. For example, models designed to simulate storm runoff from particular rainfall events may safely ignore evaporation, or long-term models of water yield may be able to ignore the detailed dynamics of surface water flow.

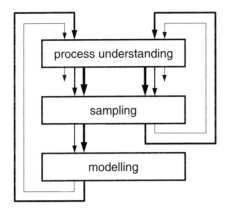

Figure 3.2. Schematic diagram of the iterative process of using understanding along with data to develop models which improve our understanding and so on.

Basic approaches to modelling

The modelling literature is replete with different ways of classifying models. Refsgaard (1996) presents an excellent description of model types and definitions relevant to modelling. Singh (1995) discusses classifications in terms of how processes are represented, the time and space scales that are used and what methods of solution to equations are used. Here we focus on three basic features, useful for distinguishing approaches to modelling in catchment hydrology – these are (i) the nature of the basic algorithms (empirical, conceptual or process-based), (ii) whether a stochastic or deterministic approach is taken to input or parameter specification, and (iii) whether the spatial representation is lumped or distributed. The first question is whether the model makes any attempt to conceptualise the basic processes or simply operates as a calibrated relationship between inputs and outputs. Empirical, regression or "black-box" models are based on input–output relationships without any attempt to describe the behaviour caused by individual processes. An example is $runoff = a \cdot (rainfall)^b$, where we derive a and b via a regression between measured rainfall and runoff. The next step up in complexity would be conceptual–empirical models wherein the basic processes such as interception, infiltration, evaporation, surface and subsurface runoff etc. are separated to some extent, but the algorithms that are used to describe the processes are essentially calibrated input–output relationships, formulated to mimic the functional behaviour of the process in question. The classical example is the STANFORD watershed model (Crawford and Linsley, 1966), and derivatives of this modelling genre are still in use all over the world. As the quest for deeper understanding of hydrological processes has progressed, models based as much as possible on the fundamental physics and governing equations of water flow over and through soil and vegetation have been developed. These are often called physically-based models and they are intended to minimise the need for calibration (i.e. model parameter optimisation) by using relationships in which the parameters are, in principle, measurable physical quantities. In practice, these parameters can be difficult to determine so these models are best thought of as complex conceptual models (Beven, 1989).

Another basic distinction between models is whether stochastic or deterministic representations and inputs are to be used. Most models are deterministic

– i.e. a single set of input values and parameters is used to generate a single set of output. The term 'stochastic' in the hydrological literature tends to be used synonymously with 'statistical', and implies some random component in the model. In stochastic models, some or all of the inputs and parameters are represented by statistical distributions, rather than single values. There is then a range of output sets, each derived from different combinations of the inputs and parameters and each of them associated with a certain probability of occurrence. It is also possible to have stochastic representations for some model components so that a given set of inputs can yield a range of output responses. The stochastic approaches are often used in model sensitivity and uncertainty analyses (see Section 3.4 and Chapter 11). Their advantage is that they provide a conceptually simple framework for representing heterogeneity when the explicit spatial or temporal detail is either not known (although at least the relevant statistical properties need to be known) or is not important (Jensen and Mantoglou, 1993).

Irrespective of these approaches to process conceptualisation, if we are concerned with spatial patterns in landscapes, our models must represent these processes in a spatially explicit manner. This is done by dividing the area to be modelled into elements, within which the processes are represented. The nature of these elements is discussed in the following section. The resulting models are termed *distributed models* to distinguish them from *lumped* models that are not spatially explicit – i.e. which treat catchments as a single unit and so average the effects of variability of processes in space. These notions of *lumped* or *distributed* do not indicate anything particular about the methods used for representing individual processes, but simply indicate the approach to spatial representation.

In the next section we discuss details of spatial model structure and in reading these details, it is useful to keep in mind the following ideas. Models of catchment hydrology must represent complex systems made up of interactions between many components, most of which vary in space and time. There is little point in representing one component in great detail while greatly simplifying another on which it depends. For example, soil erosion or water quality models may contain great detail in the representation of soil detachment or chemical decay of pollutants in transport, but if the basic runoff model that drives the hydrology is simplistic and inaccurate, the benefits of the model detail cannot be realised. Modellers need to balance the complexity of model components, recognising that the model accuracy will be limited by the weakest important component – be it process representation, spatial detail or difficulty in getting the data needed to determine parameters or to test the model. There will also be practical limitations imposed by things such as software availability and budgets. Although there is no single answer to "what is the right model?" this chapter seeks to give guidance on this issue. As shown in Figure 3.2, modelling is an iterative process and the point at which we "break the loop" and actually use a model in a practical application is dependent on factors that will be different for every modelling application. The following section is a discussion of key aspects of spatial model structure that will help a modeller decide when to "break the loop" and the consequences of doing so.

3.3 SPATIAL MODEL STRUCTURE

This section presents some generic features of spatial hydrological models and the pros and cons of the fundamental choices that a modeller makes when either choosing (or building) a model for a particular application. It is not intended to be a review of the myriad of models presented in the literature. Readers are directed to books such as Singh (1995) or Abbott and Refsgaard (1996) for collected summaries of models.

3.3.1 Spatial Discretisation

The fundamental building block of a spatial model is the model element. This is the minimum unit within which we can explicitly represent spatial heterogeneity and so defines the scale to which we must interpolate input data and represent the fundamental processes, as well as setting a minimum scale at which model testing can occur.

In hydrology, the significance of topography (the fact that water generally flows down hill!), and the relatively widespread availability of digital terrain data, have meant that the choice of model element size and type is often dictated by the way in which (and the scale at which) we represent topography. Topographic representation does not have to define the way model elements are structured but it commonly does so.

Terrain can be represented digitally in three basic ways (Moore et al., 1991) – gridded elevation data, contour data giving x, y coordinates of points of equal elevations, and irregularly spaced x, y, z data. These three forms of elevation information translate to the four basic constructions for models of spatial hydrological response (Figure 3.3).

By far the most common form of model construction is based on rectangular (usually square) elements (e.g. Abbott et al., 1986; Wigmosta et al., 1994; also Chapters 6, 7, 12 and 13). A small number of models use contour data to construct a mesh of elements bounded by adjacent contours (equipotentials) and orthogonal trajectories (streamlines). The best known examples are TOPOG (Vertessy et al., 1993; Chapter 10) and Thales (Grayson et al., 1995; Chapter 9). Triangulated Irregular Network (TIN) based hydrological models develop a flow mesh from the triangular facets derived by joining adjacent data points (e.g. Palacios-Vélez et al., 1998). There are also some models that are not based directly on digital elevation data but rather use a subjective discretisation of a catchment. A well known example is the KINEROS model of Smith et al. (1995) who define rectangular elements that preserve the key features of average sub-catchment slope, flow path length and area (Chapter 6, Figure 3.3).

Each approach has its advantages and disadvantages as summarised in Table 3.1. As noted in the Table, it is not a trivial exercise to route flow through gridded elements. A multitude of approaches exist in the literature from the simple D8 algorithm (O'Callaghan and Mark, 1984) which sends all the water to the downslope neighbouring element that has the greatest elevation

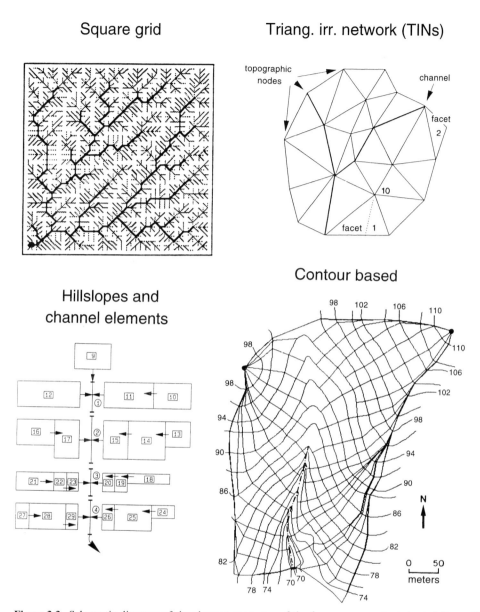

Figure 3.3. Schematic diagram of the element geometry of the four process-oriented rainfall runoff models. The contour based model (bottom right hand corner) shows the topography of the R5 catchment in Oklahoma (contour interval is 0.6 m).

drop, to multiple flow direction algorithms (e.g. Quinn et al., 1991) and more sophisticated approaches that approximate the flow tubes of contour based models (Costa-Cabral and Burges, 1994; Tarboton, 1997). While the more sophisticated approaches are more realistic than D8, the value of using them depends on the quality of the original DEM (see Chapter 2, p. 39). Poor quality DEMs are not magically improved by using more sophisticated terrain analysis algorithms. Similarly, models that are rudimentary with respect to the way they

Table 3.1. Approaches to explicit terrain representation

Approach	Advantages	Disadvantages
Gridded elements	DEMs often available as grids Computationally simple to set up dynamic models Many models available for use Simple to overlay other spatial information	Flow directions not straightforward to define (see text) Uniform density of points means inefficiently large number of elements if detail is to be maintained in key areas of the terrain
Contours and streamlines	More naturally suited to the routing of surface flow Able to assume 1-D flow in each element	Setting up of flow mesh requires specialised software – software must also be designed to avoid the inefficiency of large elements in gullies and small elements in divergent areas Does not allow flow to cross streamlines (see text) Few models are designed for this structure
TIN facets	Most efficient form of surface definition – least number of elements for most terrain detail	Flow routing is not trivial Data are not common (except direct from field survey) Few models are designed for this structure
Conceptual elements of hillslopes and stream segments	Based on the assumption that it is only those features preserved in the conceptual elements (e.g. average slope, flow path length, area) that are really important to model response Able to assume 1-D flow in each element Lead to a small number of elements - faster model running times	Discretisation generally done manually Uncertainty about the validity of the main assumption in some applications Few models are designed for this structure

use terrain information may not benefit from better terrain analyses. These more sophisticated methods, however, are more realistic and will not make such problems worse, so are to be recommended. Contour/streamline approaches have less ambiguity with the way in which terrain analyses are applied. The flow direction is defined by the "flow mesh" and one-dimensional flow is assumed in each flowtube. The problem is that some level of diffusion across the streamline is likely in nature but cannot be easily represented. Diffusion is due to surface roughness effects causing flow across the main slope and because the discrete nature of the elements causes significant differences in flow depths between adjacent elements, particularly in convergent areas. This is illustrated in Figure 3.4 which shows a contour based discretisation of a micro-catchment (Figure 3.4a) used by Moore and Grayson (1991), along with simulated and observed patterns of saturation (Figure 3.4b), and the depth of saturated flow simulated in a number of elements across the valley (Figure 3.4c). In reality, the differences in flow depth of the elements across the valley were smoothed out due to lateral gradients in flow depths. While this problem is well recognised in contour/streamline approaches, it has not been overcome in the existing models. TINs are very efficient methods for representing terrain because the density of points can be varied to suit the complexity of the surface, being dense where elevation is changing rapidly and sparse in flatter areas. The problem with TINs for distributed modelling is that flow paths are difficult to represent. Usually flow is represented as a network of channels defined by the edges of the facets being "fed" by the area of each facet. The subjective discretisation of a catchment into conceptual rectangular elements used in KINEROS is a simple approach that results in many fewer elements than is common in the previous methods. Goodrich (1990) showed that the simple representations are a sound basis for hydrological modelling and it is perhaps surprising that this method has not been more widely used.

The four methods of terrain discretisation described above are spatially explicit methods as they directly define the terrain as it appears in reality. *Distribution* modelling is a fundamentally different way of defining model elements. TOPMODEL (e.g. Beven and Kirkby, 1979; Chapter 11) is the best known distribution model but others exist such as RHESSys (Band et al., 1993) and Macaque (Watson et al., 1998). In this conceptual approach, it is assumed that particular parts of a hillslope will behave identically, from a hydrological point of view, if they have the same value of a carefully defined parameter – in the case of TOPMODEL, the wetness index of Beven and Kirkby (1979), $\ln(a/\tan\beta)$ where a is the upslope area and β is the slope. The index is computed across the hillslope (via analysis of a DEM for example) and the distribution of the index is then discretised into intervals. The hydrological response of each interval is simulated and then combined to give the response of the hillslope. Here the "model elements" are not contiguous parts of the landscape, but conceptual locations on a hillslope. In their original forms, these models were not intended to be spatially explicit. Indeed the whole idea of distribution models was to overcome the computational burden of spatial representation. Nevertheless, it is possible for the simulated response to be "mapped" back onto the landscape via the pattern of, say, wetness index values (Quinn and Beven, 1993; Quinn et al., 1995). This gives the impression of a spatially explicit model output, although the patterns are a direct function of the underlying index. While TOPMODEL is the best known model of this type, the distribution function approach can be used with any index and is, for example, widely applied in snow modelling, where terrain elevation is used as the index.

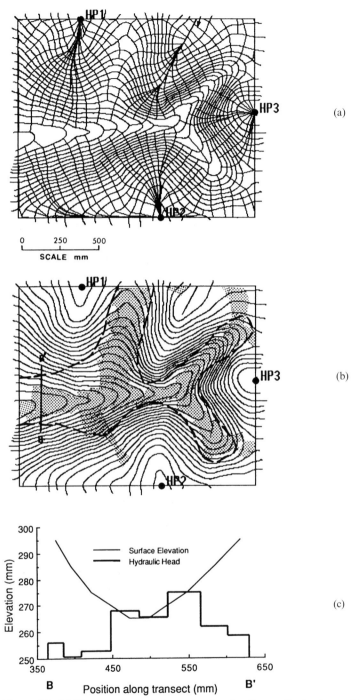

Figure 3.4. (a) Topography and element network of a sandbed micro-catchment simulated using Thales; (b) observed (lines) and simulated (stippled) saturated areas; (c) predicted flow depths and surface elevation of elements in section B–B′ marked in (b). (From Moore and Grayson, 1991; reproduced with permission.)

Another suite of models base their representation of spatial response on the concept of a "hydrological response unit" (HRU) (e.g. Leavesley and Stannard, 1995). It is assumed that a catchment can be subdivided into elements (HRUs) which are a particular combination of elevation, land cover, slope, aspect, soils and precipitation distribution. Each HRU is treated as a model element. These approaches arise from the use of lumped conceptual models and provide a mechanism for improving the representation of spatial heterogeneity in such models. The extent of spatial linking between HRUs depends on the time steps used in the models. For example, PRMS (Leavesley and Stannard, 1995) can be run in a daily mode (where outputs from each HRU are simply summed to give overall response) or a "storm mode" where HRUs are conceptualised as flow planes emptying into a channel network down which flow is routed. Models using the HRU structure provide spatial patterns of catchment response but the resulting patterns are heavily influenced by the original choice of HRUs and it is common to have relatively few HRUs compared to the number of elements in a terrain-based element discretisation.

In general, models based on elements defined by the terrain are more flexible in their representation of spatial patterns because they tend to have many more elements than HRU-style representations and are not constrained by the form of a distribution function.

3.3.2 Element Size

The element size (or model discretisation) in grid-based catchment models is often set to the resolution of the DEM that is available. As discussed in Chapter 2, DEMs are almost always interpolated from topographic maps or spot surveys and are therefore associated with a range of potential interpolation problems. These will carry through to the catchment model. It is common for "interpolation artefacts" such as flats and pits to occur and it is necessary to carefully assess the credibility of a derived DEM (Quinn and Anthony, in review) and remove these artefacts based on a subjective interpretation of the landscape and additional information (e.g. Rieger, 1998). Also, the choice of element size sets a limit to the level of terrain (and other) detail that can be explicitly represented in the model.

If terrain detail at a scale finer than the element size is important to the hydrological response of the system, this needs to be accounted for by the approaches discussed in Section 3.3.4. A common example of this problem is the presence of an incised stream or terrace on a flood plain. The details of the feature and the terrain immediately around it are likely to be very important, yet are unlikely to be properly represented. Furthermore, many of the characteristics derived from DEMs such as slope and wetness index vary with the scale of the DEM, and so parameters in the models that utilise these characteristics become scale dependent (e.g. Wolock and Price, 1994; Zhang and Montgomery, 1994; Band and Moore, 1995; Quinn et al., 1995; Gyasi-Agyei et al., 1995; Bruneau et al., 1995; Saulnier et al., 1997c). If we could define a particular length scale that represented the natural small-scale variability of various catchment properties, we could define an ideal scale for DEMs, but this is not possible so we must accept the fact of model dependence on DEM resolution. This has important implications for model calibration and validation, as model parameter values are likely to change with the size of the elements.

In 1988, Wood et al. introduced the notion of the Representative Elementary Area (REA), at least in part as a framework for thinking about what are appropriate element sizes for distributed models. They considered the REA to be an area beyond which explicit representation of spatial variability was needed but within which relatively simple, and spatially non-explicit approaches could be used. In other words, the idea was to select model elements that are sufficiently large to average out all the small-scale variability and the element size where this occurs was the REA. Wood et al. (1988) have been interpreted as implying that the REA may have a universal value. While the concept of a universal REA is enticing, it is clear that the ideal size for a model element will be entirely dependent on the processes being represented and the nature of the climate, terrain and vegetation where the model is being applied (Blöschl et al., 1995; Woods et al., 1995; Fan and Bras, 1995). Blöschl et al. (1995) illustrate the lack of a universal size for the REA, via simulations with different dominant sources and scales of variability in precipitation, soil and surface properties. The concept of the REA is, however, important and highlights a critical issue – that variability can be explicitly represented only at scales larger than the element size, while variability at the sub-element scale must be represented in a lumped way – the total variability being the sum of the explicit and the sub-element variability.

The choice of resolution in spatial models therefore determines what variability can be *explicitly* represented (i.e. representing differences from element to element) and what must be represented *implicitly* (i.e. within an element – see Section 3.3.4). This will depend on:

(i) the scale and nature of variability of the dominant processes,
(ii) the structure of the model itself (in terms of how explicit and implicit variability are represented),
(iii) what information is available to characterise the processes and variability,
(iv) the purpose for which the model is being developed, and
(v) in addition there may be computational considerations.

To satisfy (i) we need to match the model element size to the scale of variability of the dominant processes. Recall from Chapter 2 that scale incompatibilities can occur when the model spacing, extent or support does not match that of the process scale variability. Ideally we would choose an element size that captures the variability of the main processes we are wanting to model, but this is not always straightforward. In the time domain there are obvious considerations for the choice of model resolution. For example, a resolution of less than six hours is needed to properly resolve the diurnal energy cycle, but in the space domain often the choice is not so obvious as the "best" scale is likely to be either unknown, or will be quite different for equally important processes. In some cases, a process is so dominant that this choice is made easier. In Chapter 7, wind drift was such a dominant process controlling snow depth that so long as this was represented well, other processes could be ignored – the model discretisation only had to match the needs of the snow drift representation. In most practical applications, however, the choice of model element size will be a compromise and will depend on the considerations discussed below.

The way in which we conceptualise the processes in a model also affects what might be the "ideal" element size, and it is important to recognise that "finer" is not necessarily "better". Watson et al. (1998), for example, showed that, despite the obviously more realistic terrain representation of high-resolution DEM data, simulations of runoff using a distribution function model were no better than with a coarser scale DEM. This was because the basic structure of the model could not make use of the additional terrain information in the detailed data. This highlights the general point in (ii) that there will be interaction between the structure of a model and the ideal size for a model element. The choice of element size also depends on how much information is available on the variability that is known (or assumed) to be present. We may have a model that can have very small elements and, in principle, can represent great spatial detail, but if we do not have the data to define the spatial variability of the model parameters, there is unlikely to be value in using very small elements. Just as with the earlier discussion related to Figure 3.2 and the broader modelling endeavour, we could keep iterating for ever trying to define the "perfect" element size but ultimately it is a pragmatic decision, specific to a particular problem. In particular, when distributed catchment models are used for predictions in practical applications there may be a need for an output at a particular resolution. The challenge then is to deal with the inconsistencies between the model and reality imposed by that pragmatic choice. This is discussed further in Section 3.3.5.

It is likely that in the near future, GIS and computer technology will substantially relax constraints imposed by computer run times, although it is clear, as mentioned above, that improved computational resources will not make models any more accurate representations of reality. But interesting new possibilities are likely to arise, such as the use of nested models that have large elements in some areas and small elements where detail is needed (something already done in fluid dynamics). This may help overcome some of the problems identified here by enabling areas requiring detailed simulations to be separated from those less hydrologically active. The problem then will be to reconcile the way in which processes are represented and parameters are determined at the different scales.

3.3.3   Linking Elements – Explicit Representation of Spatial Variability

The approaches to spatial discretisation provide a way of representing surface topography in a computationally manageable form, but once we want to start routing flow, we must consider not only the (static) surface but also the dynamics of water flow on the surface and in the subsurface. While full 3-D hydrological models have been developed (e.g. Binley et al., 1989), their computational burden, numerical instability and the fact that the level of information needed to run them is rarely available, mean that they are not suitable for most modelling applications. The vast majority of spatial hydrological models can be conceptualised to have two primary flux directions, vertical and lateral, each of which can be represented with varying degrees of complexity. The vertical fluxes are precipitation, evapotranspiration and infiltration into the soil, while the lateral fluxes are the residual of the vertical fluxes and are generally separated into surface and subsurface flow. A fundamental choice for the modeller is the extent to which these fluxes are explicitly routed. The modeller must balance the complexity of representation of each of these fluxes with the purpose to which the model will be put. At one extreme is the case of land surface schemes in General Circulation Models (GCMs) that generally contain a great deal of detail in the controls on vertical fluxes, yet provide little or no lateral linking. This is because their primary purpose is to "feed" the atmosphere with the appropriate amounts of energy and water with little interest in lateral components. In contrast, catchment hydrological response is all about the lateral components.

Laterally, both surface and subsurface flow must be routed in some way, but the timescale of these processes is very different. Surface flow velocities will be of the order of 0.1 to 1 m s^{-1} while subsurface flow velocities (unless soil pipes or large macro pores are present) are likely to be 1×10^{-5} to 1×10^{-8} m s^{-1} (or even slower). Obviously the choice of routing method will depend on the timestep and overall timescale of the modelling.

For models designed to simulate storm event response, the timestep is small (of the order of minutes) and it is necessary to dynamically route surface flow (e.g. Chapter 10 and Chapter 6). Both overland flow and channel flow are often represented by the continuity equation (3.1); the kinematic approximation to the momentum equation (3.2); and Manning's equation (3.3) for the roughness term:

$$\frac{\partial Q(A)}{\partial s} + \frac{\partial A}{\partial t} = q(t) \tag{3.1}$$

$$S_0 = S_f \tag{3.2}$$

$$Q = S_f^{1/2} \cdot n^{-1} \cdot R^{2/3} \cdot A \tag{3.3}$$

where Q is the discharge, A is the cross-sectional area of flow, s is the distance in downslope direction, t is time, q is the lateral inflow (or outflow) rate per unit length, S_0 is the surface slope, S_f is the friction slope, n is Manning's roughness and R is the hydraulic radius. These are standard equations for which many solution schemes have been presented (e.g. Moore and Foster, 1990). The approach to spatial discretisation becomes important to the question of surface flow routing because modelled flow path lengths vary depending on the different types of terrain representation. The distributions of surface flow path lengths from a contour based, and grid based (using the D8 algorithm) terrain representation are shown schematically in Figure 3.5. The contour based representation results in shorter path lengths than the grid-based approach where, particularly with the D8 algorithm, zig-zag flow paths are defined. The differences in flow path lengths will directly translate into differences in the hydrograph shape. In practice, the flow resistance parameter n can be used to compensate for these differences in model structure – i.e. to get the same runoff response from each model, one would need lower n values in the grid-based terrain representation. This is a clear example of how the choice of model structure interacts with the parameter values.

In distributed models where the interest is on the seasonal water balance rather than on the shape of the runoff hydrograph and the detailed within-event processes, significantly larger time steps are used. If the model timestep is large relative to the expected flow times for surface runoff, simple mass accounting can be used without regard for dynamic routing. Daily models of small catchment runoff are an example where this approach may be perfectly adequate. This approach is used in Chapters 6, 9 and 11.

For subsurface flow, however, the response can be so slow that some sort of explicit routing is needed. It is usual to assume that lateral flow occurs only under saturated conditions – i.e. virtually all spatial hydrological models used at the catchment scale ignore unsaturated lateral flow. Saturated flow is commonly modelled by combining the continuity equation (3.1) with Darcy's law:

$$Q_s = A_s \cdot S_0 \cdot K_{\mathrm{sat}} \tag{3.4}$$

Where Q_s is the subsurface discharge, A_s is the cross-sectional area of subsurface flow, S_0 is the slope of the impermeable layer which is often set to the surface slope, and K_{sat} is the saturated hydraulic conductivity. Note that (3.4) also makes the kinematic assumption as the gradient of the head potential has been replaced by S_0. Some models (e.g. Thales, Chapter 9 and TOPOG, Chapter 10) route subsurface flow along with surface flow using the same flow mesh. They therefore assume an impermeable layer that is parallel to the surface topography and over which lateral flow occurs. Where there is interaction with local groundwater systems, a more appropriate modelling approach is to use the downward "outflow" from the vertical flux balance to feed a saturated groundwater model. This approach is used in MIKE SHE (Chapter 13) which also uses the same element

Figure 3.5. Schematic diagram of the distribution of flow path lengths for grid and contour based terrain representations of a catchment.

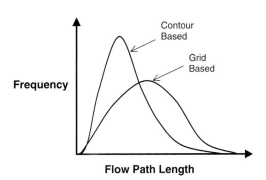

grid for the surface and subsurface components, although other combinations are possible (Chapter 12).

Spatial linking in *distribution* models is a different proposition again. In TOPMODEL and its variants (Chapter 11), it is usual to use large timesteps that allow for simple mass accumulation of surface flow, but the lateral linking of subsurface flow is more difficult to conceptualise. As discussed earlier, these models treat the "elements" as parts of the hillslope that have equivalent values of, for example, a wetness index. These "elements" are not contiguous in space so there can be no explicit routing of flow down a hillslope, as would be used in fully distributed models. Subsurface flow is instead simulated in two different ways – one for determining baseflow from a hillslope and the other for determining the change in soil moisture storage for any hillslope interval. For each timestep, the downward residual flux for each interval is added together to give a total input to a conceptual store. In TOPMODEL, baseflow is then simulated as a function of the store and the remaining water is redistributed across the hillslope according to the wetness index distribution. This redistribution can be conceptualised as an "infinite lateral conductivity" whereby any differential recharge (in different intervals or "elements") is immediately redistributed across the whole hillslope every timestep. This imposes a large amount of spatial structure on results from models such as TOPMODEL and may constrain their ability to represent spatial patterns of water tables and soil moisture because the shape of the spatial pattern is always directly linked to the wetness index pattern. Woods et al. (1997) and Barling et al. (1994) introduce a dynamic nature to the standard wetness index pattern but these ideas have yet to be incorporated into a distribution function based hydrological model. A step towards this idea has been made by Watson et al. (1998) who addressed the problem of representing lateral subsurface flow in distribution function models in a different way. They consider that, at the other extreme to the "infinite lateral conductivity" concept, there is a "zero lateral conductivity" analogy, whereby any differential recharge is maintained within the particular interval ("element") and not redistributed at all. Under this condition, the intervals can be thought of as a series of disconnected "buckets". Watson et al. (1998) introduced a "redistribution factor" which adjusted the level of redistribution between these two extremes. In testing on a catchment in south east Australia, it was found that the model worked best when the factor was very low, indicating that the model was behaving close to the "zero conductivity" extreme with very limited lateral redistribution.

The same general considerations that were discussed for the choice of model element size (p. 61) apply to issues of linking elements in distributed models. Firstly it is necessary to consider the processes in relation to the time and space scales of the modelling, e.g. surface runoff routing is important in storm models but not in long-term yield (water balance) models. This gives an indication of the appropriate model structure to use. In *distributed* models, spatial linking is conceptually straightforward while this is not so in *distribution* models, although some form of spatial linking can still be represented. The problem is generally with obtaining the information to characterise the variability and set the parameters, such as the flow resistance factor n in Manning's equation (3.3) or K_{sat} in Darcy's equation (3.4). Here the way the processes are represented within a grid element becomes important.

3.3.4 "Sub-Grid" Variability – Lumped Representation of Spatial Variability

In Section 3.3.2 we concluded that the choice of element size is ultimately a pragmatic decision and there will always be some processes and variability that are important *within* an element. How then can this "sub-grid" or sub-element variability be represented? Clearly, any representation will not be spatially explicit, or else we would just be defining a smaller element size. Rather, these approaches need to represent the effects of variability that the model cannot resolve.

Consider Figure 3.6, which illustrates some of the issues in conceptualising sub-element variability related to overland flow. In reality, flow patterns at the micro-topographic scale (much smaller than elements in a distributed model) will be highly complex (e.g. Emmett, 1970; Abrahams et al., 1989) which will cause a highly variable depth distribution across the hillslopes (Figure 3.6a). However, the way surface runoff is commonly conceptualised in an element is as a uniform depth of flow defined by solving equations (3.1–3.3) (Figure 3.6b). This means that the flow resistance factor in (3.3) must somehow be related to the sub-element scale microtopographic roughness. If the element contains a channel, a stream or a rill (which can be considered as a very small-scale channel), a further consideration is how to represent the "non-

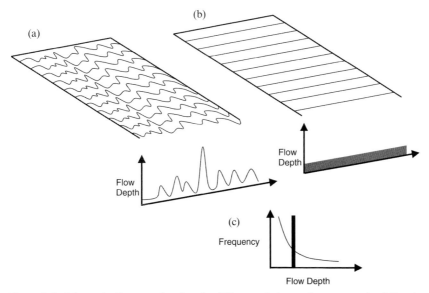

Figure 3.6. Schematic diagram showing the difference between the way overland flow is conceptualised in an element and how flow depths may be influenced by microtopography.

stream" parts of the element. These can be thought of as small catchments which "feed" the channel (Figure 3.7b) or they could be thought of as separate overland flow planes that produce shallow surface flow to be routed into the next element (Figure 3.7a). The most common approach is the former, but this is another example of the need to conceptualise the processes *within* an element.

Clearly there is a strong interaction between the way we conceptualise the process, the model structure we need to use, and therefore the parameters that we need. In practice there are a number of general approaches to representing sub-grid (or sub-element) variability of model parameters in distributed models of catchment hydrological response.

Assuming that a *point value* is valid for a whole element is the simplest approach – i.e. assuming the variance of the parameter *within* an element is zero. This assumption is commonly made for parameters such as saturated hydraulic conductivity (K_s) where a measurement from a field or laboratory test (with a small support) is used to represent an entire element. The assumption of zero variance is highly unlikely to be valid for parameters such as K_s because blocks of soil the size of a typical model element ($10s–100s\,m^2$) behave very differently to soil blocks of the size that are used in laboratories. This leads to the likelihood that the point-value approach will result in a poor model. When one attempts to improve the model through, for example, calibration, the calibrated parameter no longer represents the point value but rather some sort of average for the element – i.e. an "effective" parameter.

The *effective parameter* approach attempts to overcome the conceptual problems with point values. Effective parameters are single values that reproduce the bulk behaviour of a finite area or a finite volume. They therefore cannot be measured at a point and do not necessarily relate to point measurements at all. Figure 3.8 illustrates the notion of effective parameters for saturated flow through a porous medium where Θ_{eff} is a value of saturated hydraulic conductivity that, when used in equation (3.4), simulates a discharge that is equal to what would be expected from the heterogeneous (more realistic) block. In the overland flow example of Figure 3.6, an effective Manning n would be a single value that is able to reproduce, say, the overall hydrograph from the element (although clearly it would not be able to also reproduce the flow depth and velocity distributions). A great deal of work has been done on effective parameters, particularly for hydraulic conductivity in subsurface flow systems (e.g.

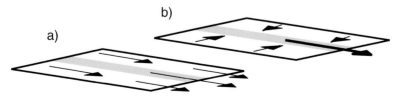

Figure 3.7. Different conceptualisations of flow within an element containing a channel (e.g. a stream on a floodplain or a rill on a hillslope).

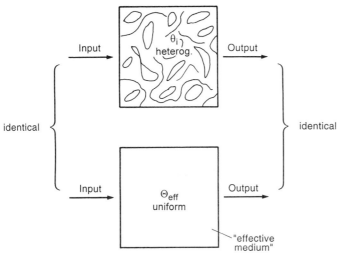

Figure 3.8. Schematic description of an effective parameter Θ_{eff} (such as hydraulic conductivity) for flow through a heterogeneous porous medium.

Gelhar, 1993; Wen and Gómez-Hernández, 1996), infiltration and surface flow parameters (Farajalla and Vieux, 1995; Willgoose and Kuczera, 1995) and soil hydraulic characteristics for setting parameters in land surface schemes (e.g. Kabat et al., 1997) – see review in Blöschl (1996). For some well-defined types of heterogeneity, equations can be derived to calculate effective values from statistics or descriptions of the heterogeneous field, but more generally, effective parameters are defined empirically via calibration. An example from surface flow is the work of Engman (1986) who derived effective values of Manning's n for a range of surface roughness types by analysing the outflow hydrographs from many runoff plots with different surface treatments. While the simplicity of effective parameters is attractive, there are problems with their use. For example, it is possible to reproduce the effects on lateral flow of macropores by using Darcy's equation (3.4) and a very high value of hydraulic conductivity. While the lateral flux values may then be correct, the distribution of flow velocities is not (because macropores will flow quickly compared to flow through the soil matrix) so simulations of, say, solute breakthrough curves will be incorrect. In a more extreme example, Grayson et al. (1992a) obtained acceptable runoff simulations using an effective value of infiltration to simulate infiltration excess surface runoff from a catchment where saturated source area runoff occurred (i.e. the physical process representation was wrong). It is also not guaranteed that effective parameter values can be found, i.e. it may be that no single effective parameter exists that produces the same response (discharge output in the example of Figure 3.8) as the heterogeneous (more realistic) block for the process and conditions considered. Binley et al. (1989), for example, were unable to define effective parameter values for soil hydraulic properties in a coupled surface–subsurface flow model. As modellers, we are often left with little choice but to use the effective parameter approach, but we must recognise that effective parameters may have a narrow range of application and an effective parameter value that "works" for one process may not be valid for another process.

Sub-element variability can also be represented by *distribution functions*. In this approach, the variability within an element is represented by a distribution of values and this distribution is used in the equations rather than a single value. It is common to represent the distribution as a number of classes (i.e. discretise the distribution) and apply the model equations to each class. The element response is then the sum of the responses of the classes. In some very simple models, it is possible to use continuous functions, e.g. Entekhabi and Eagleson's runoff model (1989) uses a distribution function for rainfall intensity and one for infiltration rate which are then convoluted to give runoff (those parts where rainfall intensity is greater than infiltration rate). Goodrich (1990, Chapter 6) uses the more common discrete approach and represents the sub-element variability of infiltration rates by conceptualising the planar elements of KINEROS as a series of parallel strips to which he applied saturated hydraulic conductivity values according to a log-normal distribution. Famiglietti and Wood (1994) used the distribution of Beven and Kirkby's wetness index to break up elements into intervals to which was applied a local model simulating evaporation. One can even think of models such as TOPMODEL being sub-element representations where the "element" is a whole hillslope and the distribution function of wetness index represents the sub-grid variability.

A third approach is to *parameterise* sub-grid variability directly. While this approach has a long history in other geosciences such as meteorology, it has so far rarely been used in hydrology. One example is the representation of rill overland flow within model elements by Moore and Burch (1986). The effect of rills in an element is represented by a lumped equation:

$$R = F \cdot A^m \tag{3.5}$$

where R is the hydraulic radius, A is the cross-sectional area of flow and F and m are parameters. These two parameters represent the detailed geometry of the rills in a lumped manner. It can be shown that m equals 0.5 for trapezoidal or parabolic geometries compared to 1 for sheet flow (Moore and Burch, 1986), and other studies have shown that in practice, m varies between these values (e.g. Parsons et al., 1990; Willgoose and Kuczera, 1995). The parameters F and m can be calculated from a graph of R versus A which can be obtained from a microtopographic survey of a small plot. These parameters can then be applied to the whole area where that type of surface occurs. This type of parameterisation does represent sub-grid processes in a realistic manner and it appears that there is opportunity for further use of parameterisation in subgrid representations of catchment models.

3.3.5 What Complexity Is Warranted?

As we have argued throughout this book so far, process understanding, data and modelling are linked in a potentially "infinite loop" where we get more complete representations and understanding in each iteration (Figure 3.2). At some point we have to "break out" of the loop so we can actually use a model for practical applications. When we do so, we have to accept the remaining inconsistencies between reality and the model. So when do we "break out"? What level of complexity do we need?

There are two fundamental approaches to these questions. The first is to begin with the simplest model that is able to reproduce the *measured* system behaviour and introduce added complexity *only* when it consistently improves the fit to observations and satisfies our understanding of the system response (i.e. implies hydrological behaviour that we know to occur). This approach (the "downward" approach) places a high priority on field data, parameter identifiability (the notion that a set of optimum parameters can be well established) and testability. It implies that we might ignore processes that we know to occur if their representation does not improve measured model performance. This is the only approach that strictly follows Popper's (1977) notion of the scientific method and the need for falsifiability, but in a practical application, it may result in a model that is too simple to address the problem of interest.

The second approach is to model all the processes thought to be important, and assume that because the conceptualisations of individual processes are "right", the overall model is "right". This is the "upward" approach. Something closer to the second approach has been common practice in distributed catchment modelling, but leads to a model that is probably too complex and cannot be properly tested. The large number of model parameters that result from this approach leads to numerous combinations of parameter values giving predictions of similar quality, i.e. the parameters are not identifiable (see Beven, 1989; Grayson et al., 1992a, b). In other words, when simulations based on different parameter sets are compared to observations, it is not clear which set of the parameters gives a better fit to the data.

This difficulty arises because the complex model is an "ill posed system" (in a mathematical sense). Groundwater modellers have been aware of this for decades and term it "non-uniqueness", while other terms are also used, e.g. "equifinality" (Beven, 1996). Many combinations of parameter values can lead to similar simulations of observed behaviour such as runoff at the catchment outlet or water level in groundwater bores. This implies that the observations are insufficient to properly test the model structure or parameters, i.e. it is not possible to *identify* the model structure or parameters. It also means that even if a model is able to simulate a particular type of observation, it *does not* indicate that other predictions made by the model are correct. For example, a model may give good fits to streamflow at a catchment outlet, but this does not indicate accurate simulation of streamflow at internal gauging stations or correct spatial patterns of saturation deficit. This has been clearly shown by many researchers, yet is commonly ignored by model users who confidently display spatial predictions based on an implicit assumption that because the outflow hydrograph was correct, the internal predictions are also accurate (see Chapter 13 for further discussion). This problem of "extrapolation into the unknown"

(Bergström, 1991), becomes acute when, for example, untested spatial runoff predictions are used as input to sediment or contaminant transport equations that require accurate flow depth and velocity estimates. It is also a problem for using these models to predict the effects of changing land use or climate, because, although some model parameters can be altered to simulate the expected effects of such changes, the uncertainty in the predictions is unknown.

The fundamental problem is that many of the equations we use to represent processes require calibration – their parameters cannot be directly measured. This is true even of "physically based" equations because they are invariably applied at a scale different to that at which they were derived. They then become conceptual representations with scale dependent parameters (see Philip, 1975; Klemeš, 1983; Hillel, 1986; Anderson and Rogers, 1987; Beven, 1989, 1996; Grayson et al., 1992b; Refsgaard et al., 1996). There is also a strong argument that some of these equations are wrong at different scales (e.g. Beven, 1996 argues that Darcian assumptions are simply wrong at the field scale). In any case, every time a new (or more complex) process description is included in a model, more parameters are added, each of which must either be calibrated or have a value assigned. The complexity and scale-dependent behaviour of nature is such that it is generally difficult to do this with any precision. The nonlinearity of processes such as surface and unsaturated flow, or evaporation, exacerbates the problem. Therefore, each new process introduces more "degrees of freedom", making testing even more difficult.

Ultimately the simple "downward" and complex "upward" approaches should converge since we want models that represent important spatial processes but that can be tested for both their internal and lumped predictions. They need to be tested well enough to know they are producing the "right" results for the "right" reasons. This will be possible only when there are sufficient observations to enable each process representation and the interaction between them to be tested. The use of new and different data is a vital step towards this convergence. Examples of new data types include geochemical tracers that provide information about flow pathways (Kendall and McDonnell, 1998), using multiple outputs such as simulated streamflow *and* salinity simultaneously (e.g. Mroczkowski et al., 1997) and of course, the motivator for this book, spatial patterns of hydro-logical behaviour.

Each of the case study chapters is an illustration of where pragmatic choices have been made regarding the appropriate level of model complexity and the consequences of those choices. The choices can be viewed within the context of Figure 3.2, i.e. deciding on the dominant processes, perhaps with the help of some initial measurements, and then building (or modifying) a model structure to represent those processes and collecting more data (and especially spatial patterns) both to develop understanding of the processes and to better represent them in the model. The general philosophy illustrated in the chapters varies but is generally more akin to the "upward approach" although the approach used in Chapter 9 would be considered "downward" with complexity being added where needed to better explain the spatial data. An "upward" example is Chapter 6 where Houser et al. used a relatively complex model (TOPLATS) but found that many of the components could be ignored (in fact needed to be for good simulations), so long as the spatial variability of precipitation was properly represented. Again this was made possible by the extensive spatial data available.

Several of the case study authors conclude that their models probably need to be more complex to deal with remaining inconsistencies between data and simulations. For example: in Chapter 7, Tarboton et al. discuss the need for better treatment of radiation and mass movement of snow in steep terrain; in Chapter 9, Western and Grayson indicate that modelling of soil crack dynamics during seasonal wetting and drying is likely to result in better simulations; in Chapter 10, Vertessy et al. indicate that further improvement may need pipeflow to be explicitly represented; in Chapter 11, Lamb et al. discuss the possibilities of changes that relax some of the constraints imposed by the TOPMODEL structure. In most cases, further data would be needed to test whether these added complexities actually helped reconcile simulation and reality.

These case studies illustrate the "infinite loop" of Figure 3.2. At some point we break out and leave some "loose ends" that can be resolved only by re-entering the loop. When we face a modelling problem in practice, we need a starting point in terms of model complexity. Figure 3.9 illustrates the conceptual relationship between model complexity, the availability of data for model testing, and predictive performance of the model. We use the term "data availability" to imply both the amount and quality of the data in terms of its use for model testing. Having pattern data is equivalent to "large" availability while just runoff data would imply "small" availability. We use the term "model complexity" to mean detail of process representation. Complex models include more processes and so are likely to have more parameters.

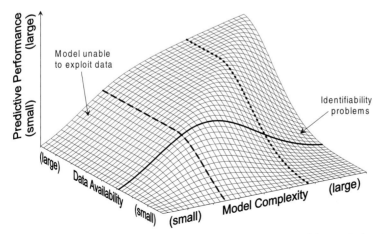

Figure 3.9. Schematic diagram of the relationship between model complexity, data availability and predictive performance.

If we have a certain availability of data (e.g. solid line in Figure 3.9), there is an "optimum model complexity" beyond which the problems of non-uniqueness described previously become important and reduce the predictive performance. There are too many model parameters and not enough data to test whether the model is working, or is working for the right reasons, which means that both the model structure and the model parameters cannot be identified properly. We can use a simpler model than the optimum, but then we will not fully exploit the information in the data (e.g. intersection of solid and dashed lines). For given model complexities (e.g. dashed and dotted lines), increasing data availability leads to better predictive performance up to a point, after which the data contains no more "information" to improve predictions; i.e. we have reached the best a particular model can do and more data does not help improve performance (the dashed and dotted lines flatten out as data availability increases). In this case, we could consider a more complex model to better exploit the information in the data, and this is something considered in some of the case studies in this book. The more common situation for practical applications of distributed modelling is represented by the intersection of the dotted and solid lines, where we are using too complex a model with limited data and so have identifiability problems. Increased data availability is needed to significantly improve the predictive performance (Chapter 13).

Ultimately the answer to "what complexity is warranted" depends on the objectives of the modelling exercise and knowledge of the system. The key point is that it is not useful to add complexity when we have no way of testing whether this improves a model or makes it worse. The topic of model complexity has been a source of stimulating discussion in the literature and interested readers may wish to consult some of the following (Bair, 1994; Bathurst and O'Connell, 1992; Beck, 1987; Beven, 1987, 1989, 1996; Refsgaard et al., 1996; Bergström, 1991; Grayson et al., 1992b; Jakeman and Hornberger, 1993; Hillel, 1986; Konikow and Bredehoeft, 1992; Klemeš, 1983, 1986a; James and Burges, 1982; Morton, 1993; Oreskes et al., 1994; Philip, 1975; Smith et al., 1994).

In order to know if a model is "improved", criteria need to be set that reflect how the model will be used, i.e. that test the components of a model that are critical to the output of interest. For example, if we want to simulate patterns of soil erosion on hillslopes, we would need to test model predictions against flow depth, velocity and sediment concentration from various places on hillslopes, rather than just against catchment runoff and sediment load. Hence there is a direct link between the purpose of the model, the level of complexity we can justify, the details of the data we have for calibration and testing, and the criteria we use to quantify performance.

The following sections summarise the conceptual underpinnings of model calibration and testing, approaches to the representation of predictive uncertainty, and how the value of added complexity, particularly in spatial models, might be assessed.

3.4   MODEL CALIBRATION AND TESTING

Model calibration is the process of choosing the "best" set of parameter values while model testing is the process we go through in an attempt to define the uncertainties in predictions, or at least get a feeling for their magnitude. Calibration is a process of optimising a model's parameter values to improve the "fit" between simulated and observed catchment responses. The "fit" can simply be a qualitative assessment of how close two time series appear (e.g. plots of observed and predicted hydrographs, or observed and predicted spatial soil moisture) but will generally be more quantitative.

An "objective function" provides a quantitative measure of the error between simulations and observations. We seek to optimise parameter values so the objective function is minimised. The choice of objective function determines what part of a model's performance is being tested. For example, when comparing observed and predicted streamflow, a range of objective functions are commonly used, e.g.:

1. Sum of the squares of the difference between simulated and observed flows (this emphasises large flows since larger absolute errors are likely to occur compared to low flows, i.e. mainly tests whether the peaks match).
2. Sum of the absolute values of the differences between observed and simulated flows (as for 1 but less sensitive because errors are not squared, i.e. tests whether the water balance matches).
3. Sum of the ratio between observed and predicted flows (places more equal emphasis on high and low flows since ratios are used but could result in large absolute errors in extreme flows).

Note the different emphasis of each function on different flow components. In a spatial context, we could use similar types of functions; e.g. we could apply 1 to a point-to-point comparison of observed versus predicted patterns. It is likely, however, that such simple functions would not exploit all the information in the patterns and that more sophisticated approaches would be better tests of model performance. Some more sophisticated approaches are presented in Section 3.5.

Once an objective function is chosen (noting that this is a subjective task), it can be minimised by either trial and error or an automated optimisation procedure. Sorooshian and Gupta (1995) and McLaughlin and Townley (1996) present reviews of approaches to calibration and automated optimisation in surface and groundwater models respectively. They highlight a range of problems such as the presence of multiple optima and strong correlations between parameters, which can lead to the identification of a non-optimum set of parameters. These problems are the result of the large number of parameters in distributed models. It is therefore common to try to reduce the number of parameters to be optimised, i.e. reduce the "degrees of freedom". This is done by choosing a *pattern* for a spatially distributed parameter and then calibrating usually a single parameter that sets the values for the whole pattern. A common example relates to soil hydraulic properties. Because of their spatial variability, they are sometimes calibrated, but in a distributed model, this may mean many thousands of "free" parameters (one or more for every model element). Clearly this is not tractable, so a *pattern* may be imposed, for example by the soil type, which sets the *relative* hydraulic properties for every element. A single parameter is then optimised that defines the absolute values from the pattern of relative values. This reduction in degrees of freedom makes the optimisation procedure tractable and significantly reduces the identifiability problems discussed above. But of course it imposes a pattern on the distribution of soil properties that may or may not be realistic (see examples in Chapters 6, 9 and 10). Wetness indices can be used in the same way to define, for example, the pattern of initial saturation deficit for a runoff model, leaving a single parameter to be calibrated that sets the absolute value of the pattern (e.g. Grayson et al., 1992a). Clearly this step of reducing degrees of freedom comes at a cost and will undermine model performance if the choice of pattern is poor.

Once a model is calibrated and optimised, it needs to be tested. There is some confusion in the literature about terminology related to testing (see Oreskes et al., 1994; Konikow and Bredehoeft, 1992; de Marsily et al., 1992). These are discussed in detail in Chapter 13 in the context of using distributed models for prediction in practical applications. Suffice to say here that the term *verification* should be reserved for strict tests such as of model code where analytical and numerical simulations are compared and should not be used in relation to model testing against real data. On the other hand, *validation* is where observations and simulations are compared using data that were not part of the calibration. As discussed in Chapter 13, a model is *validated* for a particular application and a successful validation in one example does not imply that the model is validated for universal use. Again, objective functions can be used to provide a quantitative assessment of the validation step.

The extent to which the validation step really tests the model structure and parameters is obviously dependent on the type of data available for testing. Klemeš (1986b) presented some standard approaches for lumped rainfall runoff modelling, based on variations of "split sample" tests where the observed period of record is split with some data used for calibration and the other for validation and these two sets should be independent. These approaches and their application to spatial modelling are described by Refsgaard in Chapter 13. The important question to consider when assessing both the calibration and validation steps is "just what is being tested?" i.e., how well do the comparisons test the model structure, what can be inferred about the identifiability of parameters or the confidence (predictive uncertainty) in simulations?

These questions can be answered qualitatively, but the complexity of distributed models has led to the development of more formal approaches. Uncertainty in prediction of hydrological response can arise from four sources: uncertainty in the input data due to sampling or interpolation error; uncertainty in simulated responses due to model parameter errors; and uncertainty related to the hypotheses that underlie the model itself and the model structure. Note also that there may be uncertainty in the data against which the model output is calibrated and tested. This will affect the certainty we can have in model structure or parameters.

Quantifying these various sources of uncertainty and their effect on overall predictive capability is an active area of research (e.g. Sorooshian and Dracup, 1980; van Straten and Keesman, 1991; Beven and Binley, 1992; Mroczkowski et al., 1997; Kuczera and Parent, 1998; Gupta et al., 1998; Franks et al., 1998). The more recent approaches make use of Monte Carlo simulation (i.e. running a model many times using parameters or inputs drawn randomly from a particular distribution (Tanner, 1992; Gelman et al., 1997). In principle, these methods enable quantification of the different errors, have the ability to assess the value of additional information in constraining model uncertainty and can compare different model structures in terms of their ability to reproduce observed behaviour (see Chapter 11). But they are still being developed and are presently limited by large computational demands, differing amounts of subjectivity and, in some cases, methods that lump all errors into parameter uncertainty. Once these limitations are overcome, these methods should gain wide use and will have a major impact on modelling.

The stage of development and computational requirements of these sophisticated methods means that most studies undertake only simple assessments of errors and uncertainty such as "sensitivity analyses". In sensitivity analyses, parameters are varied over particular ranges and an assessment is made of the extent to which the output varies. This approach can give the modeller an idea of which parameters dominate the model response – i.e. which are the crucial ones to "get right", but allows for only limited consideration of the complex inter-relationships between model parameters that generally exist, particularly in distributed models.

Another simple approach to assessing model quality is to carefully analyse the residuals between observed and simulated responses. Unbiased and randomly distributed residuals imply that the model does not contain structural problems that lead to systematic errors, and that it exploits all of the information present in the data. Small residuals indicate good model accuracy, and similarity between the size of the residuals from the calibration and validation steps is another sign that the model structure is good. Analysis of residuals was illustrated in Western et al. (1999a), who used spatial plots of differences between measured and simulated soil moisture (using a terrain index model) to illustrate that all of the terrain "signal" in the soil moisture measurements was represented in their model. The remaining errors were not related to terrain and could have been either sampling noise or due to randomly distributed soil or vegetation effects. Mroczkowski et al. (1997) apply this approach in a more rigorous manner using multiple response time series data to explore different structures for a streamflow and salinity model and illustrate the value of different types of information for improving the rigour of model testing.

The hydrological community still has a long way to go in model testing to achieve comprehensive assessments of predictive uncertainty. Even the most basic of testing approaches such as those discussed by Klemeš (1983) are often not applied, and application of the more complex approaches is rare. But it is important to recognise that even the most sophisticated assessment of uncertainty is useful only if we have detailed data available of a type that really tests the basic hypotheses of the model. This is where the importance of patterns becomes clear, yet "goodness of fit" measures for patterns are not well developed. Indeed comparisons of observed versus simulated patterns rarely go beyond simple visual comparison of the basic patterns, or at best, residual patterns. The following section looks in more detail at approaches to pattern comparison. It is likely that one day these may become part of comprehensive assessments of predictive uncertainty based on spatial patterns, with the error analyses between observations and simulations being used to inform model development.

3.5   SOME PATTERN COMPARISON METHODS

One of the basic motivations for this book is the belief that more refined and insightful use of spatial patterns is vital for progressing the science of hydrology and for better constraining the uncertainty in our predictions of the future. Each of the case studies presented in this book makes comparisons between observed and simulated patterns of various types to assist modelling and process understanding. In this section we introduce a number of possible methods for comparing spatial patterns that have the potential for assessing the quality of hydrological predictions. These methods have been selected because the results from them can be interpreted hydrologically, thus enabling identification of model components that are performing poorly and perhaps producing improvements in hydrological understanding. Some of the methods have been used in the case studies presented later, while others are logical extensions, which may be used in the future. This is a very new area for hydrology and techniques are still being trialed and developed.

3.5.1   Methods Used in This Book

Visual comparison of simulated and observed spatial patterns provides qualitative information about the predictive ability of a model and is used throughout this book. Albeit simple and subjective, it is an extremely useful method as areas of consistent error can be identified and it may be possible to make qualitative associations with the model components causing the errors. The disadvantage of this method is that it does not provide a quantitative measure of model performance. Thus, it is not possible to extend this method to automated optimisation techniques which makes inter-model comparisons rather subjective. Nevertheless, it is easy to implement, and, as will be illustrated in the chapters that follow, enables a lot of insight into the model performance and limitations.

Point-by-point comparison methods include both direct comparison of simulated and observed "pixel" values as well as the *mapping* of residuals between observed and simulated patterns (see Chapter 9). These techniques can provide information about bias (mean error), random simulation errors (error variance) and any organisation that may be present. The mean error and the error variance are similar to statistics used in traditional model evaluation using time-series; however, they can be applied in a spatial context to test internal model predictions. This method can also provide the spatial correlation structure (variogram) of the residuals, which gives information about the spatial scale or correlation length of the errors (e.g. Western et al., 1999a). If the correlation length of the errors is smaller than the model element scale, it can be concluded that the errors are due to either measurement error or to small-scale variability not resolved by the model. Since the model does not aim to represent these features, it can be concluded that the model is performing as well as can be expected (assuming there is no bias and a sufficiently small error variance). If the correlation length of the errors is significantly longer than the model grid scale, it can be concluded that there are patches where the errors are similar; i.e. there is some problem with the structure of the model that causes certain parts of the landscape to be better represented than others. A careful analysis of the simulated response and an understanding of the model structure gives guidance on potential model improvements. The *point-by-point* approach is used in Chapters 7 and 9, 11 and 12.

An extension of the point-by-point approach, which accounts for measurement error and sub-element variability that is not represented by the model, is to smooth the observed pattern and then compare it with simulations on a point-by-point basis. For the smoothing, geostatistical techniques such as kriging and other interpolation/estimation methods can be used as illustrated in Chapter 2, Figure 2.7. In a geostatistical approach, the nugget can be set to the sum of measurement error and sub-element variability that is not represented by the model. Point-by-point comparison between the "smooth" observed and simulated patterns then allows the separation of the effects of apparent small-scale variability (see Chapter 2, p. 32) and large-scale variability which should be represented in the distributed catchment model. Chapter 9 makes use of this approach (p. 230).

To gain further insight into which hydrologic process representations may or should be improved in the model, errors can be analysed to ascertain whether there is any *relationship with topographic variables*. Such relationships may provide hydrological insight into the cause of the errors. Relationships of this type have been examined in Chapters 7, 8, 9, 11 and 12. For example, a relationship between aspect and the error in simulated soil moisture probably indicates a problem in simulating the spatial pattern of evapotranspiration. Likewise, a relationship between a wetness index representing the control of topography on soil moisture and the error may indicate problems with the simulation of one of the lateral redistribution processes; e.g. if the gullies are simulated as being too dry, it may indicate that subsurface flow is too small, perhaps because macropore flow is occurring or because the conceptualisation of saturated/unsaturated fluxes is not correct.

3.5.2 Opportunities for Future Pattern Comparison

A limitation of the methods above (except visual comparison) is that they do not provide any information on lateral shifts – i.e. where the basic shape of patterns is correct but their location is shifted. *Optimal local alignment* (e.g. Bow, 1992) is a method used in pattern recognition and provides information on space shifts between two patterns. A field of shift vectors is calculated in the following way. Initially the whole domain is divided into subareas. Then correlation coefficients between point-by-point comparisons of the observed vs. simulated patterns are calculated for corresponding subareas. The relative position of the two corresponding subareas is then changed (i.e. shifted) and correlations are again calculated (i.e. a cross-correlation analysis). The optimum shift (i.e. optimum alignment) is where the correlation is best. This approach has the potential to identify model mismatches in the direction of the hillslopes as well as other shifts such as those associated with biases due to aspect or the way soil parameters were imposed, or georeferencing problems. To use this method successfully, it would be necessary to have small subareas but there is a tradeoff between having sufficient points in each subarea for accurate computation of correlations and obtaining detailed spatial information (high resolution) in the resulting vector field. While appearing to be a sound approach, this method has yet to be illustrated in the hydrological literature and a large number of points in space are needed.

Because water flows through the landscape along pathways that are dominated by the topography, pattern shifts may be associated with particular terrain features. This could be explicitly taken into account when comparing simulated and observed patterns. Transects of simulated and observed variables such as soil moisture along and across topographically defined flow pathways could be examined to search for shifts between the simulated and observed patterns. Cross-correlations between the simulated and observed patterns could also be calculated, but using a coordinate system consisting of topographic streamlines and contour lines. This approach may be useful in identifying problems with the lateral flow redistribution component of the models and may need fewer points in space than the optimal local alignment method, but has yet to be illustrated in the literature.

The preceding methods are applicable to patterns at specific points in time. However, as illustrated in the case study chapters, changes in the patterns over time provide additional information about the dynamics of the hydrological system. By calculating differences between two observed patterns and comparing the resulting spatial pattern of temporal differences with the equivalent information from a model, insight into the dynamics of a landscape and the ability of a model to capture those dynamics may be obtained. These dynamics will be spatially variable. All the above methods can be applied to spatial patterns of temporal differences. In Chapter 8, Troch et al. use temporal comparisons of patterns from remotely sensed data using the point-by-point method and principal component analysis. In Chapters 7 and 9, the differences between patterns of snow cover and soil moisture respectively are compared to simulated changes in model 'storages'. There is clearly plenty of scope for development and application of pattern comparison methods. We believe that pattern comparisons will ultimately become a routine part of distributed model calibration and testing. They will enable much more definitive tests of model performance to be made and significantly improve the confidence with which we can claim our models do indeed represent the *right* processes and get the *right* answers for the *right* reasons.

3.6 FINAL REMARKS

In these introductory chapters, we have illustrated that natural spatial variability is omnipresent and must be better understood if we are to advance our knowledge of hydrological processes and improve hydrological prediction. We have introduced some basic concepts to deal with spatial variability and the patterns it manifests. Some specific tools that can quantify variability and allow it to be represented in models were described in Chapter 2. We have proposed that better utilisation of spatial patterns is necessary to properly test and develop the numerous spatially explicit hydrological models that are presently available, and to further our understanding of hydrological processes. The next two chapters are reviews of spatial patterns in precipitation and evaporation. These are followed by the case study chapters which are examples that will test whether our assertions are sound.

4

Patterns and Organisation in Precipitation

Efi Foufoula-Georgiou and Venugopal Vuruputur

4.1 INTRODUCTION

Rainfall possesses a complex spatio-temporal structure which has been the subject of many studies. Traditionally, applied hydrologists have recognised the importance of this structure on runoff production, and have tried to analyse and model it using simple descriptions, such as the depth–area–duration (DAD) curves, the area-reduction curves, and the hyetographs or normalised hyetographs (see Chow et al., 1988; Linsley et al., 1982; Viessman and Lewis, 1996). The DAD curves depict, for a given duration, the area of the storm over which a given depth is equalled or exceeded. The area-reduction curves depict, for a given duration again, the decrease of the maximum storm depth (measured at a point) when it is averaged over increasing areal extent around that point. The normalised hyetograph (or mass curve) depicts, at a particular location or as average over an areal extent, the percentage of total storm depth (normalised depth) versus the percentage of storm duration (normalised time).

The above are of course simple measures of the complex spatio-temporal variability of the observed storm patterns, but still provide a means of comparing observed patterns to each other or to extremes. They also provide the means of parameterising design storms and reconstructing their spatio-temporal patterns to be used in predicting design hydrographs through rainfall-runoff modelling. For example, a 12-hour design storm depth at a point (computed from a frequency analysis or from the Probable Maximum Precipitation methodology) can be converted to "a design storm pattern" by assigning it a DAD curve and a design mass curve. That is, the spatial internal structure of the design storm could be reconstructed using the selected design DAD curves and an assumed shape for isohyets, and the temporal distribution of the total rainfall depth over its duration could be obtained using the selected design mass curve. All these attempts to reconstruct the space-time variability of a storm were based on the recognition that this variability plays an important role in runoff production (e.g., see Kiefer

Rodger Grayson and Günter Blöschl, eds. *Spatial Patterns in Catchment Hydrology: Observations and Modelling* © 2000 Cambridge University Press. All rights reserved. Printed in the United Kingdom.

and Chu, 1957; Huff, 1967; Pilgrim and Cordery, 1975; and Milly and Eagleson, 1988; among others). It should be noted that in the early hydrology days, due to data limitations (typically one or a few raingauges over the basin) more detailed descriptions of the storm spatiotemporal patterns were not possible.

The recognition of the importance of the small-scale space-time rainfall variability on runoff modelling led to considerable research efforts in developing stochastic point process models or phenomenological spatiotemporal models of rainfall which could be used to simulate precipitation patterns, conditional on preserving a desired total depth (e.g., see Gupta and Waymire, 1979; Kavvas and Delleur, 1981; Waymire et al., 1984; Valdes et al., 1985; Seed et al., 1999; see also the review of Foufoula-Georgiou and Georgakakos, 1991). In more recent years, the wide availability (at least in the United States) of NEXRAD or other radars that continually monitor rainfall at high spatial and temporal resolution (typically pixels of 2 or 4 km and at intervals of 6–10 min) have provided unique opportunities to better understand and quantify the small-scale rainfall variability. These efforts have provided considerable insight on the effect of rainfall resolution on the accuracy of runoff production estimates (e.g., see Kouwen and Garland, 1989; Krajewski et al., 1991; Ogden and Julien, 1993, 1994; Michaud and Sorooshian, 1994a; Obled et al., 1994; Faurès et al., 1995; and Winchell et al., 1998) despite the still unresolved problems related to the accuracy of the NEXRAD estimates of precipitation (e.g., see Collier and Knowles, 1986; Pessoa et al., 1993; among others). Also, the need to unify descriptions over scales (i.e., rainfall variability over a small area with rainfall variability over a larger scale) and to parameterise subgrid-scale rainfall variability parsimoniously, has prompted the introduction of new ideas and tools for analysing and modelling space-time rainfall patterns, namely, the ideas of scale-invariance (see Schertzer and Lovejoy, 1987; Gupta and Waymire, 1990; Kumar and Foufoula-Georgiou, 1993a,b; among others).

Scale invariance implies that variability of rainfall, or another quantity such as rainfall fluctuations, exhibits a statistically similar structure under proper renormalisation of space and/or time coordinates. The scale-invariance concept for rainfall has been conceptually justified based on the idea that, after all, raindrops are tracers in the turbulent atmosphere and thus the documented presence of scale-invariance in fully-developed turbulence might also be manifested in the rainfall fields produced. Irrespective of the theoretical rationale and foundation of the scale-invariance ideas of rainfall, compelling evidence has been accumulated over the past decade that indeed some features of rainfall exhibit scale invariance (for example, see Foufoula-Georgiou and Krajewski, 1995; and Foufoula-Georgiou and Tsonis, 1996). This evidence has paved the way for the development of new classes of space-time rainfall models which are applicable over a large range of scales. Models based on scale invariance are considerably more attractive compared to the previous generation of stochastic point process phenomenological models in that they: (a) are much more parsimonious, i.e., 2 or 3 parameters versus 10 to 12 parameters in the previous models; and (b) are scale-independent, i.e., applicable to a wide range of scales without changes in the model parameterisation or parameter values.

Another approach to the generation of high-resolution space-time patterns of precipitation is via numerical weather prediction models. These models have been considerably advanced in recent years in terms of physical parameterisations, data assimilation and numerical schemes, but still require extensive computational resources to run at high resolutions (e.g., see Droegemeier, 1997). Typically, they run in a nested mode where higher resolution domains are nested within a larger lower resolution domain which provides boundary fluxes and larger-scale environmental forcings to the smaller domain. The very high resolution domain involves sophisticated microphysics suitable to explicitly resolve cumulus-scale convection. The outer domain typically uses convective parameterisation schemes which depend on the model resolution. The accuracy of quantitative precipitation forecasts (QPFs) produced by these models has not been fully evaluated yet and is an issue of ongoing research (e.g., see Fritsch et al., 1998). Moreover, several studies have demonstrated the sensitivity of the resulting precipitation patterns to the chosen convective parameterisation schemes (even if these schemes are only used at the lower-resolution outside domain), to the chosen model resolution, to the degree of prescribed heterogeneities, and to the land–atmosphere exchange mechanisms embedded in the model (e.g., see Kain and Fritsch, 1992; Sivapalan and Woods, 1995; Avissar and Liu, 1996; Brubaker and Entekhabi, 1996; Wang and Seaman, 1997; Warner and Hsu, 2000; among others).

This chapter concentrates on stochastic descriptions of space-time rainfall variability based on scale-invariant parameterisations. It reviews some of the recent research describing and modelling spatiotemporal rainfall patterns and demonstrates that indeed the seemingly complex patterns of rainfall exhibit simple underlying statistical structures which can be unravelled with proper methodologies. We focus only on spatiotemporal descriptions based on scaling of rainfall fluctuations. Other descriptions based on multiscaling of rainfall intensities or multiplicative cascades (e.g., Schertzer and Lovejoy, 1987; Tessier et al., 1993; Over and Gupta, 1994; Seed et al., 1999) are not reviewed herein and the reader is referred to the original publications for such developments or to Foufoula-Georgiou and Krajewski (1995) for a brief review and further references.

Our presentation in this chapter evolves around three major questions:

1. If spatio-temporal patterns of rainfall exhibit an organised structure, how can this be used in building parsimonious models which are applicable over a range of scales and can also be used for the purpose of downscaling (i.e., reconstructing small-scale variability from large-scale averages)? This last issue is especially important given the increasing availability of remotely-sensed data whose reliability is often considered adequate only if observations are averaged over large scales, and also given the need to interpret the results of global or continental scale studies to the hydrologic basin scale.

2. If rainfall spatiotemporal patterns exhibit organisation, how can this help us to understand the still open question of relating the physics of the atmosphere with the statistics of the produced precipitation patterns?

3. If documentable space-time organisation exists in observed precipitation patterns, is the same structure reproduced by state-of-the-art numerical weather prediction models which eventually will be used to predict future storms and the resulting floods? And, moreover, what does the success or lack of success of the model in reproducing this structure tell us about the physics and parameterisations we currently use in atmospheric models?

4.2 SPATIAL RAINFALL ORGANISATION

To explain the idea of multiscale rainfall variability, consider the left-most field of Figure 4.1. It represents the radar-depicted rainfall intensity field of an extreme squall line storm of June 27, 1985 over Kansas and Oklahoma at 0300 UTC at a resolution of 4 km. If this field is degraded to different scales e.g., 8, 16, 32, 64 km by averaging, then the question arises as to whether the spatial rainfall variability of this field changes with scale in a systematic way. Moreover, notice that when we average or filter the process at a small scale to go to a larger scale, some "information" about its spatial variability is lost. This lost "information" when going from one scale to a coarser one by averaging can be preserved by keeping the so-called rainfall "fluctuations" (i.e. the difference in intensities in adjacent pixels in terms of space and/or time). If the variability of the rainfall intensities themselves does not exhibit a simple statistical structure over scales, could it be that such a structure exists in the rainfall fluctuations?

Kumar and Foufoula-Georgiou (1993a,b) and later Perica and Foufoula-Georgiou (1996a) performed a multiscale analysis of spatial rainfall fields using a discrete orthogonal wavelet transform. This transform within the multiresolution framework of Mallat (1989) provides a filter which simultaneously keeps the "averages" (smoothed fields) and "fluctuations" (details lost) as the scale changes. The two-dimensional orthogonal wavelet transform is a directional filter, so it can handle anisotropy, and is a reconstructive filter so that from the average field at 64 km and the fluctuations at scales 8, 16, 32 and 64 km, the field at the smaller scale (higher resolution) of 4 km can be reconstructed. Of course, statistical parameterisation of the details will result in a statistical reconstruction of the high-resolution field. These ideas give rise to three important questions: (a) does the spatial variability of the rainfall intensity fields at a range of scales, exhibit a simple structure which can be parameterised with a few coefficients?; (b) what about the structure of rainfall fluctuations or details as a function of scale?; and (c) what physical parameters might control the way this variability changes over scales?

These three questions were considered by Perica and Foufoula-Georgiou (1996a,b) who analysed a large number of mid-latitude mesoscale convective

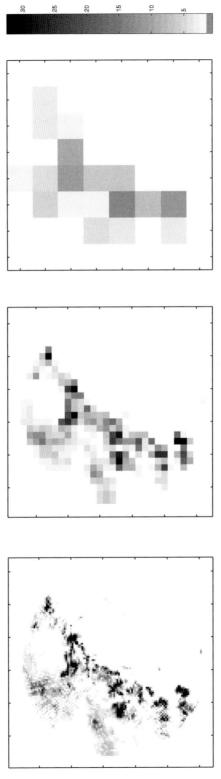

Figure 4.1. The June 27, 1985 storm over Kansas–Oklahoma at 0300 UTC. The left panel shows the observed radar fields at 4 km resolution and the middle and right panels show the degraded fields at 16 and 64 km resolution, respectively. The intensities have been mapped onto the same 32 colours for display purposes.

systems monitored by the Oklahoma–Kansas radar during the May–June, 1985 period of the Preliminary Regional Experiment for Storm-Central (PRESTORM) field program (see Cunning, 1986 for details of this experiment). It was found that although rainfall intensities themselves did not have a simple well-structured way of changing their variance as a function of scale, the "standardised rainfall fluctuations" i.e., rainfall fluctuations normalised by their corresponding-scale local rainfall averages, exhibited Normality and simple scaling, i.e., their variance changed with scale in a log–log linear way. This is not surprising since the multiscale rainfall averages carry in them the signature of deterministic background features which are particular to the rainfall-producing mechanism, making it unlikely for them to exhibit simple scaling relationships over a significant range of scales. However, once these underlying deterministic features are removed by filtering, the remaining features (e.g., deviations from local means or spatial rainfall gradients) do not seem to have a characteristic length scale and are more amenable to stochastic parameterisations which might present similarities over scales.

Specifically, let $\overline{X_\lambda}$ denote the value of the rainfall average at scale λ at a particular pixel, and $X_{\lambda,i}$, the value of the rainfall fluctuation (difference of values at adjacent pixels) at the same scale λ and direction i (e.g., latitude, longitude and diagonal). Standardised rainfall fluctuations at scale λ and direction i are defined as $\xi_{i,\lambda} = X'_{\lambda,i} \sqrt{\overline{X_\lambda}}$. Perica and Foufoula-Georgiou (1996a) computed these standardised fluctuations using an orthogonal Haar wavelet transform and found that between scales of 4 km and 64 km, for which data were available, $\xi_{i,\lambda}$ exhibited Gaussianity and simple scaling implying that

$$\frac{\sigma_{\xi,L_1}}{\sigma_{\xi,L_2}} = \left(\frac{L_1}{L_2}\right)^H \tag{4.1}$$

where $\sigma_{\xi,L}$ is the standard deviation of ξ at scale L km and H is a scale-independent parameter. The values of H varied between 0.2 and 0.5 for several mid-latitude mesoscale convective systems from the PRESTORM data set. The dependence of H on direction was not very pronounced but H was found to be strongly dependent on the convective instability of the prestorm environment, as measured by the convective available potential energy, CAPE (in $m^2 s^{-2}$)

$$H = 0.0516 + 0.9646 \cdot (\text{CAPE} \times 10^{-4}) \tag{4.2}$$

(see Figure 4.2). CAPE is the buoyant energy available to a parcel rising vertically through an undisturbed environment and is a measure of the potential instability at middle to upper atmosphere. It is defined as

$$\text{CAPE} = \int_{\text{LFC}}^{\text{EL}} g \cdot \left(\frac{\Theta_c - \Theta_{\text{env}}}{\Theta_{\text{env}}}\right) dz \tag{4.3}$$

where Θ_c is the potential temperature of an air parcel lifted from the surface to level z, Θ_{env} is the potential temperature of the unsaturated environment at the

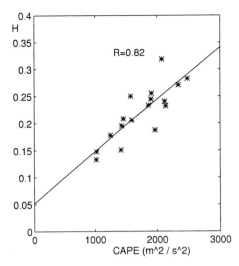

Figure 4.2. Scattergram indicating the relationship between the scaling parameter H of the standardised rainfall fluctuations and the Convective Available Potential Energy (CAPE) in m^2/s^2 of their prestorm environment. Data were used from several midlatitude mesoscale convective systems of the PRESTORM field program (after Perica and Foufoula-Georgiou, 1996a).

same level, LFC is the level of free convection and EL is the equilibrium level (e.g., see Air Weather Service, 1979; General Meteorological Package (GEMPAK), 1992).

The empirical relationship between H and CAPE is useful since CAPE can be computed from observed sounding data or by a numerical weather prediction model and then H can be estimated from (4.2) and used to infer the variability of rainfall fluctuations at any scale, given the variability at a reference scale (see also Perica and Foufoula-Georgiou, 1996b). For more information on the linkage of CAPE, and some other meteorological parameters, to statistical parameterisations of rainfall fluctuations, the reader is referred to Perica and Foufoula-Georgiou (1996a). Application examples are given in Section 4.4.1.

4.3 SPATIOTEMPORAL DYNAMICS

Spatial and temporal features of rainfall are not independent of each other but relate in ways specific to the physics of the storm-generating mechanisms. Thus, a lot of insight may be gained by studying simultaneously the spatial and temporal patterns of rainfall. Recently, Venugopal et al. (1999a) developed a methodology under which space and time structures of rainfall can be studied simultaneously at a multitude of scales. Using this methodology, they demonstrated that there exists a simple scale-invariant spatiotemporal organisation in rainfall patterns which can be unravelled by proper renormalisation of the space and time coordinates.

Let $I_{i,j}^L(\tau)$ and $I_{i,j}^L(\tau + t)$ represent rainfall intensity values averaged over a box of size L centered around spatial location (i, j) of the precipitation field at two instants of time τ and $\tau + t$, respectively (Figure 4.3). The evolution of the field at scale L and a time period t was characterised in Venugopal et al. (1999a) by the differences in the logs of the rainfall intensities $\Delta \ln I$, i.e.,

$$\Delta \ln I_{i,j,\tau}(L, t) = \ln\big(I_{i,j}^L(\tau + t)\big) - \ln\big(I_{i,j}^L(\tau)\big) \tag{4.4}$$

computed at all spatial locations (i, j). This selection (as opposed to the simpler selection of intensity differences, i.e., $\big[I_{i,j}^L(\tau + t) - I_{i,j}^L(\tau)\big]$) was made since there is evidence that the rainfall process is not additive but rather multiplicative, that is, normalised rainfall fluctuations, $\Delta I/I \equiv (I - I')/(I + I')/2$ in the terminology of Figure 4.3, and not fluctuations ΔI themselves, are independent random variables and spatially uncorrelated (see Venugopal et al., 1999a). This implies that fluctuations of ln(rainfall), i.e., $\Delta \ln I = \ln I - \ln I'$ are independent and also spatially uncorrelated random variables and can be easily characterised statistically.

The measure described above was evaluated for all locations (i, j) and all time instants τ, and for various spatial and temporal scales, L and t, respectively. Then assuming stationarity in space (i.e., independence of the specific (i, j) position) and selecting stationary regions in time (i.e., regions where the statistics of $\Delta \ln I$ do not fluctuate significantly around their mean value for the region), the probability density functions (PDFs) of $\Delta \ln I(L, t)$ were used to characterise the evolution of rainfall at several spatial scales L and temporal scales t. Notice that homogeneity in space is a reasonable assumption given that the radar frame can be seen as a fixed window within which the moving storm is observed. Thus, unless there is a specific reason to believe that a portion of the radar frame receives statistically different rainfall than the rest of the frame, intensities (or $\Delta \ln I$'s) at all positions (i, j) are considered to come from the same probability distribution.

Venugopal et al. (1999a) and Venugopal (1999) found, by analysis of several storms in different geographical regions of the world (the tropical region in Darwin, Australia; the forested region of Northern Saskatchewan, Canada; and the Oklahoma region in the midwestern United States), that the PDFs of $\Delta \ln I(L, t)$ remain statistically invariant if space and time are renormalised with the transformation $t/L^z = constant$. That is, the evolution of the rainfall field at scale L_1 and during a time lag t_1 is statistically identical to the evolution of the rainfall field at spatial scale L_2 and time lag t_2, as long as

$$t_1/t_2 = (L_1/L_2)^z \tag{4.5}$$

where z is the socalled dynamic scaling exponent.

Figure 4.3. Schematic illustrating the change in intensity of a field (rainfall in this case) over a box of size $L \times L$ centred around the location (i, j) during a time interval t.

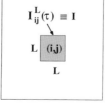

Table 4.1. Standard deviations of $\Delta \ln I$ ($\Sigma(\Delta \ln I)$) with time lag (left to right) and aggregation level (top to bottom), for a stationary region of the storm of Dec. 28, 1993, in Darwin, Australia

		Time Lag t (min)							
		10	20	30	40	50	60	70	80
L (km)	2	0.58	0.77	0.89	0.97	1.05	1.11	1.17	1.21
	4	0.47	0.67	0.79	0.87	0.95	1.01	1.07	1.12
	8	0.35	0.55	0.67	0.76	0.84	0.90	0.96	1.01
	16	0.23	0.40	0.53	0.63	0.71	0.77	0.83	0.88

Table 4.1 shows the matrix of standard deviations of these PDFs of $\Delta \ln I(L, t)$ for different spatial scales, L, and temporal scales, t, for a stationary portion of duration five hours of the December 28, 1993 storm in Darwin, Australia (see Venugopal et al., 1999a for a description of this storm). Notice that these PDFs are well approximated by normal distributions centred around zero at all space and time scales (see Figure 4.4) and thus their standard deviation completely parameterises them.

From Table 4.1, one can find, by interpolation, pairs of (L, t) such that a chosen value of the standard deviation $\Sigma(\Delta \ln I)$ remains constant. For example, Table 4.2 displays pairs (L, t) for which $\Sigma(\Delta \ln I)$ remains constant and equal to 0.6, 0.7 and 0.8, respectively. If these pairs satisfy $t \sim L^z$ i.e., if the iso-standard deviation lines plot linear on a log–log plot of L versus t, then the process is said to exhibit dynamic scaling. Figure 4.5 (left panel) shows the log–log plot of the values of Table 4.2. The fitted lines are for $\Sigma(\Delta \ln I) = 0.8$, 0.7 and 0.6 (top to bottom) and give estimates of z equal to 0.51, 0.54 and 0.58, respectively. Obviously, these iso-standard deviation lines are well approximated by straight lines implying that rainfall evolution (as characterised by the PDF of $\Delta \ln I$) exhibits dynamic scaling. Note that since the PDF of $\Delta \ln I(L, t)$ is well approximated by a Normal distribution, scale-invariance of $\Sigma(\Delta \ln I)$ implies scale invariance in the whole PDF. This was verified by computing (via interpolation) the PDFs of $\Delta \ln I$ at several (L, t) pairs which satisfy $t \sim L^z$ (for example, the pairs in Table 4.2) and checking that indeed these PDFs remained statistically invar-

Table 4.2. Time (in min) to "reach" different values of the standard deviation of $\Delta \ln I$ (left to right: 0.6, 0.7, 0.8) for various aggregation levels (top to bottom), for a stationary region of the storm of Dec. 28, 1993, in Darwin, Australia. This table is formed by linear interpolation of the values in Table 4.1

		$\Sigma(\Delta \ln I)$		
		0.6	0.7	0.8
L (km)	2	11.1	16.3	22.4
	4	16.6	22.8	31.6
	8	24.4	33.3	45.1
	16	37.4	49.4	64.7

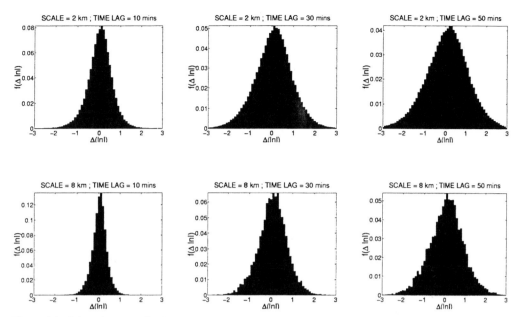

Figure 4.4. Selected PDFs of $\Delta \ln I$ (top: spatial scale of 2 km and time lags of 10, 30 and 50 min; bottom: spatial scale of 8 km, and time lags of 10, 30 and 50 min) for a stationary region of the storm of Dec. 28, 1993, in Darwin, Australia.

iant (see Figure 4.6). Similar results were found for several other storms in Darwin, Australia (see Figure 4.5 for the storm of January 4, 1994), the BOREAS region in northern Saskatchewan, and the midwestern United States region (see Venugopal, 1999).

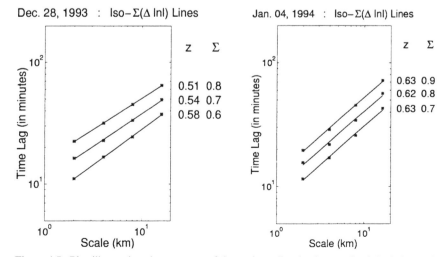

Figure 4.5. Plot illustrating the presence of dynamic scaling in the standard deviations of $\Delta \ln I$ for (a) a stationary region of the storm of Dec. 28, 1993 (left) and (b) a stationary region of the storm of Jan. 4, 1994 (right) in Darwin, Australia.

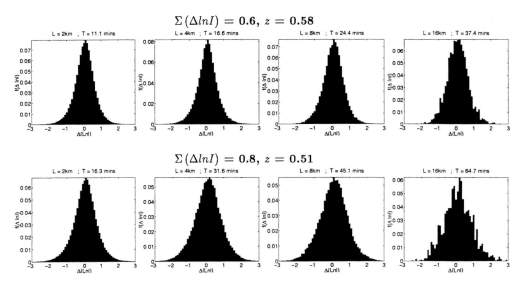

Figure 4.6. For a stationary region of the storm of Dec. 28, 1993: Confirmation that the PDFs remain statistically invariant under the transformation t/L^z = constant. The top row shows PDFs for $\Sigma(\Delta \ln I) = 0.6$, $z = 0.58$ and pairs of (t, L) which satisfy t/L^z = constant (see Table 4.2). The bottom row shows the same for $\Sigma(\Delta \ln I) = 0.8$, $z = 0.51$. Similar result holds for $\Sigma(\Delta \ln I) = 0.7$, $z = 0.54$.

The above results imply that, given the statistical structure of rainfall evolution at large scales, the structure at small scales can be statistically predicted. Application examples are given in Section 4.4.2.

4.4   RAINFALL DOWNSCALING FOR HYDROLOGIC APPLICATIONS

4.4.1   Spatial Downscaling

Recall from Section 4.2, that knowing the spatial variability of rainfall fluctuations at several intermediate scales permits the statistical reconstruction of the rainfall intensities themselves from a larger scale to a smaller scale, e.g., from 64 km down to 4 km. Based on the scale invariance of standardised rainfall fluctuations (equation (4.1)) and the relation of the scaling parameter H to CAPE (equation (4.2)), a spatial rainfall downscaling scheme was developed and implemented to several mid-latitude mesoscale convective systems with considerable success (Perica and Foufoula-Georgiou, 1996b). This scheme is able to statistically reconstruct the small-scale spatial rainfall variability and the fraction of area covered by the storm, given the large-scale rainfall field and value of CAPE in the prestorm environment i.e., ahead of the storm (see Zhang and Foufoula-Georgiou, 1997 for a numerical study of the spatial variability of CAPE and definition of a representative value of CAPE for use in (4.2)).

This spatial downscaling scheme has the advantage that its parameterisation is scale-independent and thus offers the capability of resolving the subgrid-scale spatial rainfall variability at any desired scale (at least between 4 km and 64 km, that the relationships were developed) without the need to consider a separate parameterisation scheme at each scale. Figure 4.7 shows an example where the downscaling scheme was used to disaggregate rainfall from the scale of 64 km to the scale of 4 km. It is seen that the disaggregated field compares well to the actual field. More details and a formal statistical comparison of these fields can be found in Perica and Foufoula-Georgiou (1996b) and application to other mesoscale complexes in Perica (1995). It is noted that recent evidence by Zhang and Foufoula-Georgiou (unpublished manuscript) using multi-radar data of the COMET project, suggests that the predictive relationships between H and CAPE may hold up to scales of 256 km.

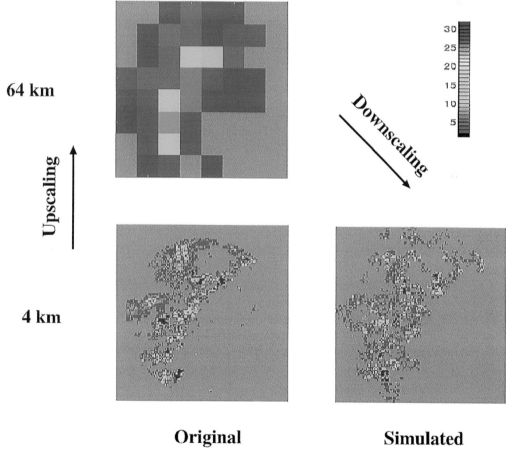

Figure 4.7. Spatial downscaling of the June 27, 1995 storm over Kansas–Oklahoma at 0300 UTC from 64 km to 4 km scale. A good agreement is found between the downscaled field (bottom right panel) and the observed field at the same resolution (bottom left panel). The intensities have been mapped onto the same 32 colours for display purposes.

4.4.2 Spatiotemporal Downscaling

Note that in the above spatial downscaling scheme, the spatial correlation of the small-scale rainfall is well preserved (see Perica and Foufoula-Georgiou, 1996b). However, if the scheme is applied independently at different instants of time, there is no guarantee that the temporal correlation (persistence) at the subgrid scales will also be preserved. In fact, a wet pixel at one instant might become dry at the next, still preserving the statistical spatial structure of the field. This might be a problem if the downscaled values were to be used in a continuous rainfall/runoff model where the "memory" of the system (e.g., soil moisture accumulation) must be well captured for accurate runoff prediction.

Recently, Venugopal et al. (1999b) proposed a new downscaling scheme which explicitly attempts the statistical preservation of both the spatial and temporal correlation of rainfall at the subgrid scales. This scheme takes advantage of the presence of dynamic scaling in rainfall evolution discussed in Section 4.3. There are several technicalities in the implementation of the spatio-temporal downscaling scheme, but the essence remains the following: small-scale space-time structures relate to larger-scale ones (in fact the PDFs of $\Delta \ln I$ remain statistically invariant) if an appropriate transformation of space and time, namely, $t \sim L^z$, is applied. Thus, given the statistical structure of rainfall at large scale, the small-scale space-time features can be statistically reconstructed based on dynamic scaling. This model is schematically depicted in Figure 4.8 and extensively discussed in Venugopal et al. (1999b).

Figure 4.9 shows the results of applying this space-time downscaling scheme to the storm of January 4, 1994 in Darwin, Australia. We started with the large-

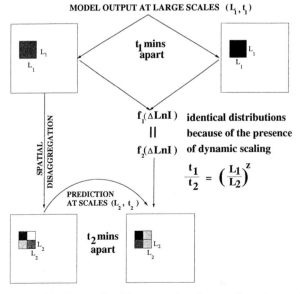

Figure 4.8. Schematic of the space-time downscaling scheme of Venugopal et al. (1999b) illustrating how the framework of dynamic scaling is coupled with an existing spatial disaggregation scheme to predict rainfall evolution at smaller space-time scales.

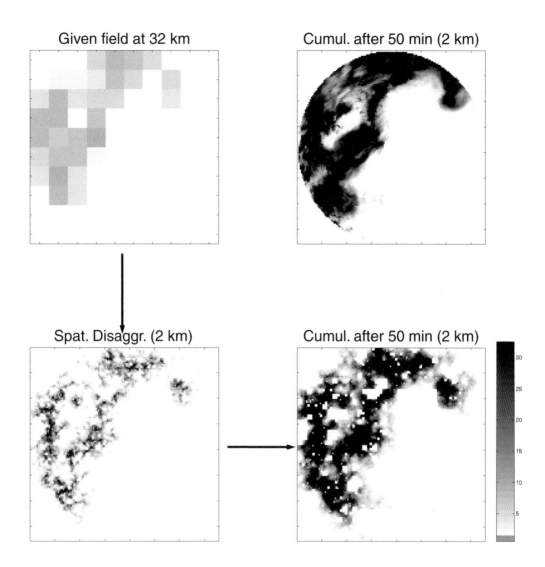

Figure 4.9. Storm of Jan. 4, 1994, in Darwin, Australia: Validation of the proposed space-time downscaling scheme by comparing the predicted 50-min cumulative rainfall patterns (bottom right panel) to the observed ones (top right panel).

scale field at 32 km (top left panel) at an instant of time (1831 UTC). This field was spatially downscaled to 2 km (see bottom left panel) using $H = 0.4$ in the scheme of Perica and Foufoula-Georgiou (1996b) discussed in the previous section. Then, the space-time scheme of Venugopal et al. (1999b) was used to evolve the 2 km field over time. It was assumed that large-scale (32 km) fields were available to us every 10 minutes. These data were used to update the distribution of changes at the large scale (32 km) which is identical to the distribution of changes at the small scale (2 km) according to the dynamic scaling hypothesis we put forth in Section 4.3.

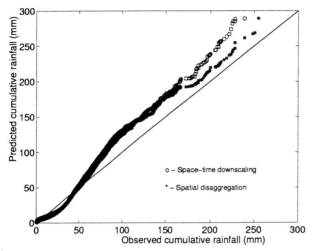

Figure 4.10. Storm of Jan. 4, 1994 in Darwin, Australia: Comparison of the observed 50-min cumulative rainfall amounts (in mm) to those predicted by space-time downscaling (circles) and spatial downscaling only (asterisks).

Figure 4.9 (bottom right panel) shows the 50-minute cumulative rainfall fields at a scale of 2 km predicted from the space-time downscaling scheme. In the same figure, the observed 50 minute cumulative field is also shown for comparison (top right panel). Visually, the space-time downscaled field compares well with the observed field (see Venugopal et al., 1999b for an extensive quantitative comparison). It is noted that, although for this storm the space-time downscaling scheme seems to overestimate the extreme 50-minute accumulations slightly more than the spatial downscaling scheme applied independently in time (see Figure 4.10), the space-time scheme is able to reproduce the temporal correlation at the sub-grid scales much better than the spatial scheme alone (see Figure 4.11).

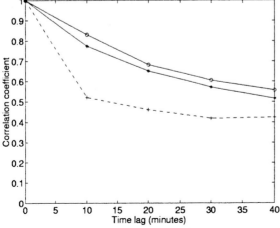

Figure 4.11. Storm of Jan. 4, 1994, in Darwin, Australia: Comparison of the temporal correlation at the 2-km scale computed from the observed fields (asterisks), from the predicted fields by the space-time downscaling scheme (circles) and by the spatial downscaling scheme only (dashed line).

Overall it is concluded that the space-time downscaling scheme discussed above is able to successfully reproduce the spatial and temporal correlation of rainfall at the subgrid scales given large-scale averages of precipitation and the downscaling parameters, H and z (see also Venugopal et al., 1999b for other applications and extensive quantitative evaluation of model performance). Providing downscaled 2 km precipitation fields in a rainfall-runoff model is expected to yield more accurate estimates of runoff and other fluxes as compared to those that would be obtained if the 32 km precipitation fields were used. Also, preserving temporal persistence in the downscaled rainfall is important in many hydrologic studies since the time-history of rainfall intensities is known to affect soil-moisture storage and runoff production from a basin. The exact degree to which the predicted runoff is affected by including or ignoring temporal persistence of rainfall at the subgrid scale remains yet to be quantified. Such a study is currently in progress in our group and the results will be reported in the near future.

4.5 CAN ATMOSPHERIC MODELS REPRODUCE THE OBSERVED PRECIPITATION PATTERNS?

Accurate forecasting of the onset, duration, location and intensity of precipitation via numerical weather prediction models, remains still a difficult challenge. Generally efforts in model development (in terms of physical and numerical advances) have outpaced efforts in detailed model validation. Specifically, methods that compare the forecasted high-resolution precipitation patterns to the observed ones, such that deficiencies in microphysical parameterisations and other small-scale structure representations can be depicted, lag behind. Traditional measures of forecast performance are too coarse for this purpose and provide only limited information about the ability of the numerical weather prediction model to mimic the dynamical environment of the storm which created the observed complex spatiotemporal rainfall pattern. Consequently, they also provide limited feedback as to how to go about improving the model (in terms of microphysical parameterisation, data assimilation, increased resolution, etc.) since they cannot quantitatively assess the detailed effect of these improvements on the forecasted precipitation pattern.

In a recent paper (Zepeda-Arce et al., 2000) several new multiscale statistical measures which can depict how well the small scale-to-scale variability and organisation of the forecasted fields matches that of the observed fields were proposed. It was demonstrated that indeed these measures are very informative compared to traditional measures of forecast verification and may lead to useful feedback for atmospheric model improvement. Some of these results are briefly discussed below.

In Zepeda-Arce et al. (2000), the multi-squall line of May 7–8, 1995 over Oklahoma was modelled by a state-of-the-art three-dimensional nonhydrostatic storm-scale prediction model (the Advanced Regional Prediction System – ARPS; see Droegemeier et al., 1996a,b; Xue et al., 1995; and Xue et al.,

2000a,b). The simulated rainfall patterns (available at the model resolution of 6 km) were compared to the observed ones (available at the radar resolution of 4 km) over a range of scales. Specifically, the predicted and observed fields were analysed for the presence of spatial and spatiotemporal scale invariances and the results compared.

The 18 minute rainfall accumulations from the observed radar patterns and 15 minute accumulations from the predicted patterns were analysed for scaling in spatial rainfall fluctuations. Good scaling was judged by a correlation coefficient of $R \geq 0.95$ in the log–log plots of the standard deviations of normalised spatial fluctuations with scale in the latitudinal and longitudinal directions and $R \geq 0.9$ in the diagonal direction (since there is greater uncertainty in estimating these values – see Perica and Foufoula-Georgiou, 1996b for their interpretation). Figure 4.12 shows the results of the analysis. When scaling was not present the estimated values of H are given, but the lack of scaling is marked on the plot by a dark square. As can be seen from Figure 4.12, the model was not able to reproduce the pronounced temporal variation of H_1, H_2, and H_3 during the storm evolution. Moreover, no directionality seemed to be present in the observed patterns ($H_1 \approx H_2 \approx H_3$) while the diagonal component (H_3) was significantly higher than H_1 and H_2 (by approximately 0.2) in the model. Generally, it was found that when scaling was present, $H_{\text{model}} < H_{\text{obs}}$. However, the standard deviations of the normalised fluctuations in all directions were higher in the model than in the observations, i.e., $\sigma_{\xi,\text{model}} > \sigma_{\xi,\text{obs}}$. These findings indicate that the model-predicted normalised rainfall fluctuations are in general more variable than the observed ones. However, the growth of this variability with scale is slower in the model than in the observations, at least for the longitudinal and latitudinal components. The drastic change in the values of H during the period of $t = 11$ to 13 hours (see Figure 4.12, top) was caused by the fact that at around $t = 11$ hours, a new squall-line started entering the domain of observation. This squall-line stayed in the domain until the end of the simulation period while the original squall-line moved out of the domain at around $t = 13$ hours. During the transition period ($t = 11$ to 13 hours) the statistical structure of the precipitation field within the domain of observation was different than before or after the transition. This, however, was not reproduced by the model-predicted patterns which had constant to only slightly increasing values of H during that period.

The radar-observed and model-predicted rainfall accumulations were also analysed for the presence of dynamic scaling within a stationary region of a few hours over which the mean and standard deviation of $\Delta \ln I$ did not vary significantly (see Zepeda-Arce et al., 2000 for details). For this region, the PDFs of $\Delta \ln I$ for spatial scales $L = 4, 8, 16$, and 32 km and temporal scales $t = 6, 12, 18, 24, 30, 36, 42, 48, 54$ and 60 min for the observed patterns were computed. For the predicted patterns the spatial scales were $L = 6, 12, 24$ and 48 km and temporal scales were $t = 15, 30, 45, 60, 75$ and 90 min. Then, the standard deviations of these PDFs were computed and (by interpolation in a tabular format) pairs of (L, t) were found which resulted in the same values of the standard deviation of $\Delta \ln I$ (see discussion in Section 4.3). This was done for

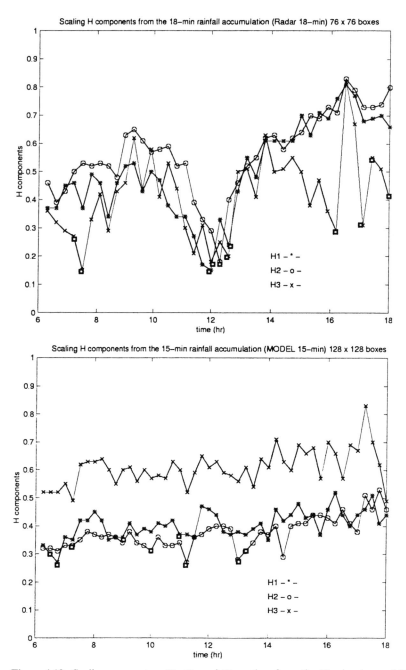

Figure 4.12. Scaling parameters H_1, H_2 and H_3 vs time from the 18-min observed (top) and 15-min simulated (bottom) rainfall accumulation patterns for the May 7–8, 1995 storm. Dark squares indicate the lack of scaling in normalised fluctuations.

both the observed and predicted patterns. These pairs (L, t) were plotted on a log–log plot as shown in Figure 4.13. It was observed that, to a good approximation, the (t, L) transformation under which the standard deviation of $\Delta \ln I$ remained constant was of the form $t/L^z = $ constant (i.e., straight-line relation-

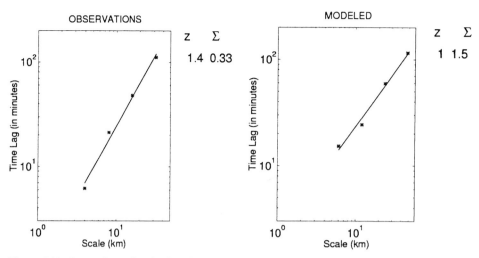

Figure 4.13. Dynamic scaling in the observed and predicted patterns of the May 7–8, 1995 storm over Oklahoma. For the observations, the iso-standard deviation line is for $\Sigma(\Delta \ln I) = 0.33$ and the estimate of $z = 1.4$; for the modelled field, $\Sigma(\Delta \ln I) = 1.5$ and $z = 1$.

ships on the log–log plots) for both the observed and predicted patterns. However, the values of z (estimated from the slopes of the log–log plots) were significantly different: $z = 1.4$ for the observed patterns and $z = 1.0$ for the predicted patterns. Also, the values of the standard deviations of $\Delta \ln I$ were much higher in the model than in observations. For example, in Figure 4.13, $\Sigma(\Delta \ln I)$ for the observations and the model-predicted fields was 0.33 and 1.5, respectively. This implies that for the same spatial scale and same time lag, the model-predicted fields change much more drastically over time than in the observations. This is consistent with the visual comparison of the fields and also with the differences in the estimates of z which as discussed below imply a faster temporal decorrelation in the predicted than the observed fields.

To understand the significance of the z value consider $L_2 = 2 \times L_1$ in the relationship $t_1/t_2 = (L_1/L_2)^z$. Then, for $z = 1$ (predictions) we get $t_2 = 2t_1$ and for $z = 1.4$ (observations) we get $t_2 = 2^{1.4} \times t_1 \simeq 3.4 \times t_1$. This implies that features twice as large will evolve two times slower in the predicted fields while they will evolve approximately 3.4 times slower in the observed fields. In other words, the predicted fields seem to have a faster decorrelation than the observed fields. It is interesting to compare the results of Figure 4.13 (and 4.5) with the schematic space-time diagram of hydrological processes in Figure 1.4 which shows a qualitatively very similar behaviour.

Another interesting comparison between the observed and predicted patterns resulted from comparison of their Depth–Area–Duration (DAD) curves for 1-hour duration. These curves plot depths of rainfall (here 1-hour accumulations) versus area of the storm over which these depths are equalled or exceeded. DAD curves compare the predicted and observed patterns in terms of their internal spatial structure irrespective of their locations. It was found that while the DAD curves tended to increase in the observed patterns from $t = 13$ to 15 hrs, they

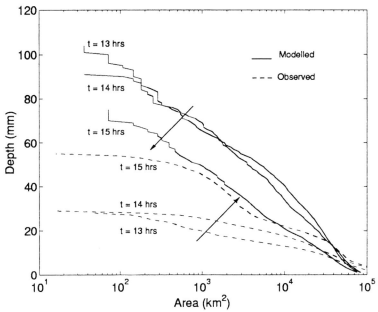

Figure 4.14. The solid lines show the DAD curves for $t = 13$, 14 and 15 hrs computed from the model-predicted 1-hour accumulation patterns; the broken lines show the same computed from the observed patterns.

tended to decrease in the predicted patterns (see Figure 4.14). Obviously, the tendency of the DAD curves to increase or decrease is related to the build-up or dissipation of the storm which might not be well reproduced in the model at least for the selected period of this particular storm. Such discrepancies would have significant effects on the predicted runoff and careful investigation and further study is warranted.

Overall, it was concluded that the above measures provided significant insight into subtle differences between the space-time structure of the predicted and observed patterns at a range of scales. These differences, although not fully interpreted yet, are much more informative than typical measures of forecast performance, such as threat scores and root-mean-square errors (see Wilks, 1995; or Fritsch et al., 1998). In addition, the scaling measures are normalised measures and are not influenced by how well the exact magnitudes and exact locations of rainfall intensities are predicted. Thus, they provide a direct assessment of how well the internal spatial structure and dynamics of the observed and predicted patterns compare to each other at a range of scales. It is hoped that these measures can provide useful feedback and guidance for improving numerical weather prediction models and this is an issue of current investigation.

4.6   CONCLUSIONS AND DISCUSSION

Many studies (for example, see Winchell et al., 1998 and references therein) have documented that the small-scale space-time variability of rainfall patterns has a significant effect on the infiltration-excess runoff volume produced by a basin. Thus, severe underestimation of the basin runoff may result when the small-scale precipitation variability is ignored and a low-resolution precipitation input is supplied into a distributed rainfall-runoff model. Similarly, if the small-scale variability of rainfall is erroneously reproduced by an atmospheric model and these predictions are used for hydrologic studies, a severe impact on the predicted runoff hydrograph may result. Moreover, concerns about the effects of climate change on water resources at the basin scale require the ability to downscale the large-scale general circulation model (GCM) predictions and to reconstruct the small-scale space-time rainfall variability to be used as an input to a hydrologic model.

In this paper, a review of a class of current approaches in parameterising the space-time variability of rainfall patterns at a range of space-time scales was presented. Although we concentrated more on methodologies developed in our group, other approaches such as Over and Gupta (1996), Seed et al. (1999) and Marsan et al. (1996) could be used in a similar way. We favour approaches based on scale-invariance because their parameterisations are scale-independent (for example, parameters H and z in the schemes presented earlier). Such parameterisations are attractive because they are parsimonious and may be related to physical properties of the storm environment as compared with parameters which depend on scale.

It is reminded that if high-resolution precipitation patterns are available, H can be estimated directly from them via a multiscaling analysis. However, if only large-scale patterns are available and need to be downscaled, the parameter H needed in the downscaling scheme must be externally predicted or assumed. The same applies for the parameter z. As discussed, the variability in the parameter H has been found empirically to be well explained by the variability in the convective available potential energy (CAPE) of the prestorm environment (Perica and Foufoula-Georgiou, 1996a) and thus CAPE can be used to predict H. Prediction of the parameter z, which parameterises the space-time evolution of rainfall at a range of scales from a similar physical observable quantity, has not been studied yet. It is anticipated that z might be related to the temporal evolution of a vertical instability measure, for example, the change of CAPE over time ($\partial CAPE/\partial t$) or to parameters describing parcel buoyancy and vertical wind shear (for example, see Weisman and Klemp, 1982). Empirical confirmation of the above assertions would require extensive meteorological observations not typically available (for instance, radiosonde observations are sparsely available in space and time). It could also well be that z can be empirically related to the standard deviation of $\Delta \ln I$ of the evolving fields. Although some preliminary evidence suggests such a possible relationship, a few cases deviated from this pattern. Analysis of more storms from different regions of the world and different climates must be done to provide an insight into the variability of the parameter z and its dependence on statistical or physical parameters of the atmosphere. Also, controlled experiments, via a state-of-the-art numerical weather prediction model which can simulate precipitation fields together with other physically-consistent parameters of the atmosphere, might provide a way of overcoming the lack-of-dense-meteorological-observations gap and at least point to possible predictive relationships of z which can be further verified from observations.

Once the parameters H and z are available, we presented evidence that the space-time downscaling scheme developed by Venugopal et al. (1999b) is fairly successful in reproducing not only the spatial correlation of the rainfall fields at the subgrid scale, but also the temporal correlation. Preserving the temporal correlation might be important when the downscaled precipitation fields are to be used in a rainfall-runoff model or a coupled atmospheric-hydrologic model to predict soil-moisture fluxes and water and energy partitioning over a basin. This remains to be demonstrated via simulation studies.

For prediction of severe flood events several hours ahead of time, we rely on state-of-the-art three-dimensional non-hydrostatic storm prediction models which can predict rainfall patterns to be used in a hydrologic model. Although numerical weather prediction models have advanced impressively over the last decade, quantitative evaluation of their performance as far as their ability to accurately reproduce the space-time structure of precipitation patterns at a range of scales, lags behind. In this chapter, we have presented some recent efforts to develop new multiscale measures for quantitative precipitation forecast (QPF) assessment. Preliminary results suggest that numerical weather prediction models might not always perform well in capturing the space-time organisation structure of the observed rainfall fields and, in particular, they might have a tendency to produce patterns with less scale-to-scale spatial variability and faster temporal decorrelation. Analysis of more cases is needed to fully quantify discrepancies between the statistical structure of predicted and observed precipitation patterns and understand the sources (such as, physical parameterisations, data assimilation methods, model resolution etc.) of these discrepancies. The increased availability of high-resolution data from NEXRAD and increased computational resources available for such studies offer unique opportunities for scientific breakthroughs and advances in atmospheric/hydrologic research, quantitative precipitation forecasting, and flood forecasting over small to large-size basins.

ACKNOWLEDGEMENTS

The material presented in this paper draws upon the research of several past and current students. Special thanks are due to Jesus Zepeda-Arce whose M.S thesis formed the basis for the materials in Section 4.5. This research was supported over the years by NSF, NOAA and NASA. Specifically, we acknowledge the support of the joint GCIP-NOAA/NASA program (grant NAG8-1519), the NASA-TRMM Program Grant (NAG5-7715), a NASA Earth Systems Science Fellowship (NASA/NGTS-5014) to Venugopal Vuruputur, and a grant jointly supported by NSF and NOAA under the U.S. Weather Research Program (ATM-9714387). We also thank Robert Houze of the University of Washington for providing us with the Darwin radar data, Jim Smith of Princeton University for providing us with the PRE-STORM data, and Kelvin Droegemeier of the University of Oklahoma for his collaboration on the storm-prediction research. Statistical analyses were performed at the Minnesota Supercomputer Institute, University of Minnesota.

5

Patterns and Organisation in Evaporation

Lawrence Hipps and William Kustas

5.1 INTRODUCTION

The evaporation of water is a crucial process in hydrology and climate. When the whole planet surface is considered, most of the available radiation energy is consumed in this process. However, a global view alone is insufficient to explain the codependence of surface hydrology and climate. Recent findings indicate that spatial variations in surface water and energy balance at various scales play a large role in the interactions between the surface and atmosphere. Advances in remote sensing have hastened the awareness of the spatial variability of the surface, and also offer some promise to quantify such variability. A point has been reached where the quantification of spatial patterns of evaporation is required in order to address current issues in hydrology and climate.

The evaporation of water at the surface and subsequent exchange with the lower atmosphere is a complex process – even for local scales and simple surfaces. When larger scales and spatial variations are considered, nonlinear processes may become pronounced, and further difficulties arise. Because of its great importance to hydrology and climate, considerable effort has been extended towards understanding and quantifying the evaporation process. Much is known about the process for uniform surfaces at local scales. However, current issues in hydrology and climate involve larger scales and non-uniform surfaces. Here there remains much to be learned. Note that evaporation can follow several avenues, including free water surfaces, soil surfaces, and transpiration by vegetation. Here we use the term evaporation in a generic sense, so that it is inclusive of any of these pathways.

5.2 GOVERNING FACTORS AND MODELS

5.2.1 Governing Factors

Before contending with spatial patterns there must be clear understanding of the processes important to a local surface. Because of the variety of ecosystems

Rodger Grayson and Günter Blöschl, eds. *Spatial Patterns in Catchment Hydrology: Observations and Modelling* © 2000 Cambridge University Press. All rights reserved. Printed in the United Kingdom.

and environmental conditions, the importance of various factors on evaporation differs from case to case. This can lead to confusion and improper generalisations about how to approach the process. We commence with a brief overview of the governing factors and subsequent interactions.

Water Supply

For land surfaces, the upper soil profile or root zone is the storage medium for water. The depth of soil in which water content must be considered must be commensurate with the root zone. Knowledge of surface water content alone is insufficient. Although soil water availability is a necessary condition for evaporation, the rate is not only a function of soil water. However, spatial variations in soil water play a direct role in spatial patterns of evaporation.

Available Energy

When water is sufficiently available, evaporation often proceeds at a rate that is proportional to available energy, usually defined by $R_n - G$, where R_n is net radiation and G is energy flowing into the soil. The large value of latent heat causes a great deal of energy to be consumed when water is available. This has led many models to treat evaporation as proportional to the available energy, and reflects the historical bias of research towards surfaces with relatively large water supplies.

Saturation Deficit

The very large negative values for water potential in the atmosphere require more useful variables such as vapour pressure or specific humidity. The gradient in humidity between the surface and the air has historically been replaced by the saturation deficit of the air, in order to linearise equations and avoid explicit dealings with surface temperature. When surface humidity values are large enough, saturation deficit effectively represents the gradient in water potential.

Turbulence Transport

Supply of water, energy, and a gradient of humidity are not enough to maintain the process, however. The water vapour must be transported away from the surface into the atmosphere, or the humidity gradient would soon decay and reduce the evaporation. So wind and turbulence play a critical role in maintaining values of saturation deficit. Unfortunately, turbulence is a very complex process without an analytical solution. As a result, it is inevitably parameterised in any treatment of evaporation.

Stomatal Conductance

Finally, when plants are considered, the situation becomes much more complex. Plants are *living* things, which limits the use of physical laws and mathematics to describe the processes. The response has been to focus on the behaviour of the stomates, since water vapour must pass through these structures. Indeed, stomatal conductance is a key mechanism by which we account for the role of the vegetation in this process.

Although stomatal conductance of plants has been studied for many years, predicting the exact behaviour remains somewhat elusive. We know that there are connections between stomatal conductance, transpiration, and several atmospheric variables such as saturation deficit. The connections between the processes are examined at scales from the sub-leaf to canopy by Jarvis and McNaughton (1986). However, the concepts of cause and effect are tenuous. Historically, stomatal conductance was assumed to respond to saturation deficit, and thereby affect transpiration. However, Mott and Parkhurst (1992) showed that transpiration may respond directly to saturation deficit, and stomatal conductance adjusts in response to transpiration. Monteith (1995a) reanalysed 52 data sets, and concluded that they support this hypothesis. Monteith (1995b) discusses the implications of this issue on approaches to model evaporation. Clearly there are complex and nonlinear interactions between plant water status, stomatal conductance, transpiration, and various atmospheric factors. The role of living vegetation in the process is not treated very directly at present.

5.2.2   Problems of Nonlinearity

A major difficulty in modelling evaporation is the strong dependency among the variables. In fact, there are no independent variables as such. Changes in any of the critical factors in principle induce changes in all others, until a new equilibrium can be reached. At small spatial scales the nonlinearities are not always very evident. Hence, many of these have historically been ignored or hidden inside the definitions of various parameters. Indeed, the common consideration of a very shallow layer of atmosphere above the surface does not allow for many of the critical feedbacks. The solution to this problem will be discussed shortly. It involves examination of the entire atmospheric boundary layer.

5.2.3   Models Describing Evaporation

Penman–Monteith Equation

This expression is the most fundamental equation available to examine the evaporation process. It is strictly valid for a leaf, but is generally considered at the scale of a canopy. A uniform surface is implicitly assumed. The equation is developed by linearising the vapour pressure gradient term, to remove any explicit dependence on surface temperature. The final equation is:

$$E = \frac{s \cdot (R_n - G) + \rho \cdot c_p \cdot D / r_a}{s + \gamma \cdot (1 + r_c / r_a)} \tag{5.1}$$

Here s is the slope of the saturation specific humidity versus temperature relation, ρ is density of air, c_p is specific heat of air, γ is c_p / L where L is latent heat of vaporisation, D is saturation deficit or saturation minus actual specific humidity, r_a is aerodynamic resistance, and r_c is stomatal resistance.

The role of turbulence and stomatal behaviour are both collapsed into resistance terms. Also note that for scales larger than a single leaf, the stomatal resistance term represents some bulk or effective value for the surface. The value of D is generally specified near the surface. Thus, there is no explicit allowance for connections and exchanges with a deeper layer of atmosphere. This equation is a diagnostic equation describing the relationships between key factors of the system. It represents a tool to examine interactions between evaporation and critical factors in the soil, vegetation, and atmosphere.

Simplifications for Special Cases

For extensive surfaces covered with vegetation, the evaporation is large and convection is small. This leads to poor coupling between the surface and atmosphere, and evaporation becomes energy limited. The evaporation flux by definition must approach the value of available energy. This value is called *equilibrium evaporation* (E_{eq}). For extensive vegetated surfaces the actual evaporation is strongly proportional to E_{eq}. This led Priestley and Taylor (1972) to propose that:

$$E = \alpha \cdot E_{eq} \tag{5.2}$$

where α is a parameter, originally defined as 1.26, although McNaughton and Spriggs (1989) demonstrate that α is not constant and depends on dynamic interactions between the surface and atmospheric boundary layer. Nevertheless, this equation is a useful tool for the special case of large and uniform regions with complete vegetation.

Use of Surface Temperature to Estimate Evaporation by Residual

If the entire energy balance equation is considered, E can be estimated by the residual if the other terms are calculated and measured. This involves determination of sensible heat flux (H). Remote sensing methods allow estimation of the surface temperature, which can be used with air temperature to estimate H using similarity theory, as described later. Since remote sensing techniques can sometimes retrieve spatial fields of surface temperature, such an approach can estimate spatial distribution of evaporation. Examples of this approach will be discussed in Section 5.6.

Coupling of Surface Energy Balance to the Atmospheric Boundary Layer

Most of the historical study of evaporation has been conducted at local scales, and considered a layer of atmosphere only a few metres above the surface. This ignores the role of large-scale atmospheric properties and the feedback between the surface and the atmosphere.

Recently, several studies have demonstrated the need to consider a continuous and interactive system that often includes the atmospheric boundary layer (ABL) as well as the air above it. McNaughton and Jarvis (1983) and McNaughton and Spriggs (1986) demonstrate how a growing ABL can entrain warm, dry air from aloft which mixes down to the surface. This can raise the value of saturation deficit, and enhance evaporation rates. The system is coupled, in that changes in the surface heat and evaporation rates affect the growth of the ABL, which in turn can feed back to alter the surface fluxes. These processes were combined into an elegant model posed by McNaughton and Spriggs (1986). These connections between the surface energy balance and the ABL must be considered in the process. They become especially important for regional scales, or to consider spatial variations in surface fluxes.

5.3   ESTIMATION OF EVAPORATION RATES USING MEASUREMENTS

There are several approaches either to measure evaporation directly, or to estimate it from other measurements. We will cover the most common and reliable approaches.

5.3.1   Local Scales

Eddy Covariance

This is the most direct approach, and attempts to actually measure the flux. The flux of water vapour can be described as:

$$E = \overline{w} \cdot \overline{\rho_v} + \overline{w' \rho_v'}$$ (5.3)

where ρ_v is water vapour density, and w is the vertical wind velocity. The primes indicate instantaneous deviations from the temporal mean. The first term represents flux due to the mean vertical wind, while the second term is the turbulence flux. In many conditions over flat surfaces with a suitable averaging period, the mean vertical velocity should be zero. The first term then vanishes, leaving:

$$E = \overline{w' \rho_v'}$$ (5.4)

The turbulence flux is equal to the covariance of the vertical wind velocity and a scalar such as water vapour density. In practice, it is not as simple as it appears. Determination of the "suitable" averaging period, presence of non-stationary conditions, non-zero mean vertical velocities, and other issues, pose challenges to making quality flux measurements. These problems are discussed in Mahrt (1998) and Vickers and Mahrt (1997). Some of these issues are also denoted in Baldocchi et al. (1988).

Bowen Ratio

If the evaporation and sensible heat fluxes are expressed in terms of turbulence diffusivities and gradients, then the ratio of sensible to latent heat flux, or Bowen Ratio, can be approximated as:

$$B = \frac{c_p \cdot \Delta \overline{T}}{L \cdot \Delta \overline{q}}$$ (5.5)

Critical assumptions made here include equality of turbulence diffusivities for heat and water vapour, and replacing finite differences for differential values of gradients. The energy balance equation can be used with (5.5) to obtain:

$$E = \frac{R_n - G}{1 + B} \tag{5.6}$$

If measurements of available energy and vertical changes in temperature and humidity are made, E can be calculated. This assumes that available energy can be measured *without error*. For uniform surfaces with large values of vertical gradients, the Bowen Ratio technique works well. However, for heterogeneous surfaces, the assumption of equality in heat and water vapour diffusivities is likely to be violated.

Flux Gradient Approach – Monin–Obukhov Similarity Theory

Monin-Obukhov Similarity theory (MOS) can be used to estimate the vertical profiles of wind speed as well as momentum, heat and water vapour fluxes with only a few parameters. It is based on an assumption that the turbulent transport of a quantity is proportional to the product of the turbulence diffusivity, K, and the vertical gradient in mean concentration C. The height-dependent eddy diffusivity is assumed to be a function of the momentum transport and atmospheric stability. For momentum, heat and water vapour, the gradients are related to the fluxes using similarity parameters. Integrated forms of the resulting expressions have been derived (Brutsaert, 1982).

The fact that stability functions continue to be modified, raises concern about the reliability of using gradient type approaches for estimating fluxes. Large Eddy Simulation (LES) suggests that boundary layer depth has an indirect influence on MOS scaling for wind (Khanna and Brasseur, 1997). Williams and Hacker (1993) show that mixed-layer convective processes influence MOS and support the refinements made by Kader and Yaglom (1990). Clearly there are still considerable uncertainties as to the exact forms of the mean profiles as both surface heterogeneity as well as mixed-layer convective processes affect the idealised MOS profiles.

When surface values of temperature and humidity are determined, only values at one height in the surface layer are needed, along with an estimate of the surface roughness for momentum, z_{Om}, and heat, z_{Oh}, and water z_{Ow}, and surface humidity. For heterogeneous surfaces, z_{Oh} has little physical meaning, but there has been more progress in relating z_{Om} to physical properties of the surface (e.g., Brutsaert, 1982).

5.3.2 Regional Scales

Aircraft-based Eddy Covariance

Aircraft-based flux systems can in theory provide large-area flux estimation both in the surface layer and throughout the ABL. However, in a number of field programs, the latent and sensible heat fluxes measured by aircraft tend to be smaller than those measured by towers several metres above the surface (Shuttleworth, 1991). Sampling errors for both tower and aircraft-based systems are discussed by Mahrt (1998). Under nonstationary conditions, procedures for estimating sampling errors are invalid. Moreover the flux estimate is sensitive to the choice of averaging length. Vickers and Mahrt (1997) and Mahrt (1998) describe the use of a quantity called the nonstationarity ratio, to define when significant errors may exist in the measurements. Processing of aircraft measurements is considerably more involved than tower data, and collection of the data is quite expensive. However, it is the only method to directly estimate fluxes and their spatial variations at regional scales.

Regional Fluxes and Properties of the ABL

Since the atmospheric boundary layer is connected to surface processes at a regional scale, there must be a relationship between the regional surface fluxes and properties of the ABL. One approach to this issue has been to use a similarity theory for the ABL to estimate fluxes from vertical profiles of wind, temperature, and humidity in the ABL (Sugita and Brutsaert, 1991) measured using soundings from radiosondes.

A different approach presented by Munley and Hipps (1991), Swiatek (1992), and Hipps et al. (1994), related temporal changes in ABL properties to surface fluxes using fundamental governing equations for temperature and humidity. The latter two studies suggested that horizontal advection in the ABL was an important process affecting the ability to recover reasonable surface flux values. When a crude estimate of this process was made, agreement of ABL estimates with measured surface fluxes was reasonably good for two semi-arid ecosystems. However, in the application of this approach over other semi-arid landscapes containing significant variability in surface fluxes, greater discrepancies with flux observations, especially in evaporation, have been found (Kustas et al., 1995; Lhomme et al., 1997). One of the reasons for this scatter is footprint issues.

5.3.3 Footprint Issues

In order to interpret an estimate of a surface flux of mass or energy, one must know from where the flux originated. A source area or region upwind of the surface contributes to a measured flux at a given height. This source area is called the "footprint" and is the area over which measurements are being influenced (see Chapter 2, p. 19) The contribution from each surface element varies according to upwind distance from the location of the measurement, and atmospheric diffusion properties. In order to determine the region associated with a flux value or the footprint, some type of model must be used.

There are two main approaches in footprint models: analytical solutions to the diffusion equation, and Lagrangian models. The analytical approaches derive solutions to the diffusion equation using parameterisations such as similarity theory for turbulence diffusion. There are also other critical assumptions made, such as no spatial variation in the surface flux. This results in equations that require only a few inputs, and are relatively easy to implement. Lagrangian models are more complex and numerically simulate the trajectories of many thousands of individual particles. Knowledge of the turbulence field is needed to allow the trajectories to be computed. When the results of many particle journeys are compiled, the relative contribution of various upwind distances to the flux can be determined. Examples of the analytical category are Schuepp et al. (1990), Horst and Weil (1992), and Schmid (1994). Lagrangian approaches are presented in Leclerc and Thurtell (1990) and Finn et al. (1996).

For heterogeneous surfaces, knowledge of the footprint of any flux measurement is absolutely necessary, in order to interpret spatial variations in fluxes. A current limitation is that present footprint models generally assume a spatially constant flux at the surface. In reality, fluxes will vary in space. The effects of spatial variations in surface properties and fluxes on the resulting footprints remain to be determined, i.e. the measurements represent the bulk effects but we cannot use them to easily define detail of the spatial patterns.

5.4   SPATIAL VARIATIONS OF EVAPORATION

It is of great importance in hydrology to be able to quantify the spatial distribution of evaporation. It certainly has some connections to the traditional hydrologic outputs at the catchment scale, such as streamflow. However, the spatial distribution of water balance, especially at larger scales, has strong connections with the atmospheric conditions and hydroclimatology of a region. Qualitatively, the important surface properties that relate to spatial variations in evaporation are understood rather well. Spatial changes in water balance are connected to those of the root zone soil moisture, vegetation density, stomatal conductance, net radiation, saturation deficit, and turbulence intensity.

There have been some advances in determination of spatial fields of some of the above properties using remote sensing information. In particular, net radiation, *surface* soil moisture, and vegetation density can be estimated spatially with remote sensing and auxiliary data (Kustas and Humes, 1996; Carlson et al., 1994).

We can define several issues that pose difficulties in assessing the spatial patterns in water balance, including difficulties associated with the definition and description of heterogeneous surfaces, and the effects of such surfaces on fluxes and the aggregation of fluxes over the landscape. These must be resolved in order to develop the ability to quantify spatial variations in the surface fluxes.

5.5 DIFFICULTIES POSED BY HETEROGENEOUS SURFACES

When surfaces are heterogeneous, several issues arise. First, most models and measurement approaches either explicitly or implicitly assume a uniform surface. Second, the spatial variability in critical properties can cause nonlinear processes to become important.

5.5.1 The Notion of Heterogeneity

Heterogeneity is a rather descriptive term, and is often used somewhat ambiguously. Unfortunately, there is at present no universal approach to quantify the degree of heterogeneity. This is partly because the importance or effects of nonuniformity seem to depend upon the process that is being considered. The difficulty in quantifying what we mean by heterogeneity is indicative of the complexity of the entire issue of water and energy balance of inhomogeneous surfaces. Here we discuss some of the recent approaches to this problem.

Heterogeneity exists at all spatial scales, from variations within individual leaves (Monteith and Unsworth, 1990), to the canopy level where evaporation and sensible heat may originate from significantly different sources (Shuttleworth and Wallace, 1985), to larger scales where nonuniformity can affect atmospheric flow (Giorgi and Avissar, 1997). Besides scale, the type of heterogeneity may also be important. For example, de Bruin et al. (1991) showed that variations in temperature and humidity fields have a different effect on Monin–Obukhov similarity than variations in the wind field.

For purposes of estimating evaporation either directly via measurement of eddies, or indirectly using flux–gradient relationships, heterogeneity at the canopy scale and larger is of primary concern. At smaller scales, physically-based methods which consider both biological and fluid dynamics have been developed for scaling from the leaf to canopy scale (Norman, 1993; Baldocchi, 1993). However, they can be quite complicated and may only be applicable under ideal conditions, such as a canopy that is horizontally homogeneous (Baldocchi, 1993). The issue is how to define when the surface can no longer be treated as homogeneous.

5.5.2 Determining when a Surface is Heterogeneous

No exact methodology or theory exists to determine a priori when a surface can no longer be considered uniform. Measurement of turbulent fluxes and statistics is one indirect method, where deviation of the Monin–Obukhov similarity functions from those determined over uniform surfaces has been shown to be an indicator of heterogeneity (e.g., Chen, 1990a,b; de Bruin et al., 1991; Roth and Oke, 1995; Katul et al., 1995). Similarity theory requires the correlation between temperature and humidity to be near unity. This is not true for non-uniform surfaces (Katul et al., 1995; Roth and Oke, 1995), due to the source and/

or sink of evaporation differing from that of sensible heat flux. Unfortunately, these approaches do not provide a measure of the degree of heterogeneity.

Remote sensing may hold potential as a means of quantifying surface spatial variability by calculating spatial power spectra for surface radiance or reflectance values (Hipps et al., 1996). This requires pixel resolution fine enough to discriminate between plant and soil, which is often not available from satellites. Moreover, the shape of spatial power spectra depends upon the spatial resolution of the surface data (Hipps et al., 1995). This brings forward a critical issue. The degree of heterogeneity or spatial variability may be dependent upon the spatial resolution at which the surface is observed (see Chapter 2, p. 19).

Another indirect approach suggested by Blyth and Harding (1995) uses remotely sensed surface temperature along with wind and temperature profiles in the surface layer, to derive the roughness lengths of heat and momentum. The relationship between these values is related to heterogeneity of the surface. Both theory and observations indicate that transfer of momentum is more efficient than heat (Brutsaert, 1982). For homogeneous surfaces the ratio of roughness length for momentum, z_{OM}, and heat, z_{OH}, is essentially a constant, usually expressed as the natural logarithm $\ln(z_{OM}/z_{OH}) = kB^{-1}$ where $kB^{-1} \sim 2$. Many studies, especially for partial canopy cover surfaces, have found kB^{-1} significantly larger than 2 with values generally falling between permeable-rough, $kB^{-1} \sim 2$, and bluff-rough, $kB^{-1} \sim 10$ (Verhoef et al., 1997). So the ratio of the roughness lengths is an indirect indicator of the degree of departure from a uniform surface. This result is caused by several factors which include effects of the soil/substrate on the remotely sensed surface temperature observation, canopy architecture and the amount of cover (McNaughton and Van den Hurk, 1995).

5.5.3 Application of Single and Dual-source Approaches to Heterogeneous Surfaces

There is a fundamental problem in representing a heterogeneous surface as a single layer or source, which is implicit in the application of, for example, the Penman–Monteith equation, because of the significant influence of the soil/substrate on the total surface energy balance. Thus, the surface resistance to evaporation has lost physical meaning because it represents an unknown combination of stomatal resistance of the vegetation and resistance to soil evaporation (Blyth and Harding, 1995). This has prompted the development of two-source approaches, whereby the energy exchanges of the soil/substrate and vegetation are evaluated separately (e.g., Shuttleworth and Wallace, 1985). Nevertheless, some studies reported the Penman–Monteith equation to be useful for evaporation estimation over heterogeneous surfaces (e.g., Stewart and Verma, 1992; Huntingford et al., 1995). In fact Huntingford et al. (1995) found little difference in performance of two-source approaches versus the Penman–Monteith for a Sahelian savanna. However, these studies arrive at reliable evaporation estimates only after the stomatal response functions are optimised with the measurements from the particular site. Therefore, as a predictive tool, the Penman–Monteith approach will be tenuous for heterogeneous surfaces without a priori calibration. By performing such a priori calibration, much simpler formulations such as the Priestley–Taylor equation can yield evaporation predictions similar to two-source approaches for heterogeneous surfaces (Stannard, 1993).

5.5.4 Application of Surface-layer Similarity above Heterogeneous Surfaces

For several decades Monin–Obukhov Similarity (MOS) theory has been used to relate mean profiles of scalars and wind to the turbulent fluxes of heat and momentum (Brutsaert, 1982; Stull, 1988). However, serious limitations exist in the application close to the canopy due to roughness sublayer effects (e.g., Garratt, 1978, 1980). For heterogeneous surfaces we are presently unable to resolve the relative influence of all the mechanisms involved, and more importantly have been unable to develop a unified theory to correct MOS for effect of the roughness sublayer on mean profiles and turbulent statistics (Roth and Oke, 1995).

An example of the effect of heterogeneity on MOS profiles is shown in Figure 5.1 for a desert site containing coppice dunes and mesquite vegetation (Kustas et al., 1998). In Figure 5.1 d_0 is the zero plane displacement. This is a length to account for the fact that in tall vegetation, the source and sinks are above the ground surface, so the heights are specified as distances above a new reference value which makes the relationship between fluxes and gradients valid. While the roughness sublayer does not appear to affect the wind profile, the actual temperature profile departs significantly from the idealised MOS predicted profile. This is probably due in part to the complicated source/sink distribution of heat (Coppin et al., 1986). Over this site, the heat sources are the interdune regions and heat sinks are mesquite vegetation randomly distributed over the surface. As a result, significant scatter between predicted and measured heat fluxes has been reported using the above MOS equations (Kustas et al., 1998).

5.5.5 Effects of Heterogeneity on Surface Fluxes and Aggregation

As mentioned, determination of the spatial distribution of the critical surface properties that relate to evaporation is becoming possible at many scales with advances in remote sensing. However, there are issues about how to properly determine and interpret variables of interest from remote sensing data. For example, the interpretation of radiometric temperature in terms of the heat flux process is far from simple (Norman and Becker, 1995). Remote sensing estimates of vegetation are subject to variations in density and geometry. Only

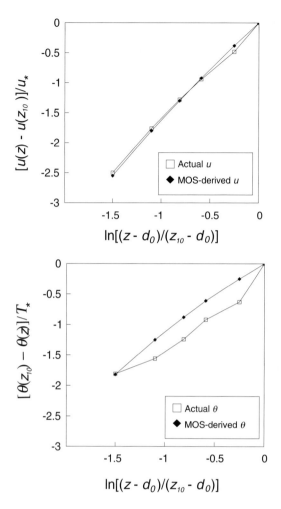

Figure 5.1. Plots of normalised wind $[u(z) - u(z_{IOm})]/u_*$ and temperature $[\theta(z_{IOm}) - \theta(z)]/T_*$ versus $\ln[(z - d_0)/(z_{IOm} - d_0)]$, with u_* and T_* estimates from the eddy covariance measurements. Actual versus MOS-derived normalised profiles of u and θ representing an average of all unstable profiles (see Kustas et al., 1998).

upper soil moisture can be estimated by remote sensing, while plants respond to water in the entire root zone.

In order to model the fluxes, the actual patches of surface types must be delimited. Identifying various patches is not trivial, as it requires determination of the properties that are of hydrological importance, as well as the magnitude of spatial changes which are significant. Also, the scales of heterogeneity must be determined so that the models can be implemented at commensurate spatial scales, i.e. the characteristic scale of the process must match the modelling scale (see Chapter 2, p. 27).

However, even if there were complete knowledge of the distribution of the critical biophysical properties of the surface, there are other issues to be addressed. At some scales of heterogeneity, nonlinear effects may become important. For example, the properties and processes at one surface may affect those of a nearby surface. Several examples can be posed here. Significant spatial changes in surface water balance, common in semi-arid regions, result in transport by the mean wind of heat and saturation deficit from drier to wetter surfaces. This can enhance the evaporation and alter the energy and water balance of the latter surfaces. This effect of advection on evaporation is detailed in Zermeño-Gonzalez and Hipps (1997). In addition, Avissar (1998) has shown results with mesoscale models that suggest secondary circulations can form between warm and cool adjacent patches. These may carry significant vertical fluxes of mass and energy, which will not be reflected in local measurements of turbulence transport, nor accounted for in models treating each spatial surface element independently.

Finally, the fluxes and governing properties do not both aggregate linearly. The actual surface fluxes can be added linearly (the flux from each spatial element can be summed, and normalised to yield average flux). However, the spatial averages of the critical properties when input into the flux equation, do not yield the correct value for the average flux (see the discussion on effective parameters in Chapter 3, p. 68). Since, we generally have available, at best, the spatial distribution of the surface properties, the aggregation up to larger regions is a problem.

Ultimately, the above factors create difficulties in properly aggregating the fluxes up to larger regions. This so-called aggregation problem remains unsolved in a general way at present. However, remote sensing may provide spatially distributed hydrologic information critical in addressing scaling issues (Beven and Fisher, 1996). There are several directions which have been posed. These include the determination of effective parameters for surface properties (Lhomme et al., 1994), and treating surface properties as probability density functions, and inputting them into mesoscale atmospheric models (Avissar, 1995). We do not directly address this issue here, but simply note that the spatial distribution of evaporation and the aggregation problem are ultimately connected.

In the meantime there have been attempts to estimate spatial patterns of evaporation using a combination of modelling and remotely sensed information. As a result of the problems discussed above, these methods can be used only under restrictive assumptions and require data that is not commonly available. Nevertheless, they provide a way forward.

5.6 EXAMPLES OF ESTIMATING SPATIAL VARIATIONS OF EVAPORATION

Surface energy balance models using remotely sensed data have been developed and used in generating spatially distributed evaporation maps (Kustas and Norman, 1996). For many of these models, surface temperature serves as a primary boundary condition (e.g., Bastiaanssen et al., 1998). Clearly, the spatial variation of surface temperature is not enough to estimate the variation in evaporation since the amount of vegetative cover, water deficit conditions, and aerodynamic roughness strongly influence the turbulent transport and thus the aerodynamic–radiometric temperature relationship (Norman et al., 1995).

Promising approaches described below, explicitly evaluate flux and temperature contributions from the soil and vegetation using the conceptual modelling philosophy of Shuttleworth and Wallace (1985). The modelling strategy is to consider the Penman–Monteith type of approach strictly for the vegetated fraction, and a similar resistance type analogue for the soil component (i.e. a two-source approach). In this case, the vapour pressure gradient term is not linearised as in equation (5.1), but is a function of the vegetation and soil temperatures which is derived from remotely sensed observations of canopy cover and surface temperatures and model inversion. Along similar lines, the approach of Norman et al. (1995) uses the Priestley–Taylor approximation for the vegetated component only, but with the extension that the alpha value can approach zero (i.e., no transpiration). This is necessary since the model is constrained by both the energy balance and radiative temperature balance between model-derived component temperatures and the remotely sensed surface temperature observations.

While the above formulations address the issue of aerodynamic-radiometric temperature relationships, determining spatially distributed heat fluxes at regional scales will invariably require incorporating surface–atmospheric feedback processes. Several approaches have made significant progress in this area. Following Price (1990), Carlson et al. (1990, 1994) combined an ABL model with a soil–vegetation–atmosphere–transfer (SVAT) scheme for mapping surface soil moisture, vegetation cover and surface fluxes based on a fundamental relationship between vegetation index (i.e., cover) and surface temperature. Using ancillary data (including a morning sounding, vegetation and soil type information), root-zone and surface soil moisture are varied, respectively, until the modelled and measured surface temperatures are closely matched for both 100% vegetative cover and bare soil conditions. Further refinements to this technique have been developed by Gillies and Carlson (1995), for potential incorporation into climate models. Comparisons between model-derived fluxes and observations have been made by Gillies et al. (1997) using high resolution aircraft-based remote sensing measurements. Approximately 90% of the variance in the fluxes was captured by the model for the conditions of their study.

The Two-Source Time-Integrated model of Anderson et al. (1997) (presently called ALEXI), provides a practical algorithm for using a combination of satellite data, synoptic weather data and ancillary information to map surface flux components on a continental scale (Mecikalski et al., 1999). The ALEXI approach builds on the earlier work with the Two-Source model (Norman et al., 1995) by using remote brightness temperature observations at two times in the morning hours, and considering planetary boundary layer processes. The methodology removes the need for a measurement of near-surface air temperature and is relatively insensitive to uncertainties in surface thermal emissivity and atmospheric corrections on the GOES brightness temperature measurements. Anderson et al. (1997) and Mecikalski et al. (1999) have shown that surface fluxes retrieved from the ALEXI approach compare well with measurements, albeit under some restrictive assumptions. The ALEXI approach is a practical means to operational estimates of surface fluxes over continental scales with 5–10 km pixel resolution. It also connects the surface properties and processes with the development of the atmospheric boundary layer, which is necessary to realistically describe the system.

A relatively simple two-source model using the framework described by Norman et al. (1995) has been used to generate surface flux maps (Kustas and Humes, 1996; Schmugge et al., 1998). The model was designed to use input data primarily from satellite observations. Several simplifying assumptions about energy partitioning between the soil and vegetation reduce both computational time and input data required to characterise surface properties. The inputs include an estimate of fractional vegetative cover, canopy height, leaf width, surface temperature, solar radiation, wind speed and air temperature. The remote sensing data from the Monsoon '90 experiment (Kustas and Goodrich, 1994), conducted in a semi-arid rangeland catchment in Arizona, have been used to evaluate the model. An example of an evaporation map generated from the two-source model is shown in Figure 5.2. A Landsat-5 TM image was used to generate a fractional vegetative cover and land use map for deriving vegetative height and roughness. A network of surface flux stations (approximate locations displayed as discs in the figure) provided spatially distributed solar radiation, wind and air temperature observations (Kustas and Humes, 1996). Aircraft surface temperature observations for a day with the largest variation in moisture conditions were used. The pixel resolution is 120 m, similar to the resolution of Landsat TM thermal band. The calculated latent heat flux field shows a wide range in values from about 50 to nearly 500 W m^{-2}. This variation is due in part to a recent precipitation gradient over the study area, with essentially no rainfall occurring in the western quarter of the image and gradually increasing to significant amounts in the north-eastern portion (Humes et al., 1997). In addition, the model computes higher evaporation rates for the areas along the ephemeral channels (the green and blue stripes) which contain more and taller vegetative cover, since there is typically more available water in these areas.

Comparison of model versus observed half-hourly latent heat flux from the flux measurement sites is illustrated in Figure 5.2 (values in W m^{-2}). There is qualitative agreement between model and observed fluxes (i.e., higher observed latent heat fluxes are in areas with higher modelled fluxes). However, it is not straightforward to determine how to weight the pixels within the source footprint of the observations. Note that patches with the highest and lowest latent heat fluxes were not within the observation network. This makes it difficult to validate regional flux models with a network of local flux measurements in heterogeneous regions (Kustas et al., 1995). Several pixels surrounding the eight surface flux stations were averaged for three days in which soil moisture conditions were different. The comparison between model and observed latent heat fluxes is illustrated in Figure 5.3. A standard error of approximately 30 W m^{-2} and $R^2 = 0.8$ is obtained. These are similar to the results found in the other modelling studies described above.

These examples illustrate that, despite the conceptual problems identified earlier in the chapter, we have made progress towards methods for estimating spatial

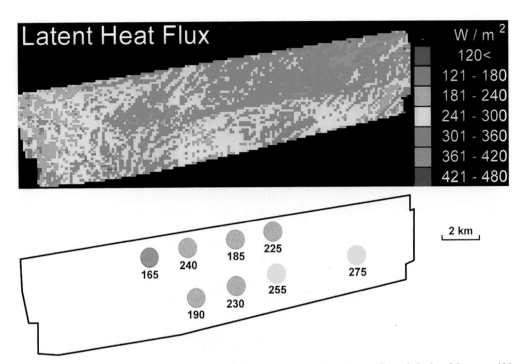

Figure 5.2. Evaporation image created from remote sensing data collected during Monsoon '90 used in a simple two-source model described in Norman et al. (1995) and estimates of evaporation from metflux stations (discs). Note that the size of the discs does not represent the measurement area. See also Kustas and Humes (1996).

variations in evaporation. Presently, these are applicable only under special circumstances, requiring detailed remote sensing data, cloud-free conditions, some limiting assumptions related to the "footprint" problem, and provide only a snapshot view of spatial variations.

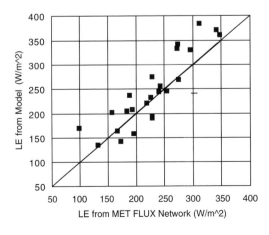

Figure 5.3. Comparison of two-source model-derived LE versus LE observations from the METFLUX network for three days of aircraft remote sensing observations during the Monsoon '90 experiment. See Kustas and Humes (1996) and Schmugge et al. (1998) for details.

5.7   CURRENT FRONTIERS IN EVAPORATION RESEARCH

There are several problems that presently limit our abilities to examine and model spatial variations in evaporation. These include capabilities of making accurate measurements of critical processes over appropriate scales, as well as missing theoretical knowledge about processes and scaling issues.

5.7.1   Measurement Issues

Available Energy

Ultimately, the energy and water balances are inextricably connected. When we consider spatial distribution of fluxes, it is necessary to measure or estimate available energy at various spatial scales. This remains a serious difficulty. Remote sensing information offers promise to allow estimates of spatially distributed net radiation (Diak et al., 1998). However, soil heat flux remains a more serious difficulty, especially for heterogeneous surfaces. In such cases, measurements of spatial averages are nearly impossible, as the number of sites required is likely prohibitive. There are some studies that have related the ratio of G/R_n to remotely sensed radiance indices (Kustas and Daughtry, 1990) and some analytical treatment of this issue (Kustas et al., 1993). However, there is as yet no general solution to this problem.

Longer Timescale Estimates Covering Seasonal and Yearly Trends

There are relatively few studies that have produced a good set of spatially distributed flux measurements to validate models. In addition, these have been generally conducted over rather short time periods, for a variety of reasons. We need to examine the seasonal changes in the fluxes themselves, as well as properties and processes that connect to evaporation and water balance at catchment scales. Little such information is presently available. Some attention is needed to acquiring more data at sites over a number of seasons.

5.7.2   Modelling Issues

Aggregation

Earlier, we briefly addressed the complex issue of aggregation, or how to scale processes and fluxes over a range of spatial scales. Because of the depth and complexity of the subject, we did not cover it in detail. Ultimately specifying spatial variations in evaporation and water balance and their implications to climate will be predicated upon reaching an adequate solution to the scaling or aggregation problem. Currently we appear to be missing fundamental ideas to allow a general theoretical solution to the problem. The atmospheric modelling community involved in Soil–Vegetation–Atmosphere Transfer (SVAT) schemes is starting to recognise the potential of remote sensing information in addressing scaling and aggregation issues in hydrology and meteorology (Avissar, 1998). Preliminary studies using remote sensing data with SVAT schemes indicate the effects of using aggregated information on large-scale evaporation estimates is relatively minor (e.g., Sellers et al., 1995; Kustas and Humes, 1996; Friedl, 1997). This result, however, depends on the scale of heterogeneity (Giorgi and Avissar, 1997) and on the sensitivity of the model parameterisations to surface properties affecting evaporation (Famiglietti and Wood, 1995). We still lack the knowledge to make any general conclusions about these issues.

Combining Surface–Atmospheric Interaction with Remote Sensing Approaches

Earlier, we pointed out current research efforts attempting to merge ABL models with SVAT schemes. The reason for doing this is that wind, temperature and humidity profiles within the fully turbulent region of ABL (i.e., mixed layer) relate to surface fluxes integrated upwind having length scales several orders of magnitude larger than the ABL depth. With ABL depth, typically on the order of 1 km during daytime convective conditions, the wind and scalar quantities should reflect integrated values of surface heterogeneities roughly 10 km upwind. Therefore, by combining spatially variable information on vegetation cover and type and surface temperature from remote sensing with ABL processes, there is the potential of creating the appropriate links between spatially variable surface fluxes and atmospheric feedbacks. The three examples discussed in Section 5.6 demonstrate possibilities of such an approach. They also indicate the issues involved in linking the ABL, SVAT models, and remote sensing data to represent heterogeneous surfaces. There are still processes not yet expressed in these approaches, such as local or mesoscale advection effects.

5.7.3   Conclusions

As our understanding of hydrology and climate has advanced, the importance of evaporation and its spatial distribution has become more evident. Although there is a wealth of theoretical and measurement information available about evaporation, most of it is confined to rather uniform surfaces, and small spatial scales. Even in these cases, all is not yet known.

The current issues in surface hydrology and climate demand attention to spatial and temporal distributions of evaporation at a range of scales. The feedbacks between the evaporation at the surface and atmospheric processes and circulations are often intricate, and cannot be generally ignored. Inevitably this involves dealing with heterogeneous surfaces, which at best stretch the limits of many of our current approaches. However, the advent of remote sensing information offers to make available the spatial variations of several critical surface properties. The key is how to properly connect this information to the actual fluxes. At this stage we have relatively few cases available where these issues can be carefully examined on the landscape, but clearly some real progress has been made in this issue.

PART TWO

CASE STUDIES

Runoff, Precipitation, and Soil Moisture at Walnut Gulch

Paul Houser, David Goodrich and Kamran Syed

6.1 INTRODUCTION

The research presented here was undertaken at the Walnut Gulch Experimental Watershed (30°43′N, 110°41′W) near Tombstone, Arizona, which is operated by the Southwest Watershed Research Center (SWRC), Agriculture Research Service (ARS), U.S. Department of Agriculture (USDA). The extremes in rainfall and temperature in this region lead to great spatial heterogeneity in soil hydrological processes. Observations from a series of nested gauging stations, a dense network of precipitation gauges, and remotely sensed soil moisture estimates, in concert with specialised remote sensing, surface characterisation, and numerical simulation have led to numerous insights into the nature, causes, and effects of hydrologic spatial patterns in this semi-arid catchment. The nature, representation, and interrelation of spatial rainfall patterns and their impact on the spatial distribution of runoff and soil moisture is described. Additionally, the representation of this spatial behaviour through the integration of observations in a distributed hydrologic model using data assimilation methods is assessed.

6.1.1 Description of Study Area

The Walnut Gulch Experimental Watershed (Figures 6.1 and 6.3) was selected as a research facility by the United States Department of Agriculture (USDA) in the mid-1950s. Prior appropriation water laws resulted in conflicts between upstream land owner conservation programs and downstream water users. Technology to quantify the influence of upland conservation on downstream water supply was not available. Thus, scientists and engineers in USDA selected Walnut Gulch for a demonstration/research area that could be used to monitor and develop technology to address the problem. In 1959, facilities needed for soil and water research in the USDA were identified in a United

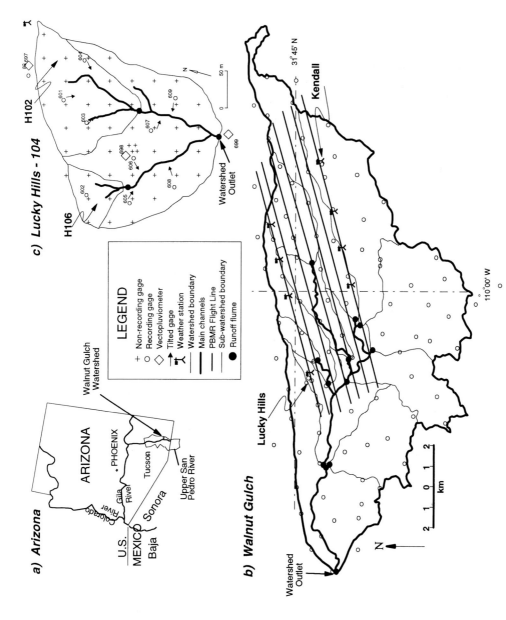

Figure 6.1. The Walnut Gulch Experimental Watershed: (a) location map; (b) catchment map including raingauges, major drainages, runoff flumes, flight lines; and (c) Lucky Hills subcatchment map.

States Senate Document (U.S. Senate Committee, 1959). The Southwest Watershed Research Center in Tucson, Arizona, USA, was created in 1961 to administer and conduct research on the Walnut Gulch watershed (Renard et al., 1993).

The Walnut Gulch Experimental Watershed (WGEW) is defined as the upper 148 km^2 of the Walnut Gulch drainage basin in an alluvial fan portion of the San Pedro catchment in southeastern Arizona (Figure 6.1). Depth to ground water varies from 45 m at the lower end to 145 m in the centre of the catchment. Soil types range from clays and silts to well-cemented boulder conglomerates, with the surface (0–5 cm) soil textures being gravelly and sandy loams containing, on average, 30 % rock and little organic matter (Renard et al., 1993). The topography can be described as gently rolling hills incised by steep drainage channels which are more pronounced at the eastern end of the catchment near the Dragoon Mountains. The mixed grass-brush rangeland vegetation, which is typical of southeastern Arizona and southwestern New Mexico, ranges from 20 to 60 % in coverage. Grasses primarily cover the eastern half of the catchment, while the western half is bush-dominated.

This rangeland region receives 250–500 mm of precipitation annually, with about two-thirds of it as convective precipitation during a summer monsoon season. The balance of precipitation falls during winter frontal storms of Pacific origin (Figure 6.2) and potential evapotranspiration is approximately ten times annual rainfall. The runoff in the ephemeral streams is of short duration and is typically near critical depth (Renard et al., 1993).

Currently, eighty-five recording rain gauges, eleven primary catchment runoff-measuring flumes, and two micrometeorological observation (Metflux) stations make the WGEW a valuable research location. During Monsoon '90 (July 23 to August 10, 1990), eight Metflux stations provided continuous measurement of local meteorological conditions and the surface energy balance, and extensive remote-sensing observations were made (Kustas and Goodrich, 1994). Figure 6.3 shows some of the monitoring equipment and gives an impression of the landscape.

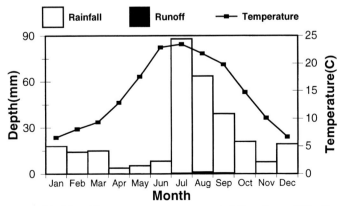

Figure 6.2. Monthly average temperature, rainfall, and runoff for the Walnut Gulch Experimental Watershed.

Figure 6.3. Mosaic of three photos of the WGEW: (a) catchment landscape, with the city of Tombstone, Arizona in the distance; (b) a large runoff measurement flume; and (c) the Lucky Hills metflux site.

6.1.2   Review

Many recent studies of hydrologic variability have shown that land surface heterogeneity has a profound impact on hydrologic phenomena (Milly and Eagleson, 1988; Pitman et al., 1990; Avissar, 1992). Spatial and temporal variability in meteorology (precipitation, wind speed, humidity, radiation, and temperature), soils (hydraulic conductivity, porosity, water retention, topography, and thermal properties), and vegetation (stomatal resistance, leaf area index, albedo, and root depth) interact in a highly nonlinear manner to produce complex heterogeneity in soil moisture, runoff, and evapotranspiration (Ghan et al., 1997). Detailed analysis of surface observations have provided valuable insights into the nature and causes of surface heterogeneity (Seyfried and Wilcox, 1995). It is well established that variability in precipitation is among the most important causes of variability in soil moisture and runoff (Ghan et al., 1997). However, because soil moisture integrates the temporal variability of precipitation, knowledge of the instantaneous precipitation distribution does not necessarily provide a complete picture of hydrologic variability. The interrelations between the complex processes causing hydrologic variability also change in time and space (Seyfried and Wilcox, 1995).

Physically-based hydrologic models have great potential for helping to unravel the complexities of hydrologic heterogeneity, by helping us to critically analyse the problem, organise our thoughts and data sets, and to test our hypotheses. Modelling the impact of meteorologic, soils, and vegetation heterogeneity on surface hydrology has taken two general directions: (1) *spatially distributed modelling* that uses spatially distributed inputs of relevant soil, vegetation, and meteorology to enable better prediction of hydrologic patterns; and (2) *statistical-dynamical modelling* approaches in which homogenous land patches are identified and modelled as a single unit. This facilitates the development of probability density functions, that when combined with the physically-based hydrologic equations, characterise the variability of the hydrologic system (Avissar, 1992).

Numerous studies have investigated the nature and prediction of hydrologic spatial variability at Walnut Gulch (Kustas and Goodrich, 1994; Schmugge et al., 1994; Goodrich et al., 1994, 1995; Humes et al., 1997; Syed, 1994; Jackson et al., 1993; Michaud and Sorooshian, 1994b; Houser et al., 1998, 2000. Generally, these studies have shown that the highly convective, and therefore spatially variable nature of precipitation, has profound impacts on the spatial distribution of soil moisture and temperature, the production of runoff, and the partitioning of the surface energy balance. The variability of soils and vegetation were generally found to have a second-order modifying effect on the spatial variability imposed by precipitation, and at high resolutions, surface temperatures and fluxes were found to be strongly correlated with topography.

This chapter summarises the work on characterisation and simulation of spatial variability of soil moisture and runoff in response to spatial precipitation patterns at the WGEW. First, a discussion of patterns and characteristics of precipitation, soil moisture and runoff based on observations is presented in Section 6.2. This is followed in Section 6.3 with a discussion of modelling and spatial inferences of precipitation, soil moisture, and runoff.

6.2 OBSERVATIONS

6.2.1 Observed Spatial and Temporal Characteristics of Walnut Gulch Precipitation

Knowledge of spatial and temporal characteristics of rainfall is crucial for better understanding this important component of the hydrologic cycle and to represent it more realistically in rainfall-runoff models. Various spatial storm characteristics which are considered important in runoff production of a catchment include, but are not limited to, areal storm coverage, its intensity patterns, direction of storm movement (Osborn, 1964), its position within the catchment (Michaud, 1992), and the extent and intensity of the runoff-producing storm core (Koterba, 1986).

There are three factors which generally limit the reliability of these computed spatial rainfall measures. First, there are inherent limitations in the data-collection procedures. The raingauge design, technician experience, and digitising methodology play important roles in establishing the accuracy of the data. Second, WGEW rainfall is observed at points scattered over a finite area. The interpolation techniques generally used to generate a continuous representation from point data have limitations. Third, interpretation of computed measures may differ depending on the interpolation method used. It is appropriate to devote some discussion to these limitations before describing the observed spatial nature of rainfall in detail because precipitation is such a dominant driver of catchment hydrology at the WGEW (see also Chapter 2 for a general discussion of interpolation issues).

In Walnut Gulch, rainfall observations from more than ninety gauges are available (Figure 6.1b). These are standard weighing type gauges that record the cumulative depth of precipitation continuously as a line trace on a revolving chart driven by an analog clock. The chart completes one revolution in 24 hours and remains in place for seven days before it is replaced with a fresh chart. These charts are manually checked and inferred for starting and ending times of rainfall events. To identify spatial rainfall patterns, the point observations must be transformed to a relatively continuous field, which is achieved by using spatial interpolation methods. Several methods of interpolation are available and each has its strengths and weaknesses which are thoroughly documented in the literature (Myers, 1994) and discussed in Chapter 2, pp. 26–45.

An analysis of Walnut Gulch rainfall data and a mathematically defined synthetic surface showed that both kriging and multiquadric interpolation methods (Shaw and Lynn, 1972) such as splines, produce similar results based on cross-validation residuals. Given these results, the multiquadric method was used here because it does not require a labor-intensive a priori definition of a variogram or correlogram.

A total of 302 summer thunderstorm events that occurred in the period from 1975 to 1991 were analysed, 85 of which produced runoff at the outlet of the catchment. An event was defined as a rainfall episode separated from other rainfall episodes by at least 1 hour. The rainfall data from all 91 gauges was discretised into 10 minute time slices and then used in the multiquadric inter-polation process to estimate rainfall values on a 100 m grid covering the entire catchment which were used to compute the storm areal coverage, position, and movement at a range of subcatchment scales.

At the Lucky Hills catchment scale (< 5 hectares), the assumption of spatially uniform rainfall was tested by making measurements with 10 recording and 49 non-recording raingauges, 9 tilted non-recording gauges, and 3 vectopluvi-ometers for a range of events during the 1990 monsoon season (Figure 6.1c). The data were analysed to: 1) assess the precipitation measurement uncertainty due to gauge type, calibration, data reduction and placement; 2) assess the impacts of wind on precipitation observations; and 3) evaluate the impact of spatial and temporal rainfall variability on the estimate of areal precipitation over the catchment.

A histogram of areal total storm coverage for the 302 events is plotted in Figure 6.4a. Slightly less than half the total number of storms cover the entire $148 \, km^2$ catchment. An event rarely delivers rain to the entire catchment instan-taneously, but may affect large portions of the catchment over its entire duration. The contrast between total storm areal coverage and within storm coverage is illustrated in Figure 6.4b. The local nature and high spatial variability of these convective storms is evident. About one-third of storms occur in the range of 30 to $50 \, km^2$ and about half are greater than $50 \, km^2$ with a maximum of $120 \, km^2$. The spatial extent of storm cores was found to be even more limited. Out of a total of 302 events, 53 events contained a high-intensity storm core (10 minute intensity $> 25 \, mm/hr$), of which about 25 % were in the range of areal coverage of 2 to $3 \, km^2$, 50 % ranged from 3 to $9 \, km^2$ and the remaining 25 % were larger than $9 \, km^2$ with a maximum of $34 \, km^2$.

It is also interesting to observe how these storms, on average, grow and decay in time. To examine this, the average of areal coverage of all storms for each 10 minute time step was computed from a common start time (Figure 6.5). The dotted line is the result of averaging every storm whether or not it reported any rain in a particular 10 minute interval (n was always 302). The rapid growth of the areal coverage of a storm and its recession are shown in Figure 6.5. When only those events that reported some rain in a particular time step were averaged, the first 1.5 hours was largely unchanged, indicating that most of the storms last more than 1.5 hours. However, beyond 1.5 hours, there is a sharp deviation, indicating that the longer dura-tion storms have substantial spatial coverage (greater than $50 \, km^2$). When the average rainfall volume in successive time steps was plotted in a similar fashion, its shape was very similar to the areal coverage plot except that

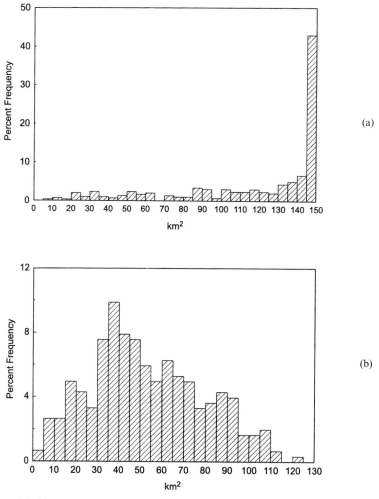

Figure 6.4. (a) Spatial storm coverage on an event basis. (b) Spatial storm coverage averaged for 10 minute time increments through storm duration.

its peak occurred at 70 minutes. This indicates that rainfall intensities typically peak in the first 1.5 hours of the storm even though the average storm event lasts for 4.75 hours.

Due to the limited extent of runoff producing storm intensities and high runoff transmission losses, the location of the storm core within the catchment is also very important. An example is shown in Figure 6.6 for a storm that occurred on July 30, 1989. The storm had two distinct periods of high-intensity rainfall (Figure 6.6a). The first one occurred near the start of the event, and was located near the catchment outlet (Figure 6.6b). This was followed by a low-intensity period of about 90 minutes. The second high-intensity burst then occurred, and was located near the head of the catchment (Figure 6.6c). This example clearly illustrates that, in this environment, it may be difficult to uniquely define a rainfall event using the arbitrary criteria currently utilised in

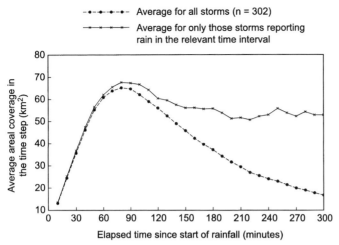

Figure 6.5. Progression of average spatial storm extent within event duration.

Walnut Gulch data processing, as these two periods of high-intensity rainfall are likely to have been two independent thunderstorm cells.

At the small-catchment scale, the intensive observations made at Lucky Hills – 104 resulted in several significant findings. It was found that the range of observed variation over the catchment was greater than the variation that would result from total measurement error (i.e. even at this scale spatial variability exists). An example of a storm (August 12, 1990) exhibiting the largest absolute variation in rainfall over the 4.4 hectare Lucky Hills – 104 catchment is shown in Figure 6.7. In addition, geostatistical analysis indicated the presence of first-order drift with corresponding rainfall gradient ranges from 0.28 to 2.48 mm/100 m with an average of 1.2 mm/100 m. These gradients represent a 4 % to 14 % variation of the mean rainfall depth over a 100 m distance indicating that raingauge location is particularly important if only one gauge is available. This suggests that the typical uniform rainfall assumption is *invalid* at the 5 hectare scale in regions where convective thunderstorm rainfall is significant. Spatial rainfall variation at this scale is attributed to localised wind effects from down drafts associated with the relatively random location of air mass thunderstorms in relation to the catchment. The overall WGEW raingauge network depicted in Figure 6.1b will not be able to resolve the rainfall spatial variations and patterns at the 5 hectare catchment scale for storms such as those illustrated in Figure 6.7. However, the density of the large area network is such that overall gradients in rainfall depth of typical air mass thunderstorms are captured. This should allow approximate estimation of the first-order drift noted for all but one of the storms observed with the small area network at Lucky Hills. While it may be possible to estimate gradients in rainfall depth with the large area network, rainfall maxima or minima occurring between the gauges will not be resolved at the 5 hectare scale. Spatial rainfall variation at this scale has important implications as testing and validation of process-based hydrologic models are often conducted on small research catchments using the spatially uniform rainfall assumption (single rain-

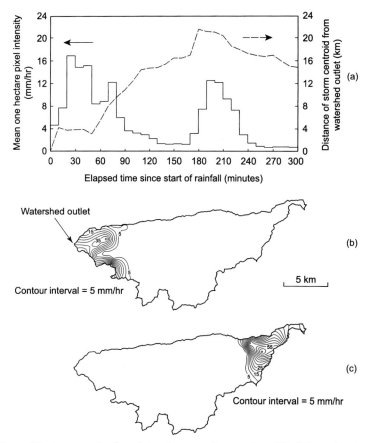

Figure 6.6. An example of spatial and temporal separation of high-intensity storm cells. (a) Spatially averaged storm intensities (solid line) and storm centroidal position (dashed line) in 10 minute time increments for the event of July 30, 1989. (b) Intensity contours during 40–50 minute interval. (c) Intensity contours during 190–200 minute interval.

gauge). The impacts of observed rainfall variability on runoff modelling at this scale are discussed in Section 6.3.3.1.

6.2.2 Observed Spatial and Temporal Characteristics of Walnut Gulch Soil Moisture

In Walnut Gulch, soil moisture observations were made by in-situ gravimetric sampling, resistance sensors, and Time Domain Reflectometry (TDR) sensors, as well as by microwave remote sensing. During Monsoon '90, three replicate gravimetric surface soil moisture samples were collected daily at the eight Metflux sites (Schmugge et al., 1994). These were converted to volumetric soil moisture using bulk density measurements made at each site. The only continuous soil moisture measurements made during Monsoon '90 were those made with resistance sensors (Kustas and Goodrich, 1994). They were placed at 2.5 cm and 5 cm below the surface at all eight Metflux sites. These sensors are generally difficult to

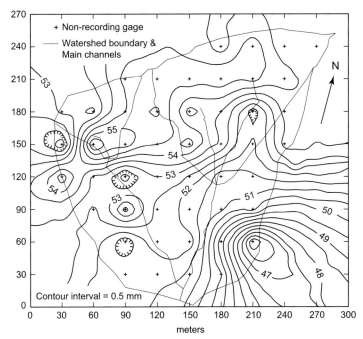

Figure 6.7. Contour map of the rainfall depth for the storm of August 12, 1990 (interpolation by isotropic kriging). Storm duration is 4 hours 42 minutes with about 75 % of the rain falling in 26 minutes.

calibrate and tend to drift. Therefore, following the recommendation of Stannard et al. (1994), the resistance data were calibrated against the gravimetric samples and then used to interpolate gravimetric data to each model time step. TDR measurements were made at daily intervals and at multiple depths down to 0.5 m at two of the Metflux sites (Kustas and Goodrich, 1994).

Engman (1991) described NASA's 21 cm wavelength (1.42 GHz), passive microwave Push Broom Microwave Radiometer (PBMR) instrument as "a mature and reliable instrument with a good history of soil moisture measurements". This approach relies on the large dielectric contrast between water and dry soil at long (> 10 cm) microwave wavelengths that causes the soil's emissivity to be a function primarily of moisture content. Vegetation can reduce the range of microwave brightness variation, totally obscuring the soil moisture signal if it is present in sufficiently large amounts. Fortunately this remote-sensing technique works well because Walnut Gulch has minimal vegetation (Schmugge et al., 1994).

During the Monsoon '90 field campaign (July 23 to August 10, 1990), the PBMR instrument was flown on board the National Aeronautics and Space Administration (NASA) C-130 aircraft. Six days (July, 31 August 2, 4, 5, 8 and 9, 1990) of microwave brightness temperature were collected over an 8 × 20 km area in the northeastern portion of the catchment (Schmugge et al., 1994). The period was very dry prior to the first flight, which was followed by 5 cm of rain falling over most of the study area on August 1, 1990. This produced a significant decrease in brightness temperature (50 to 60 K) on August 2, 1990. The successive flights on August 4, 5, 8, and 9, 1990 showed the effects of some smaller rain storms and drydown of the area. A strong east-west spatial pattern is also evident and is strongly correlated to the observed soil and vegetation gradients. The changes in brightness temperature at six of the eight Metflux sites (Figure 6.1b) were well correlated with rainfall ($R^2 > 0.9$) and in-situ soil moisture ($R^2 = 0.8$) (Schmugge et al., 1994). The linear relationships established between microwave brightness temperature and gravimetric soil moisture at each Metflux site by Schmugge et al. (1994) were used with an inverse distance weighting scheme to invert microwave brightness temperature to soil moisture (Figure 6.8) (Houser et al., 1998).

The PBMR data have been analysed using geostatistical methods (see Chapter 2). The analysis also showed that the correlation structure varies with time. The July 31, 1990 PBMR observations, taken during dry conditions, show little spatial correlation, i.e. there is only very short range correlation probably due to random pattern of surface properties. One day after the large precipitation event on August 1, 1990, the variogram changes to linear with a very small nugget and a range beyond the observation area (15 km), i.e. the storm pattern imposes a large-scale pattern on the brightness temperature. Three days after the August 1, 1990 storm, a range of about 3.5 km becomes apparent, that is, as the surface dries, the scale of correlation decreases. Eight days after the storm, some spatial structure is still evident in the PBMR variogram but there are also significant random components in the pattern (Houser et al., 1998), i.e. the brightness temperature pattern imposed by the storm is disappearing and the random pattern of surface properties is dominating again.

A multispectral scanner was also flown on NASA's C-130 aircraft during the Monsoon '90 field experiment. Using the NS001 thermal band (10.9–12.3 μm), in conjunction with a radiative transfer algorithm (LOWTRAN 7) that corrected for atmospheric effects on the signal, surface temperature distributions were derived (Figure 6.8) (Humes et al., 1997). The image on August 1, 1990 shows areas of low temperatures corresponding to isolated cumulus clouds. There are also clear discontinuities between the two flight lines, which are attributed to the time difference in data acquisition (see Humes et al., 1997). It is clear that the surface temperature is strongly influenced by the surface soil moisture since it has a high correlation with both the PBMR observations and the rainfall distribution, and also shows the effects of shading by topography and larger amounts of vegetation in drainage lines.

In addition to the Monsoon '90 PBMR remotely sensed and ground-based gravimetric measurements, a similar suite of measurements was carried out over Walnut Gulch in 1991 using ESTAR, the airborne electronically steered thinned array L-band radiometer (Jackson et al., 1993). As in 1990, a wide range of soil moisture patterns and conditions were captured as flights were made before and after several significant rainfall events. With the ground-based data as well as the patterns of soil moisture acquired in 1990 by the PBMR instrument, the viability

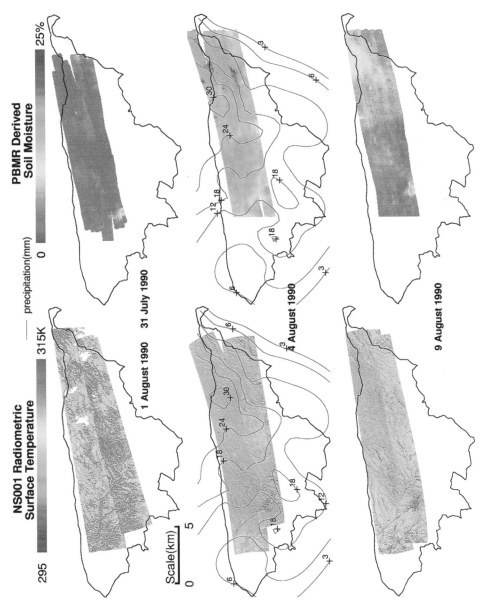

Figure 6.8. Push Broom Microwave Radiometer derived soil moisture, NS001 derived surface temerature (Humes et al., 1997), and accumulated precipitation for the WGEW during the Monsoon '90 experiment (Houser et al., 1998). The precipitation contours shown for August 4, 1990 show the precipitation that fell from multiple storms between August 1 and August 4.

137

of the ESTAR instrument for soil moisture estimation was established by this study (Jackson et al., 1993).

Soil moisture patterns in the WGEW, as observed by both in-situ and remote sensing, are complex, with large variability at all scales. However, some spatial structure is evident, arising from highly-localised convective precipitation, drydown processes, and surface characteristics such as soil and vegetation type.

6.2.3 Observed Spatial and Temporal Characteristics of Walnut Gulch Runoff

The interactions of rainfall patterns and antecedent soil moisture patterns will of course play a role in determining patterns of runoff generation. The nested structure of the runoff observation network within the WGEW affords an opportunity to examine spatial runoff patterns to some degree. In Figure 6.9 the runoff per unit area for each of the gauged catchments resulting from the August 1, 1990 storm is illustrated as a circle at the outlet of each gauged catchment whose size is proportional to the runoff magnitude. As expected, runoff was generated in regions of high rainfall. Sufficient runoff was generated from this event so that the flow was able to traverse approximately 15 km of dry ephemeral channel and reach the overall catchment outlet. While many of the catchments produced no runoff from the August 1, 1990 event, a more regular pattern of runoff distribution is observed for a ten year average (1969–1979). The general trend apparent in this data is a reduction in mean annual runoff per unit area with increasing drainage area.

These trends are consistent with the ephemeral semi-arid nature of the catchment where runoff is not augmented by ground water inflows. Without a saturated channel system, the dry loose alluvium present in the vast majority of the larger channels is able to absorb a significant volume of surface runoff. Depending on the location of rainfall, these channel transmission losses can also significantly impact peak runoff rates (Renard et al., 1993). An example

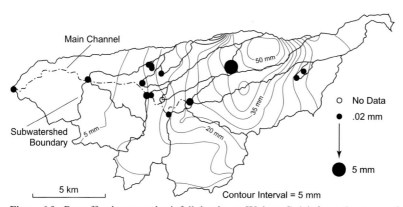

Figure 6.9. Runoff volumes and rainfall depths on Walnut Gulch from the storm of August 1, 1990.

of the impacts of channel transmission losses on runoff volume and peak runoff rate is illustrated in Figure 6.10 for the event of August 27, 1982. This figure depicts the storm isohyets and the hydrographs at flumes 6, 2, and 1. Because the rainfall is isolated above these flumes the change in hydrograph runoff volume and peak rate is solely attributed to transmission losses.

Channel transmission losses effectively decrease the correlation between rainfall and upland soil moisture patterns, and observed runoff. In the extreme case, all locally generated runoff may infiltrate into the channels. In this case any connection between rainfall and soil moisture patterns is severed downstream of the terminal location of the runoff front. Goodrich et al. (1997) concluded that explicit treatment of channel transmission losses is required for modelling catch-

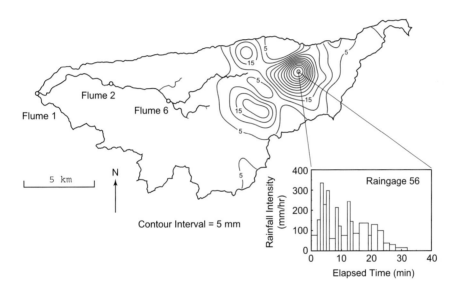

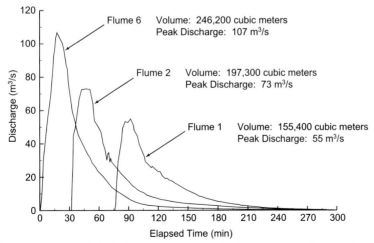

Figure 6.10. Storm total isohyets and hydrographs from flumes 6, 2, and 1 for event of August 27, 1982.

ments larger than roughly 40 hectares. In the case where two runoff events occur over the same reach of channel within a short period of time (< 3 days), runoff from the second event can be greatly enhanced as transmission losses are largely satisfied by the prior event.

6.3   MODELLING AND SPATIAL INFERENCES

6.3.1   Precipitation Modelling

Spatial precipitation modelling efforts posed and tested using WGEW data were initiated with extensions from point observations to area (Osborn, 1977); depth–area (Osborn and Lane, 1972); and, point–area frequency conversions (Osborn and Lane, 1981). These studies provide methods for areal distribution of rainfall uniformly without internal storm pattern information. Several stochastic models have also been developed to predict the spatial and temporal distribution of thunderstorm rainfall (Osborn et al., 1980; Eagleson et al., 1987; Islam et al., 1988; Jacobs et al., 1988). In general these models were able to reproduce the main statistical features of rainfall patterns. However, model stationarity assumptions limited model results as they were not able to describe observed nonstationary storm behaviour.

While these models have some utility in predicting rainfall patterns that are statistically similar to observations, the remotely sensed spatial patterns (Section 6.2.2) have the potential to estimate observed spatial rainfall patterns on an event basis. As noted in that section, Schmugge et al. (1994) found a high correlation of change in brightness temperature between flight acquisition dates and total inter-flight rainfall. The data from one set of PBMR flights in 1990 and two sets of ESTAR flights in 1991 are illustrated in Figure 6.11. As illu-

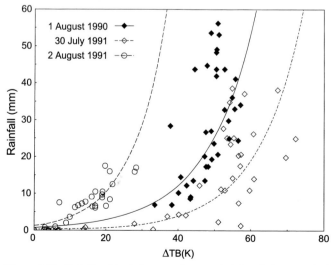

Figure 6.11. Relationship between raingauge total and decrease in brightness temperature (ΔTB) for three Walnut Gulch rainfall events. The lines are fitted exponential curves for the three events.

strated in the figure, a simple exponential model describes the relationship quite well. The respective R^2 values for this function are $R^2 = 0.68$, 0.83, and 0.79 for July 31 to August 2, 1990 (labelled August 1, 1990), July 30 to August 1, 1991 (labelled 30 July 1991) and August 2 to August 3, 1991 (labelled 2 August 1991) data. The variability in these relationships may be caused by differences in within-storm rainfall intensity patterns, infiltration, or evapotranspiration. The failure of the relationship above rainfall amounts of 30 mm is likely the result of two factors. First, for high rainfall amounts generating runoff, a portion of the rainfall is conveyed offsite and is concentrated in channels. In this case the full rainfall amount does not infiltrate and increase local soil moisture. The second factor has to do with the dynamic brightness temperature range of the L-band radiometer. For sparse vegetation the sensitivity of brightness temperature (TB) is about 2.5 K per percent of soil moisture. The dynamic range, given constant physical temperature, can be estimated by multiplying the sensitivity by the soil moisture range. Observed ranges of soil moisture vary from 18 % to almost 30 % depending on soil type. The corresponding dynamic range would vary from roughly 45 to 75 K.

The relationships illustrated in Figure 6.11 can be inverted to predict rainfall patterns. This analysis was carried out by Jackson et al. (1993) and comparison between observed rainfall patterns from the raingauge network and those predicted using the patterns of change in brightness temperature are illustrated for the two events in 1991 in Figure 6.12. The patterns predicted using changes in brightness temperature are very similar to patterns obtained from interpolating rainfall amounts from the dense raingauge network. As Jackson et al. (1993) note, for sparsely vegetated arid and semi-arid regions similar to Walnut Gulch, these results suggest the potential of using the change in brightness temperature method to estimate rainfall over large regions which do not have raingauge networks, provided a precipitation–ΔTB relationship is available.

6.3.2 Soil Moisture Modelling

6.3.2.1 Description of TOPLATS and its Application to Walnut Gulch

The TOPMODEL-based Land Atmosphere Transfer Scheme (TOPLATS) (Famiglietti and Wood, 1994) predicts spatial distributions of land surface runoff, energy fluxes, and soil moisture dynamics given atmospheric, soil, and vegetation information. It incorporates simple representations of atmospheric forcing, vertical soil moisture transport, plant-controlled transpiration, interception, evaporation, infiltration, surface runoff, and sensible and ground heat fluxes. The model incorporates a diurnal cycle and is driven with standard meteorological data with an hourly time step, this being considered sufficient to resolve the dynamics of the land–atmosphere interaction. The subsurface unsaturated soil column is partitioned into three layers, with the upper layer corresponding to the microwave remote sensing penetration depth, the underlying root zone extending from the bottom of the surface zone to the depth of

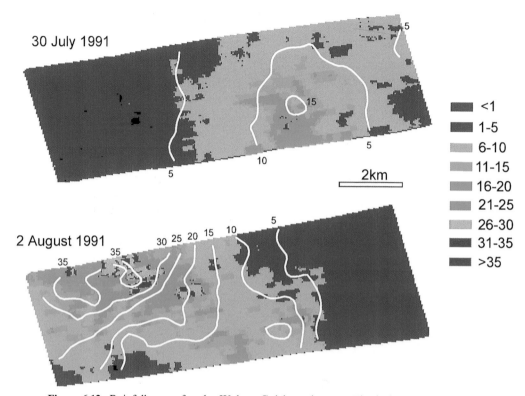

Figure 6.12. Rainfall maps for the Walnut Gulch study area. The isohyetal contour lines were derived from the observations made by the WGEW raingauge network. The images of predicted rainfall were obtained from the pre-storm to post-storm change in brightness temperature and the exponential models illustrated in Figure 6.11. All values are in mm. Top: July 30, 1991 event; Bottom: August 2, 1991 event. (Modified from Jackson et al., 1993.)

plant roots, and the transmission zone extending from the bottom of the root zone to the top of the saturated soil.

The WGEW was modelled using TOPLATS in a spatially distributed manner at a 40 m resolution from July 23, 1990 to August 16, 1990. The TOPLATS parameterisation was largely based on observations made during Monsoon '90 (Kustas and Goodrich, 1994). However, eight model parameters were not observed and had to be estimated or specified by model calibration (Houser et al., 2000).

6.3.2.2 Spatially Distributed Model Forcing and Parameters

The multiquadric precipitation interpolation algorithm (Syed, 1994) was used to produce spatially distributed precipitation values for the entire model domain from the available raingauge data. All other meteorological forcing (air temperature, wind speed, humidity, and radiation) were assumed to be spatially constant because observations from the eight Metflux stations contain insufficient information to derive spatially-variable meteorological forcing for the approximately ninety thousand TOPLATS model grid points, and an eight-site average decreases the impact of highly local meteorologic signals (i.e., solar radiation measurement errors due to vegetation and topographic shading) on larger-area simulation. Therefore, meteorological forcing was derived from averaging observations at the eight Metflux stations in place during the experiment (Kustas and Goodrich, 1994).

It is thought that optimal implementation of a distributed hydrological model requires the specification of spatial distributions of soil and vegetation parameters. Therefore, the required TOPLATS spatially variable parameters were estimated using GIS maps of Walnut Gulch vegetation and soils; several examples of these spatially variable parameters are shown in Figure 6.13 and Houser et al. (2000). Spatially distributed information on minimum stomatal resistance, root depth, leaf area index, residual soil moisture, saturated soil moisture, saturated hydraulic conductivity, percent clay, percent sand, effective porosity, and topographic index at Walnut Gulch were used as parameters in the TOPLATS to make spatially distributed predictions. All other parameters were held spatially constant. The three TOPLATS soil moisture layers were initialised on July 23, 1990 based on catchment average in-situ TDR soil moisture observations.

The simulated spatial patterns of surface soil moisture at 12:00pm on August 7, 1990, using these spatially variable parameters, are shown in Figure 6.13. For comparison, the results for a simulation using spatially constant soil and vegetation parameters are also shown. The spatially variable soil and vegetation parameters have a large impact on the spatial patterns of the simulation, which appear unrealistic because they compare poorly with observed PBMR surface soil moisture. A series of sensitivity simulations was performed to determine which subset of spatial parameters contribute most to these patterns. The use of spatially variable vegetation parameters has much less effect on predictions as compared to soil parameters; it is likely that at the WGEW soil has more control of soil moisture processes than vegetation. The parameter specifying saturated soil moisture has the most influence on simulated spatial patterns, while those which specify the percentages of sand and clay, the saturated hydraulic conductivity, and the residual soil moisture have a more moderate influence. Finally, the spatially variable topographic index has very little influence on the simulations because the process of water table interaction with the surface does not operate at the WGEW.

The artefact of enhanced spatial soil and vegetation polygons apparent in the simulations is probably not a simple mis-specification of parameter values, rather it is an artefact of discretely assigning a single set of parameters (which in reality would display high variability, see for example Chapter 10, Figure 10.4) to large areas. A more appropriate specification of spatial parameters would be continuous, as obtained with remote sensing. It might be possible to develop a smoothing algorithm that would use the soil polygon information to approximate continuously varying, spatially distributed parameters. Because the simulations using spatially constant vegetation and soil parameters compare

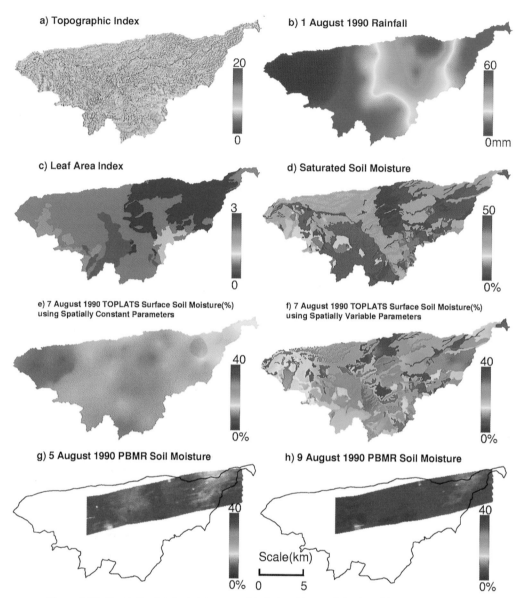

Figure 6.13. Spatially distributed topographic index (a); precipitation (b); vegetation (c); and soils (d); for the Walnut Gulch Experimental Watershed. TOPLATS spatial predictions of surface soil moisture at 12:00pm on August 7 1990 using spatially constant (e) and spatially variable (f) soils and vegetation parameters (all simulations use spatially variable topography and precipitation). Push Broom Microwave Radiometer (PBMR) derived soil moisture for August 5, 1990 and August 9, 1990 (Houser, 1996). The addition of spatially variable soils and vegetation produces unrealistic polygon artefacts in the simulation.

well to the PBMR patterns, soil and vegetation parameters were assumed spatially constant across the catchment in subsequent studies, leaving precipitation as the dominant spatially varying entity (topographic index is also variable, but with little effect).

6.3.2.3 Four-Dimensional Data Assimilation for Enhanced Soil Moisture Pattern Identification

Errors in the structure, parameters, and forcing of TOPLATS can never be fully rectified, and therefore lead to prediction errors. However, observations of model states or storages, distributed in time and space, can be used to correct the trajectory of the model, and reduce its prediction errors. Charney et al. (1969) first suggested combining current and past data in an explicit dynamical model, using the model's prognostic equations to provide time continuity and dynamic coupling amongst the fields. This concept has evolved into a family of techniques (i.e., direct insertion, Newtonian nudging, optimal interpolation, variational, Kalman filtering, etc.) known as Four-Dimensional Data Assimilation (4DDA). TOPLATS was modified to allow the assimilation of soil moisture and other state variables. The following description assumes assimilation of observed surface soil moisture, θ_o, derived from the PBMR (as shown in Figure 6.8). However, with modifications specific to the state variable, the following description can be used to assimilate other variables, such as surface temperature.

A control and a direct insertion simulation are used as the basis for evaluating the data assimilation runs. A control simulation (i.e., the simulation without data assimilation) can be considered an extreme case, in which it is assumed that the observations contain no information. The other extreme is direct insertion, where it is assumed that the model contains no information. In this case, the model prediction of surface soil moisture, θ_{sz}, is replaced with a PBMR soil moisture observation, θ_o, whenever an observation is available. With direct insertion, no data are assimilated outside the four-dimensional region (one time and three space dimensions) where observations are available; therefore, any advection of information is only accomplished via the model physics in subsequent model integrations.

In a second data assimilation technique, which is termed "statistical correction", the mean and standard deviation of the surface soil moisture states in the model are adjusted to match the mean and standard deviation of the observations. This method assumes that the statistics of the observations are perfect, which is arguably more reasonable than assuming that each observation is in itself perfect, as in direct insertion. It also assumes that the patterns predicted by the model are correct but that the predicted surface soil moisture statistics contain bias. As with direct insertion, advection of information into deeper soil layers is accomplished solely through the model physics.

A third data assimilation technique called Newtonian nudging relaxes the model state towards the observed state by adding an artificial tendency term into the prognostic equations which is proportional to the difference between the two states. These small forcing terms gradually correct the model fields which are assumed to remain in approximate equilibrium at each time step (Stauffer and Seaman, 1990). In this way, the model can be nudged toward observations within a certain distance, and during a period of time, around the observations. Newtonian nudging is implemented as follows:

$$\frac{\partial \theta}{\partial t} = F(\theta, x, t) + G_\theta \cdot W_\theta(x, t) \cdot \varepsilon_\theta(x) \cdot (\theta'_o - \theta) \tag{6.1}$$

The model's forcing terms are represented by F, θ'_o is the PBMR surface soil moisture observation at the model grid, and t is time. G_θ is the nudging factor which determines the magnitude of the nudging term relative to all other model processes, while the four-dimensional weighting function, W_θ, specifies its spatial and temporal variation. The analysis quality factor, ε_θ, varies between 0 and 1 and is based on the quality and distribution of the observations. Equation (6.1) was implemented for all three TOPLATS soil layers.

The Newtonian nudging weighting function, W, at time, t, and location, x, for each observation, I, is a combination of the horizontal weighting function, w_{xy}, the vertical weighting function, w_z, and the temporal weighting function, w_t, thus:

$$W(x, t) \equiv w_{xy} \cdot w_z \cdot w_t \tag{6.2}$$

The horizontal weighting function can be defined by a Cressman-type horizontal weighting function, as:

$$w_{xy} = \frac{R^2 - D^2}{R^2 + D^2}, \qquad 0 \le D \le R \tag{6.3}$$

$$w_{xy} = 0, \qquad D > R \tag{6.4}$$

where R is the radius of influence, and D is the distance from the ith observation to the gridpoint. The vertical weighting function, w_z, is also a distance weighting function, following Seaman (1990); thus:

$$w_z = 1 - \frac{|z_{obs} - z|}{R_z}, \qquad |z_{obs} - z| \le R_z \tag{6.5}$$

$$w_z = 0, \qquad |z_{obs} - z| > R_z \tag{6.6}$$

where R_z is the vertical radius of influence, and z_{obs} is the vertical position of the ith observation. The temporal weighting function is defined as follows:

$$w_t = 1, \qquad |t - t_0| < \frac{\tau}{4} \tag{6.7}$$

$$w_t = 0, \qquad |t - t_0| > \tau \tag{6.8}$$

$$w_t = \frac{(\tau - |t - t_0|)}{\tau/4}, \qquad \frac{\tau}{4} \le |t - t_0| \le \tau \tag{6.9}$$

where t is the model-relative time, t_0 is the model-relative time of the ith observation, and τ is the half-period of a predetermined observation influencing time window.

The final data assimilation method explored here is statistical or optimal interpolation, which is a minimum variance method that is closely related to kriging. Statistical interpolation was implemented in all three TOPLATS soil layers as follows (Daley, 1991):

$$\theta_A(r_i) = \theta_B(r_i) + \sum_{k=1}^{K} W_{ik} \cdot [\theta_O(r_k) - \theta_B(r_k)] \tag{6.10}$$

where K is the number of observation points, W_{ik} is the weight function, $\theta(r)$ is the soil moisture analysis variable, r is the three-dimensional spatial coordinates, $\theta_A(r_i)$ is the analysed value of θ at the analysis gridpoint r_i, $\theta_B(r_i)$ is the background or first-guess value of θ at r_i, and $\theta_O(r_k)$ and $\theta_B(r_k)$ are the observed and background values, respectively, at the observation station r_k.

The weight function, W_{ik}, is determined by least-squares minimisation of equation (6.10), with the assumptions that $\theta_B(r_k)$, $\theta_B(r_i)$, and $\theta_O(r_k)$ are unbiased, that there is no correlation between the model and observation error, that the error correlations are homogeneous, isotropic, and time invariant, and that the background error correlation, ρ_B, is horizontally and vertically separable (i.e., $\rho_B = \rho_{Bxy}\rho_{Bz}$) (Daley, 1991), thus:

$$\sum_{l=1}^{K} W_{il} \cdot [\rho_{Bxy} \cdot (r_l - r_k) + \varepsilon_O^2 \cdot \rho_O \cdot (r_l - r_k)] = \rho_{Bxy} \cdot (r_i - r_k) \cdot \rho_{Bz} \cdot (z_i - z_k) \tag{6.11}$$

where ρ_O is the observation error correlation matrix, ρ_{Bxy} is the background horizontal error correlation matrix, and ρ_{Bz} is the vertical error correlation matrix. ρ_O and ρ_B were estimated using PBMR observations and corresponding model predictions (Houser et al., 1998). The system of linear equations given in equation (6.11) was solved using a Cholesky Decomposition (Press et al., 1986). Each PBMR image contains over 35,000 observations, which requires solving a system of 35,000 linear equations for each model grid point each time an observation was available. Clearly the computational resources needed for this task are unreasonable; hence, a simplified method was required. This was accomplished by (a) using a random subset of 100 PBMR observations, and (b) by using 100 "super-observations", these being approximately $1\,\mathrm{km}^2$ PBMR soil moisture averages.

Catchment average time series of surface and root zone soil moisture for the various assimilations using all of the available PBMR observations are shown in Figure 6.14. All of the data assimilation methods significantly and similarly improved the simulation of surface zone soil moisture, with the exception of direct insertion, which was unable to impose an entire catchment correction and was therefore unable to adjust the model trajectory sufficiently. Nudging had the clear advantage of providing smoother temporal adjustments. All simulations produced identical surface zone soil moisture simulations after the storm on August 12, 1990 because this storm saturated the surface zone causing all past surface zone forcing to be forgotten, but this process does not occur in the model's root zone where memory of past assimilation is preserved. This sequence of events is not unrealistic; rather, it suggests a time interval at which soil moisture observations are needed for data assimilation, this interval being less than or equal to the time between storm events. In the root zone, the simulated time

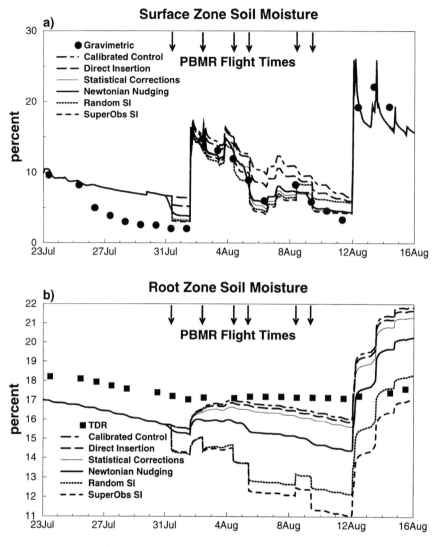

Figure 6.14. Comparison of TOPLATS catchment average surface and root zone soil moisture time series for various assimilation studies. (From Houser et al., 1998; reproduced with permission.)

series fell into two distinct groups corresponding to methods with and without the capability for vertical assimilation of information. Among the latter group, nudging assimilation performs a more conservative correction compared to statistical interpolation. None of the methods produced time series that matched the in-situ root zone observations. However, it is important to bear in mind that with only two in-situ root zone observations, the root zone spatial variability is not adequately sampled.

The control run deviates most significantly from observations near the end of the drydown on August 7, 1990, so this time is selected to demonstrate the intercomparison between assimilation methods. It should be noted that four

PBMR images were assimilated prior to this time, with the last assimilation occurring on August 5, 1990. The spatial patterns of model-predicted surface soil moisture for the different assimilation methods are shown in Figure 6.15. The best spatial patterns are considered to be those without discontinuity at the edge of the observed area, without numerical artefacts, and with a similar nature to those produced by the model without assimilation.

Simple updating is unable to advect information horizontally, giving rise to an undesirable discontinuity in the calculated soil moisture field and preserving all the observational noise. Updating also is able only to impact root zone soil moisture very slightly through model physics and preserves the discontinuity in this zone. Data assimilation via statistical corrections is able to adjust the entire surface soil moisture field to observed levels. It produces a soil moisture spatial field that does not contain discontinuities or retain the observed spatial pattern. Newtonian nudging also produces a spatial field of soil moisture without discontinuities.

Both the random and the super-observation statistical interpolation approaches result in an undesirable linear streaking feature that extends outward from the observed area that is an artefact of numerical procedures, or may be the result of a violation of the statistical interpolation unbiased background assumption. Statistical interpolation has the advantage of using error correlation functions based on the characteristics of the observations and the model predictions. However, it also has the disadvantage of being excessively demanding on computer resources when addressed as a fully posed problem with remotely sensed data, and it lacks the benefits of temporal assimilation.

There is a clear tradeoff between using a complex data assimilation technique and the ability to use all the available data due to the large computational burdens of performing data assimilation at fine resolutions using dense data sets. As the complexity of the data assimilation model increases, the size of the assimilated data set needs to decrease in order to maintain computational feasibility. Complex methods have the ability to extract more useful information from assimilated data, but simpler methods use more of the data to extract similar information. This tradeoff allows simpler assimilation techniques to perform almost as well as complex techniques. In general, this argument suggests the use of assimilation methods that are of moderate complexity, that are sound and computationally efficient, but use as much data as possible. If the information in the data can be efficiently compressed or filtered before its use in data assimilation, and if the mathematical solvers can be further optimised, it may be reasonable to use larger data sets in complex data assimilation strategies.

6.3.3 Runoff Modelling

The range of runoff models applied to, or developed with, Walnut Gulch data varies widely in both complexity and type. Early models included those based on linear regression at annual time scales (Diskin, 1970) and stochastic models for estimating the start of the runoff season, the number of runoff events per season,

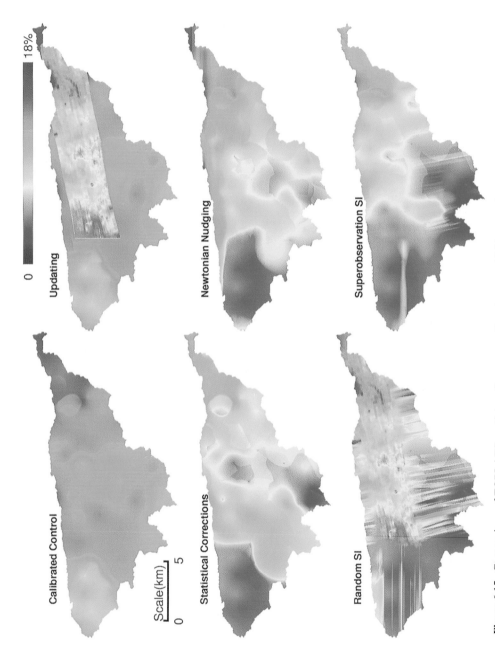

Figure 6.15. Comparison of TOPLATS spatial surface zone soil moisture at 12:00pm on August 7, 1990 from various assimilation methods. (From Houser et al., 1998; reproduced with permission.)

time interval between events, beginning event time, runoff volume, and peak discharge (Diskin and Lane, 1972; Lane and Renard, 1972). More geometrically detailed recent work by Syed (1994) further reinforced the importance of spatial distributions of both rainfall and pre-storm soil moisture availability on catchment runoff response using the 302 storm events discussed earlier with regression analyses. Simple measures of spatial characteristics of rainfall considered individually did not show a very high degree of correlation with either runoff volume or peak rate of runoff for WG1 (the whole experimental catchment). The highest correlation, R^2, was found to be between precipitation volume and runoff volume ($R^2 = 0.59$) and precipitation volume and peak rate of runoff ($R^2 = 0.53$). However, when only the precipitation volume of the storm core (intensities $> 25\,\text{mm/hr}$) was considered, the coefficients of correlation increased to 0.71 and 0.76, respectively. This clearly illustrates that the high-intensity portions of the storm core are directly related to runoff production.

A large number of other models have also utilised data from Walnut Gulch for development or validation. These include CREAMS (Knisel, 1980); SPUR (Lane, 1983a,b; Renard et al., 1993); ARDBSN (Stone et al., 1986); WEPP (Lopes et al., 1989); and CELMOD5 (Karnieli et al., 1994). A particularly well adapted model for use in arid and semi-arid regions where transmission losses are important was developed by Lane (1982).

6.3.3.1 Description of KINEROS and Its Application to Walnut Gulch

KINEROS is a physically based, event-oriented kinematic runoff and erosion model (Smith et al., 1995) that was also developed and tested using WGEW data. In this model, catchments are represented by discretising contributing areas into a cascade of one-dimensional overland flow and channel elements using topographic information. The infiltration component is based on the simplification of the Richard's equation posed by Smith and Parlange (1978):

$$f_c = K_s \frac{e^{F/B}}{(e^{F/B} - 1)}; \quad \text{and} \quad B = G \cdot \varepsilon \cdot (S_{\max} - SI) \quad (6.12)$$

where f_c is the infiltration capacity (L/T), K_s is the saturated hydraulic conductivity (L/T), F is the infiltrated water (L), B is the saturation deficit (L), G is the effective net capillary drive (L), ε is the porosity, $S_{\max}$ is the maximum relative fillable porosity, and SI is the initial relative soil saturation. Runoff generated by infiltration excess is routed interactively using the kinematic wave equations for overland flow and channel flow, respectively stated as:

$$\frac{\partial h}{\partial t} + \frac{\partial \alpha \cdot h^m}{\partial x} = r_i(t) - f_i(x, t); \quad \text{and} \quad \frac{\partial A}{\partial t} + \frac{\partial Q(A)}{\partial x} = q_l(t) - f_{c_i}(x, t) \quad (6.13)$$

where h is the mean overland flow depth (L), t is time, x is the distance along the slope (L), α is $1.49\,S^{1/2}/n$, S is the slope, n is Manning's roughness coefficient, m is $5/3$, $r_i(t)$ is the rainfall rate (L/T), $f_i(x, t)$ is the infiltration rate (L/T), A is the channel cross-sectional area of flow (L^2), $Q(A)$ is the channel discharge as a function of area (L^3/T), $q_l(t)$ is the net lateral inflow per unit length of channel (L^2/T), and $f_{ci}(x, t)$ is the net channel infiltration per unit length of channel (L^2/T). These equations, and those for erosion and sediment transport, are solved using a four-point implicit finite difference method (Smith et al., 1995).

Unlike excess routing, interactive routing implies that infiltration and runoff are computed at each finite difference node using rainfall, upstream inflow, and current degree of soil saturation. This feature is particularly important for accurate treatment of transmission losses with flow down dry channels. To explicitly account for space-time variations in rainfall patterns the model computes, for each overland flow element, the rainfall intensities at the element centroid as a linear combination of intensities at the three nearest gauges forming a piece-wise planar approximation of the rainfall field over the catchment (Goodrich, 1990). The interpolated centroid intensity is applied uniformly over that individual model element. To represent small-scale variability of infiltration that is beyond the scale of discretisation (sub-metre to metre), the model assumes the saturated hydraulic conductivity (K_s) within an overland flow element varies log-normally (Woolhiser and Goodrich, 1988; Smith et al., 1990) (i.e., it uses a sub-grid distribution function parameterisation as discussed in Chapter 3).

Validation of the KINEROS model is reported by Goodrich (1990), Goodrich et al. (1993), and Smith et al., (1995) on four Walnut Gulch subcatchments (Lucky Hills (LH)-106, 0.4 ha; LH-102, 1.4 ha; LH-104, 4.4 ha; and WG-11, 631 ha). For the Lucky Hills catchment, rainfall inputs were obtained from two raingauges, and for WG-11 ten raingauges were used. The validation process consisted of a split sample test (Chapter 3, p. 76 and Chapter 13, p. 340) with the calibration phase using approximately ten observed events on each catchment and a validation phase in which an independent set of roughly twenty runoff events were used to assess model performance using the coefficient of efficiency, E (Nash and Sutcliffe, 1970) (Table 6.1). The model was calibrated by adjusting

Table 6.1. KINEROS calibration and verification coefficient of efficiency for runoff volume and peak discharge

Basin	Calibration efficiency						Verification efficiency				Maximum no. of model elements
	Volume			Peak			Volume		Peak		
	1	2	3	1	2	3	1	2	1	2	
LH-6	0.98	0.97	0.81	0.95	0.94	0.86	0.98	0.98	0.79	0.77	30
LH-2	0.97	0.88	0.88	0.97	0.93	0.93	0.93	0.92	0.93	0.89	68
LH-4	0.97	0.96	0.89	0.98	0.88	0.88	0.99	0.99	0.92	0.96	235
WG-11	0.86			0.84			0.49		0.16		243

1 – two raingauges in Lucky Hills, ten raingauges in WG-11, using the maximum number of overland and channel flow elements
2 – two raingauges, one overland flow element, no channel elements
3 – one raingauge, maximum number of overland and channel flow elements
* – If the model predicts observed runoff with perfection, $E = 1$. If $E < 0$, the model's predictive power is worse than simply using the average of observed values.

three parameters: basin-wide multipliers on n, K_s, and CV_{Ks}. The multipliers scale the model element input parameters while maintaining relative differences based on field observations. Using this approach, the overall dimension of the adjustable parameter space remains small (see Chapter 13, pp. 342–3 for a discussion of this approach). By using the nested catchments LH-106 and LH-102 (see Figure 6.1) within catchment LH-104, internal assessment for the model's ability to reproduce runoff patterns was also possible.

As judged by the efficiency statistics, the model provides remarkably good predictions of runoff volume and peak response for the Lucky Hills catchments. An overall assessment of internal model accuracy using the nested catchments gives an E of 0.91 and 0.86 for LH-106 runoff volume and peak rate, respectively, and comparable LH-102 E values of 0.96 and 0.97. These high values of E obtained by using LH-104 multipliers for the internal catchments suggest a good deal of internal model accuracy. On WG11 the model performed reasonably well for the calibration event set, but E dropped off considerably for the verification event set due to overprediction of the two largest events in the verification set.

While we can represent a wide range of geometric catchment complexity, it is not clear just how much is required to best represent hydrological response. Does the use of a great number of model elements, and therefore a great amount of distributed input, actually improve the simulations or can simpler geometries do just as well? Which components can be simplified and which must have their spatial detail preserved? Geometric model complexity and catchment heterogeneity are closely related. More complex model representations (i.e., more overland flow and channel model elements) more closely preserve the catchment patterns of topography and channel networks. Large-scale orthophoto maps were used to discretise the catchments into a large number of elements, and a geometric simplification procedure based on stream order reduction was developed (Goodrich, 1990). Successive levels of reduction in model complexity were then carried out to assess the impacts of simplification on simulated runoff response. It was found that adequate representation of concentrated channel routing imposed a fundamental limit on simplification because concentrated channel flow can only be converted to overland flow with a distortion of the hydraulic roughness to a certain degree. For catchments greater than 1 hectare it was found that an average area for first-order channels should be roughly 10–15 % of the total catchment area (Goodrich, 1990).

The relative impact of geometric versus rainfall pattern simplification was also assessed. The error introduced when the model was simplified to a single overland flow plane with two raingauges as input was less than or equal to the error when one raingauge was used in the Lucky Hills catchments as input to a model with the maximum number of elements corresponding to the most complex geometric catchment representation (see Table 6.1). This result was even more pronounced in the larger WG-11 catchment when one versus ten raingauges was used (Goodrich, 1990). *Therefore, unless there are major differences in land use, basin discretisation should not exceed the ability to resolve input rainfall variability.*

The uncertainty in rainfall input due to small- and large-scale spatial variability suggests that the confidence in the calibration can only be equal to or less than the certainty of rainfall input data.

Faurès et al. (1995) assessed the impacts of Walnut Gulch rainfall variability on runoff simulations in the LH-104 catchment using KINEROS (Figure 6.16). Data from combinations of five recording raingauges were input into KINEROS for the event of August 3, 1990. These simulations produced a range of variation for simulated peak runoff rate and runoff volume of 15 mm/hr ($CV = 38.8\%$) and 2.6 mm ($CV = 40.0\%$), respectively. The variability in runoff model results emphasises the importance of adequately sampling the spatial distribution of rainfall in the catchment. It was also found that model output variation as a function of the number of raingauges was generally greater for small events than for large ones. This reflects the difficulty of modelling small runoff events when runoff to rainfall ratios are low and measurement error may be a larger relative percentage of the input rainfall signal. This would be expected whenever relative infiltration and rainfall rates are close, resulting in small runoff ratios. In this case, the model becomes very sensitive to both input and parameter patterns. If the uniform, single raingauge assumption were used during parameter fitting in spite of spatial variability comparable to that observed here, the variation in simulated hydrographs

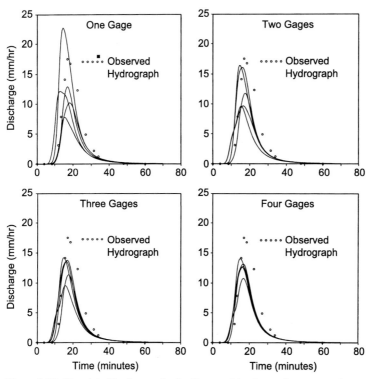

Figure 6.16. Simulated hydrographs for five combinations of one, two, three, and four raingauges in Lucky Hills-104 (August 3, 1990 event).

could be mistakenly assigned to variability of other model parameters or errors in the model structure.

Finally, the relative importance of PBMR derived remotely sensed soil moisture patterns (Section 6.2.2) on runoff simulations on the larger WG-11 catchment is examined. KINEROS is very sensitive to the estimate of the pre-storm initial relative soil saturation, SI (Goodrich, 1990). An average SI was derived for each of 256 model elements (3.4 ha mean overland flow element area) from five PBMR overflights, three of which are illustrated as volumetric soil moisture in Figure 6.8 (Goodrich et al., 1994). An increase in the variability of SI with increasing mean SI was observed; however, the highest mean SI observed was 0.45 on a zero to one scale, so the generally postulated decrease in variability as SI approaches one was not observed (Goodrich et al., 1994). Attaining a very high average SI may not be realised given the rapid drainage of the coarse soils of WG-11 and the difficulty of obtaining an aircraft overflight immediately following an intense convective thunderstorm.

In order to assess the relative importance of variability in initial soil saturation and variability in precipitation, a simulation study was performed using different combinations of observed patterns in soil saturation (from PBMR data) and precipitation (from multiple raingauges). The impact of simplifying the representation of initial soil saturation was assessed by comparing the highly complex SI pattern (256 SI values) to a single catchment average SI representation (Figure 6.17). For rainfall simplification, the case of using observations from ten raingauges in and near the catchment is compared to using rainfall from a single central raingauge (uniform rainfall). The rationale for using *average SI* but a *single* raingauge (rather than average of the ten, which would be less variable) is

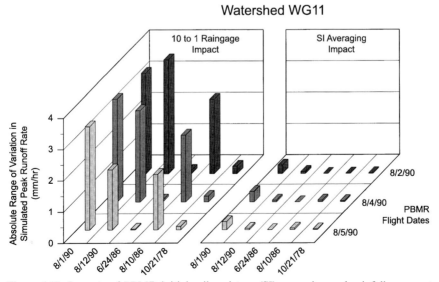

Figure 6.17. Impacts of PBMR initial soil moisture (SI) averaging and rainfall representation on simulated peak runoff rate for various storm–SI combinations.

that average *SI* data may become widely available from low spatial resolution sensors, while only single raingauges are usually available and applied uniformly to large areas.

The comparisons were made for three PBMR derived sets of *SI* (August 2, 4, and 5, 1990) and for five storms, two of which occurred during the period of PBMR overflights and three historical storms that had relatively small runoff volumes and distinct rainfall patterns. The October 21, 1978 event was relatively uniform, the June 24, 1986 event had high rainfall gradients in the upper central portion of WG11, and the August 10, 1986 event produced steep precipitation gradients in the lower portion of WG11. The magnitude of these storm/runoff events means that the influence of *SI* on runoff generation is large because the rainfall depth is of the same order as the soil water deficit (Goodrich, 1990). The absolute percentage change in peak runoff rate ranged from 0.5 % to 12.3 % for *SI* averaging and over 400 % for rainfall simplification. Based on these results, a simple basin average of remotely derived *SI* estimates at the medium catchment scale (6.31 km^2), with a greater knowledge of spatial rainfall patterns, appears to be adequate for runoff simulation. This implies that the relatively coarse resolution of potential space-based microwave instruments may be entirely adequate for defining distributed pre-storm initial soil water content conditions for rainfall-runoff modelling in semi-arid regions, provided ground truth data are available (Goodrich et al., 1994).

6.4 CONCLUSIONS

Our understanding of the complex hydrological processes active in semi-arid regions has been greatly enhanced through numerous studies of rainfall, runoff, and soil moisture patterns at the USDA-ARS Walnut Gulch Experimental Watershed in southeastern Arizona. Extremes in rainfall and temperature in this region lead to great spatial heterogeneity in soil hydrological processes. Accurate spatial and temporal knowledge of precipitation totals and intensity were found to be the most important factor in determining hydrologic catchment patterns in this region. Convective rainfall is highly localised with observations indicating that rainfall from raingauges six or more kilometres apart can be considered independent. Significant rainfall variability is also apparent over scales of several hundred metres as rainfall gradients ranging from 0.28 to 2.48 mm/100 m were observed over a 4.4 hectare catchment. This suggests that the typical uniform rainfall assumption is *invalid* at the 5 hectare scale in this or similar environments. Soil moisture patterns were profoundly impacted by precipitation. Remote sensing techniques also show potential for indirect estimation of rainfall at ungauged catchments.

The correlation structure present in the PBMR derived soil moisture changes as the surface dries following a rain storm. Storms impose large-scale correlation which decreases as the soil dries and the random (small-scale) effects of surface characteristics begin to control soil moisture variability.

The techniques of data assimilation were successfully applied to obtain better soil moisture estimates using a distributed model and remote sensing data. Overall, the Newtonian nudging method has the most desirable features for remotely sensed soil moisture data assimilation. It is the only true four-dimensional data assimilation method used in this study, and it produces relatively continuous soil moisture time series and reasonable spatial patterns. There is a clear tradeoff between using a complex data assimilation technique and the ability to use all the available data. The use of assimilation methods that are sound and computationally efficient and use as much data as possible is preferred.

The relationships between storm size, location and pattern, and the scale and geometry of the catchment are delicate and should be carefully considered when interpreting or modelling hydrological processes. A critical process in this region is ephemeral channel transmission losses. Spatially distributed models that explicitly treat runoff routing and channel abstractions are considered essential. This is supported by good calibration and verification results of models with explicit physically-based routing and infiltration components using the nested gauge data.

The conclusions described here must be considered in the context of the semi-arid Walnut Gulch environment. It should be reiterated that runoff is almost exclusively generated via an infiltration excess mechanism and annual potential evapotranspiration is roughly ten times greater than annual rainfall in this environment. In this influent environment, with annual runoff decreasing with increasing catchment size, it was found that runoff response becomes more nonlinear with increasing catchment size. Our increased understanding of this environment would not have been possible without the long-term, spatially dense observations made at the WGEW. We strongly encourage the continued operation and improvement of this exceptional outdoor laboratory.

ACKNOWLEDGEMENTS

We gratefully acknowledge the staff of the USDA-Southwest Watershed Research Center for their diligent collection, processing and interpretation of high-quality data from the Walnut Gulch Experimental Watershed. We also thank T. Jackson for providing several figures, and C. Unkrich for the preparation of many of the figures. We would also like to gratefully acknowledge support for the third author from the NASA/EOS grant NAGW2425.

7

Spatial Snow Cover Processes at Kühtai and Reynolds Creek

David Tarboton, Günter Blöschl, Keith Cooley,
Robert Kirnbauer and Charlie Luce

7.1   INTRODUCTION

In many climates, predicting and understanding the spatio-temporal variability
of snow-related quantities plays a key role in catchment hydrology. Practical
applications include the prediction of snowmelt induced floods and the estimation of water yield from snow-covered catchments for water resources management. The snow cover is also a key link in the climate system via its effect on the
surface energy and water balance, so its accurate representation is essential to a
better understanding of climate effects on the hydrological cycle. Modelling the
spatio-temporal variability of snow-related quantities is complicated by the inter-
related and multiscale nature of the processes involved. Natural snow variability
is extreme and although snow related data such as snow water equivalent is often
available in considerable temporal detail as time series (e.g. the US SNOTEL
network, NRCS, 1998), the spatial resolution of snow-related data is notoriously
poor. Often, at best, a few point measurements are available in the catchment of
interest and, because of the extreme spatial variability, point data are not very
representative of the spatial patterns and/or the spatial averages. Although runoff does provide a spatially aggregated estimate of melt water yield from a
catchment, it is not possible to infer the actual melt processes and their spatial
distribution from runoff data alone. Recently, progress in remote sensing of snow
has shown potential. Snow-covered area can be measured using a variety of
methods. However, remote sensing of snow water equivalent has not been sufficiently developed for operational observation of deep snowcover in rugged
mountain terrain (Elder et al., 1998). Therefore it has been suggested in the
literature (e.g., Blöschl et al., 1991b) to use snow cover patterns for evaluating
and improving distributed snow models. This is consistent with the general thrust
of this book of using observed patterns for assessing distributed models. As
compared to other components of the hydrologic cycle described in this book
such as rainfall, runoff and soil moisture, snow has the definite advantage that

the patterns are actually visible to the eye. However, snow cover related fluxes, state variables and model parameters are highly variable in space and time.

This chapter addresses the issues of spatial variability in snow cover and snow water equivalent, and the processes responsible for this variability. The chapter starts with a brief description of the physical processes involved in snow accumulation, redistribution and melting with an emphasis on their spatial variability. A few key modelling approaches are summarised. Two case studies in different snow environments are discussed to exemplify the range of snow processes typically encountered in catchment hydrology. The first case study is set in the Austrian Alps and is representative of the high alpine environment where snow redistribution by avalanching and differential melting caused by terrain aspect are major sources of spatial variability. The second case study is set in the Western U.S. rangelands where slopes tend to be flatter and wind drift is a major source of spatial snow variability. We conclude with a few remarks on the future of distributed snow modelling and use of spatial patterns to improve model confidence.

7.2 SPATIAL VARIABILITY OF SNOW-RELATED PROCESSES

7.2.1 Snow Accumulation and Melt at the Point Scale

Snow accumulation and melt is spatially variable due to the spatial variability in the driving processes and inputs. This spatial variability in turn results in spatially variable surface water inputs from snowmelt that affects runoff and soil moisture discussed in other chapters of this book. The spatial variability of snow-related processes has been discussed in detail by Obled and Harder (1979) and Hardy et al. (1999), and others. Here we will only give a brief review.

At a point, the accumulation and ablation of snow is a process involving fluxes of energy and mass across the snow–air and the snow–ground interfaces. Energy exchanges include shortwave solar radiation (direct solar radiation and diffuse solar radiation), terrestrial/atmospheric longwave radiation, turbulent fluxes (sensible and latent heat exchanges between the atmosphere and snow), energy fluxes associated with exchanges of mass (the energy that comes with falling rain and is carried away by meltwater), and conduction between the snow and underlying ground (i.e. ground heat flux). In alpine environments, radiation fluxes are usually larger than sensible and latent heat fluxes, but in lowlands where snowmelt tends to occur in early winter they can be much smaller (e.g., Male and Gray, 1981; Braun, 1985). Advective exchanges and the ground heat flux are usually very small, but their integrated effect over a season can be significant. Mass exchanges consist of precipitation inputs, meltwater release, and condensation/evaporation/sublimation, the latter being very small. The dynamics within a snowpack are quite complicated, involving energy and mass fluxes due to conduction, thermal radiation, vapour diffusion, meltwater movement, settling and compaction. Some of these processes lead to the formation of ice layers, which impede the downward propagation of infiltrating meltwater, resulting in concentrated finger flow and sometimes lateral flow (Colbeck, 1978, 1991).

7.2.2 Spatial Patterns of Snowmelt Processes

Energy exchanges are the main processes responsible for the differential melting of snow in a catchment. The spatial variability of direct solar radiation within a catchment is dominated by terrain slope, aspect and shading. Direct solar radiation per unit horizontal area averaged over a time interval from t to $t + \Delta t$ may be expressed as

$$Q_{si} = I_o \cdot T_f \cdot \frac{1}{\Delta t \cos S} \int_{t}^{t+\Delta t} \cos(\psi(t)) \mathrm{d}t \qquad (7.1)$$

where I_o is the solar intensity (4914 kJ m^{-2} hr^{-1} or 1367 W/m^2), T_f atmospheric transmissivity, S the local slope angle and $\psi(t)$ the time varying illumination angle, defined as the angle between the surface normal and direction to the sun. In mountain regions, terrain shading can be important in which case the integral above should only be evaluated for times when direct radiation is incident, i.e. the point is not shaded by nearby terrain. The time varying illumination angle can be accurately computed from analytical expressions and tabulated values (Dozier, 1979). Atmospheric transmissivity depends upon weather conditions and cloudiness and therefore gives rise to the largest uncertainties in estimation of incident radiation. Simple approaches to quantifying atmospheric transmissivity include those of Bristow and Campbell (1984) based upon diurnal temperature ranges and Neuwirth (1982) based on visual observations of cloudiness (i.e. the fraction cloud cover of the sky). More elaborate methods integrate radiative transfer throughout the atmosphere (e.g. "LOWTRAN 7", see Kneizys et al., 1988). The spatial distribution of atmospheric transmissivity in a catchment is random and hence essentially unpredictable but fortunately, if integrated over a period of say a few weeks, its effect on snowmelt tends to average out, so approximating transmissivity as constant over a study area is usually reasonable.

Part of the atmospheric transmissivity reduction in direct radiation is due to scattering and about one half of the scattered energy reaches the surface as diffuse radiation (the other half going out into space) (Dingman, 1994). Diffuse radiation tends to increase in cloudy conditions when more of the incident radiation is scattered. In a catchment, the spatial pattern of diffuse radiation received by the snow surface depends on the fraction of the sky dome that is visible from each point. This fraction can be quantified in terms of the sky view factor, V_d. The sky view factor is based on the assumption of isotropic radiation and is defined as the ratio of the radiation incident on a point accounting for slope, aspect and terrain obstructions, to the equivalent radiation incident on a flat and unobstructed surface. V_d only depends on terrain and does not depend on time. For any point in time, incident diffuse shortwave radiation can be estimated as

$$R_{\text{slope}} = V_d \cdot \pi \cdot I_d \qquad (7.2)$$

where I_d is the isotropic diffuse radiation intensity. In mountain regions, solar radiation reflected by surrounding terrain can also be important. It can be approximated by $(1 - V_d)$ times reflected shortwave radiation (albedo times incident radiation). Procedures for computation of horizon angles and sky and terrain view factors and discussion of their use, assumptions and limitations in estimating radiation are given by (Dozier, 1979; Dozier and Frew, 1990; Dubayah et al., 1990; Frew, 1990).

Part of the incoming solar radiation is reflected by the snow surface. The ratio of reflected and incoming radiation is termed albedo, which can vary considerably as a function of the condition and age of the snow surface. Given the magnitude of the solar radiation term in the energy balance, modest albedo changes are important to the snow surface energy balance. The albedo of snow is generally at a maximum after a fresh snowfall and decreases with time due to growth in grain sizes, and the accumulation of dust, soot and debris on the snow surface (U.S. Army Corps of Engineers, 1956). The rate of grain growth increases with snow temperature and in particular with the presence of liquid water (Wiscombe and Warren, 1981; Dozier, 1987; Marshall and Warren, 1987). The most important process controls on albedo are reflected in the parameterisations suggested by various authors. Examples include Rohrer (1992) and Dickinson et al. (1993) who proposed a parameterisation of albedo as a function of air temperature and time after snowfall, Brun et al. (1992) who parameterised albedo as a function of time after snowfall, grain size and grain type, and Marks and Dozier (1992) and Marshall and Warren (1987) who modelled grain size increase and parameterised albedo in visible and infrared bands as a function of grain size. Little is known about the spatial distribution of snow albedo in catchments; the controls mentioned above do suggest that albedo tends to be lower on south-facing slopes (in the Northern Hemisphere) due to the more rapid grain growth as a consequence of larger energy inputs as compared to other slope aspects.

Both the atmosphere and the snow surface emit black body longwave radiation that is proportional to the fourth power of absolute temperature. Incoming longwave radiation from the atmosphere is related to the vertical distribution of air mass properties (air temperature, vapour pressure) and the presence of clouds (Obled and Harder, 1979). While several parameterisations are available based on surface air temperature and vapour pressure (see e.g., Price and Dunne, 1976; Satterlund, 1979) there is considerable uncertainty in these estimates due to atmospheric variability. Radiative transfer models (e.g. "LOWTRAN 7", see Kneizys et al., 1988) overcome some of this uncertainty at the cost of more substantial data requirements. In a valley, incoming longwave radiation from the atmosphere is reduced because the adjacent mountains obscure part of the sky. Like shortwave diffuse radiation, incoming longwave radiation from the atmosphere is generally diffuse and its spatial pattern can be represented by the sky view factor analogously to equation (7.2). However, scattered and emitted longwave radiation from mountainside slopes is present and may be greater than atmospheric longwave radiation, particularly in steep valleys and cirques where the slopes are snow free. For example, Olyphant (1986) showed that the snowpack in cirques can have an additional longwave radiation input from the surrounding terrain equivalent to 500 mm melt when integrated over an entire snowmelt season, as compared to flat terrain. Similar to reflected shortwave radiation, it can be approximated by $(1 - V_d)$ times terrestrial emissions from surrounding terrain.

Outgoing longwave radiation is on average greater than incoming longwave radiation, resulting in a net loss of energy as thermal radiation from the surface. The emissivity of snow is between 0.97 and 1 (Anderson, 1976) and night time longwave radiation losses under clear skies are responsible for considerable cooling of the snow surface. However, actual heat loss is limited by the small thermal conductivity of the snow which may vary depending on snow surface properties. The spatial distribution of longwave radiation emitted by the snow in a catchment is rather complex, being controlled by the spatial pattern of surface temperature, which in turn is controlled by the overall heat budget of the snow. Cold snowpacks (prior to any melting) have low thermal conductivity which results in limited outgoing longwave radiation and large night-time depressions in surface temperature. The presence of liquid water in the snow near the surface, due to melting or rain, alters this significantly. The surface temperature remains close to freezing (0 °C) until this water refreezes. This unfrozen water near the surface represents a considerable storage of latent heat of fusion energy that may be radiated away. Melting and refreezing also results in crusts at the snow surface with altered thermal properties (conductivity and density). These processes are the compound effect of total net energy exchanges and vary spatially because of the terrain effects on incident radiation energy inputs.

Incident radiation (both shortwave and longwave) on snow beneath the vegetation canopy is limited by the radiative transmissivity of the vegetation (Verstraete, 1987, 1988; Verstraete et al., 1990) which is related to leaf area index defined as the ratio of the total surface of leaves above a ground area to that ground area, as well as leaf shape and orientation. Vegetation has a lower albedo than snow, and therefore absorbs more incident radiation and may be warmer than the surrounding snow surface. Vegetation emits longwave radiation proportional to the fourth power of its absolute temperature which results in localised melting around sparse vegetation. Vegetation also provides greater surface roughness, reducing wind speeds at the surface of snow beneath vegetation which affects the turbulent energy transfers mentioned below. In sparsely vegetated areas, the persistence of snow patches associated with patches of vegetation is a source of spatial variability, due to vegetation shading as well as wind sheltering and accumulation of snow drifts (Seyfried and Wilcox, 1995). The spatial pattern of vegetation is naturally quite variable due to temperature, radiation and moisture variability and the biological needs of different species and in most environments it is controlled by human activity which adds additional variability. This spatial variability influences the distribution of snow, and is influenced by snow distribution in a synergistic relationship. Snowmelt supplies water for vegetation. Snow also affects the environment in which vegetation species need to survive.

Turbulent energy transfers comprising sensible and latent heat fluxes are a significant component of the snow energy balance. Sensible heat fluxes depend on the temperature gradient and turbulent diffusion due to wind. Latent heat fluxes depend on the vapour pressure gradient and turbulent diffusion due to wind. Latent heat fluxes consist of evaporation and condensation of liquid water, and sublimation of ice (Male and Gray, 1981; Bras, 1990). Surface roughness and the profile of wind velocity with height control turbulent diffusion. Turbulent transfer rates also depend on atmospheric stability, which is a function of the temperature gradients (Brutsaert, 1982). Snow surfaces with surface temperature limited to remain at or below melting (0 °C) almost always have a stabilising effect on the atmosphere, tending to reduce turbulent diffusion. Spatial variability of topography and vegetation result in spatial variability in wind, wind profiles and turbulent energy fluxes which affect the spatial patterns of snow. Windspeed is higher on exposed ridgetops than in valleys. Windspeed is also higher on upwind than on downwind slopes. The variability of wind will be discussed below in the context of wind redistribution of snow where it has a greater effect. The same wind spatial variability that results in snow redistribution also has spatially variable effects on turbulent exchange.

7.2.3 Spatial Patterns of Snow Accumulation Processes

Snowfall and snow redistribution by wind are the main processes responsible for the differential accumulation of snow in a catchment.

The main control on snowfall patterns is elevation through its control on the state of precipitation. The state of precipitation (rain or snow) depends upon air temperature at the time of precipitation. The lapse rate of air temperature with elevation results in snow at higher elevations and rain at lower elevations. The snow line (the elevation separating rain from snow) varies for each precipitation event. Rain on snow may cause snowmelt at lower elevations, while at high elevation there is additional snow accumulation. The net effect of these processes is a strong dependence of snow accumulation on elevation. In addition to the effects due to the state of precipitation discussed above, topography also influences the pattern of snowfall and accumulation through orographic effects on atmospheric processes. Heavy precipitation occurs on slopes where atmospheric flow is forced over mountain ranges. Orographic lifting may also induce instability in the atmosphere, triggering convective precipitation (e.g., Dingman, 1994). On the downwind side of mountain ranges, precipitation is reduced because orographic lifting and condensation have stripped moisture from the atmosphere. Approaches to modelling orographic precipitation range from empirical correlation of precipitation with elevation, the so-called hypsometric method (e.g., Dingman, 1994; also see Chapter 2, pp. 35, 40), to models that empirically and dynamically model atmospheric flow and snowfall (Rhea, 1978; Peck and Schaake, 1990; Barros and Lettenmaier, 1993, 1994). The scale of spatial patterns associated with orographic effects is generally quite large (1 km or more) relative to the variability associated with, for example, slope and aspect effects on radiation (10 to 100 m).

In steep terrain, deposited snow frequently sloughs and avalanches, moving downslope under the influence of gravity in sometimes catastrophic fashion, coming to rest in less steep gullies and runout zones. The two main controls on sloughing and avalanching are terrain slope and the stability of the pack. Typically, hillslopes with slopes between 20° and 50° are prone to avalanching while steeper and flatter slopes are not. Avalanches usually originate in weak layers resulting from variable snow density, crystalline structure and lack of bonding between new and old snow (e.g. McClung and Schaerer, 1993; Armstrong and Williams, 1986). On flatter slopes the downslope component of gravity is insufficient to overcome the shear strength of snow.

The redistribution of snow by wind is a complex process controlled by the interaction of wind flow, topography, snow properties and surface roughness. Processes involved include scour from upwind slopes, sublimation of suspended and saltating particles, deposition on downwind slopes and especially behind terrain obstacles, where flow separation occurs. Vegetation, through its influence on surface roughness, limits the scour and enhances the deposition of blowing snow. This effect of vegetation is only present while it is not buried by snow, leading to the concept of vegetation holding capacity used in wind-blown snow modelling (e.g., Pomeroy and Gray, 1995; Liston and Sturm, 1998).

This section has shown, in a conceptual fashion, the physical processes that lead to spatial variability and spatial patterns in snow accumulation and melt. These comprise multiple processes interacting across a range of scales. McKay and Gray (1981) summarise the scales involved in various snow redistribution processes:

- Macroscale: (10^4–10^5 m) Elevation, orography, meteorological effects such as standing waves, flow of wind around barriers and lake effects.
- Mesoscale: (10^2–10^3 m) Redistribution due to wind and avalanches, deposition and accumulation related to elevation, slope, aspect, vegetative cover height and density.
- Microscale: (10–10^2 m) Primarily surface roughness and transport phenomena.

In the next section we review approaches for distributed snow modelling followed by case studies where spatial patterns of distributed measurements and model results are compared.

7.3 SPATIAL SNOW MODELLING

Spatially distributed snow models differ in terms of the degree of process representation they involve. At one end of the spectrum are empirical methods that often use statistical relationships involving temperature, radiation and terrain properties while at the other end are process based (dynamic) models (Kirnbauer et al., 1994). One example of an empirical model for estimating peak snow accumulation is the SWETREE model (Elder, 1995; Elder et al., 1995, 1998; Winstral et al., 1999). This model is based on statistical analysis of a very large number of snow water equivalent samples and uses binary decision trees to predict snow water equivalent based upon indices for radiation, wind exposure and other controls. These indices are used to subdivide a catchment into classes, starting from the most important controls and proceeding to the less important controls. A similar recent example is the model of König and Sturm (1998) which is based on topographic rules using physiographic features such as creek patterns, flat patterns, and slope patterns. These features are derived from a visual analysis of aerial photographs, and for each of them, characteristic values of snow depth and snow water equivalent are assigned. König and Sturm (1998) examined their method in the Alaskan Arctic where slopes are much flatter than in the catchments of Elder (1995) and where wind drift is the main process giving rise to differential accumulation and melting. Another contribution to mapping snow water equivalent is due to Woo et al. (1983) and Yang and Woo (1999) which, similar to König and Sturm (1998), use topographic features, but their approach is more heavily based on ground data. The advantage of this type of model is a parsimonious model structure which implies robustness and ease of use, but this comes at the cost of requiring a substantial database for calibrating the model, usually consisting of both remotely sensed images and ground data. An example of an empirical spatially distributed melt model is provided by Williams and Tarboton (1999). This model separates the energy that causes snowmelt into three components: a spatially uniform component, a component that is proportional to elevation, and one that is proportional to solar illumination (which is determined from topography). Measurements of snowmelt at several topographically unique points (called "index points") in a catchment are related to elevation and solar illumination through regression in order to factor the melt energy into the three separate components at each time step. Inputs from snowmelt measurements at the index locations are used to calibrate the regression at each time step. Then the spatial patterns of solar illumination and elevation are used to predict the spatial distribution of melt over the whole catchment.

Process-based models account for both mass and energy exchanges and keep track of state variables related to mass and energy over time. In this type of model, the catchment is usually subdivided into model elements and point snow models are applied to each element. There have been a large number of point snow models developed in the literature that range in complexity and amount of data used (e.g., Anderson, 1976; Blöschl and Kirnbauer, 1991; Jordan, 1991; Kustas et al., 1994; Tarboton et al., 1995; Tarboton and Luce, 1996; Luce et al., 1997). The main advantage of this type of model is that it allows a detailed representation of the processes giving rise to differential melting and accumulation. However, extension of point snowmelt models to catchments involves considerable problems and uncertainty, part of which is related to scale issues (e.g.

Kirnbauer et al., 1994; Beven, 1995; Blöschl, 1999). As discussed in a general sense in Chapter 3, the fundamental questions involve the selection of model elements, parameterisation of subgrid variability and nonlinearity, and distributing input data and model parameters across the catchment. Ways of selecting model elements in distributed snow models are similar to those in other hydrological models and include square grids, hillslope elements, and elevation bands (see Chapter 3).

If these elements are small enough, a detailed spatially explicit approach is possible. While this approach is conceptually simple and appealing, we must have enough detailed information to determine site parameters and inputs to each element. However, this is often not possible and element sizes are used that are relatively large as compared to the characteristic scale of the underlying variability. Often, the effective parameter approach is used where it is implicitly assumed that an average parameter over that element represents the combined effects of the processes within that element (see Chapter 3, p. 00) but as snow related processes are highly nonlinear, treatment of the variability within elements (i.e. subgrid variability) requires particular attention. Luce et al. (1997), for example, showed that this approach may yield incorrect results once the element area exceeds about 1 ha. An alternative is to use distribution functions to represent subgrid variability. The distribution function approach so far has not been widely used for representing subgrid variability in spatially distributed snow models but it has a long tradition for representing spatial variability in lumped catchment models. An early example is the areal depletion curve approach of Anderson (1973), where an empirical function is used to relate the areal extent of snow cover to mean areal water equivalent. A more recent example is Luce et al. (1999) who show that the surface water input estimated from a lumped model parameterised with a depletion curve derived from the distribution of snow at peak accumulation, compares well with the surface water input estimated from periodic measurements and from an explicitly distributed snowmelt model. A drawback of using spatial distribution functions within each model element is that one needs a minimum of two or three parameters to represent the distribution functions rather than one single parameter as in the effective parameter approach, but it is much better suited for representing the nonlinear effects of the subgrid snow processes.

Distributing input data and model parameters across the catchment draws on the understanding of the spatial variability of the processes driving snowmelt and snow accumulation. Climatic data are usually available at one or two sites within the catchment and snow courses usually provide just a few values of snow water equivalent and snow depth, so distributing this information to every model element requires assumptions to be made. This distribution procedure is essentially an interpolation problem, i.e. a problem of determining patterns from points (see Chapter 2). In the context of snow, auxiliary data for interpolation can be based on terrain features such as slope, aspect, terrain shading and view factors that can be directly computed from digital elevation models. As discussed earlier in this chapter, this approach is particularly useful for estimating detailed spatial patterns of radiation inputs to snowmelt models (Dozier, 1979; Dozier and Frew, 1990).

One of the most important issues of spatial snow models is how to represent snow redistribution by wind drift. There are three types of approaches in the literature that differ in complexity. The simplest approach is to use wind drift factors. The basic assumption of this approach is that the spatial patterns of snow and/or snowfall are similar in all years. This similarity is based on the rationale that topography is the main factor controlling wind drift and that average wind speeds and directions only differ slightly from year to year. It assumes that if snowfall is increased, the amount of accumulated snow water equivalent will be increased proportionally and the spatial pattern due to drifting will be the same. In an alpine environment, Kirnbauer and Blöschl (1994) found that this is indeed the case, with acceptable accuracy. Once the time stability is established, there are two possibilities of deriving the wind drift factors. The most accurate approach is to sample snow water equivalent exhaustively in the catchment (e.g. Cooley, 1988) and to derive snow drift factors from these data (Tarboton et al., 1995; Luce et al., 1998). For larger catchments this is not feasible, and relationships between wind drift factors and topographic attributes have been postulated (e.g. Blöschl et al., 1991b; see discussion in Moore et al., 1996). The parameters for these types of relationships can be estimated from remote sensing data and/or ground measurements. A further step up in complexity is quasi-dynamic models. A typical representative of this model genre is given in Essery et al. (1999). They applied a model of wind flow over complex terrain to arctic landscapes and used it to investigate joint probability distributions of wind speed and blowing snow occurrence. Functions that describe the joint distribution were then used to scale results up from a single-column model of blowing snow that presumes homogeneous terrain. Results are compared with results from a distributed model and spatially distributed snow surveys from the Arctic. The most complex models are dynamic models. One example is the model of Liston and Sturm (1998) which represents snow transport resulting from saltation and suspension, snow accumulation and erosion, and sublimation of the blowing and drifting snow. It is driven by a wind model that computes the flow field over the complex topography. Model inputs include climatic forcings, as well as vegetation type which is used to determine a vegetation snow-holding capacity that must be exceeded before any additional snow is available to be transported by the wind. The complex air flow in an alpine environment provides a challenge for these models, therefore in many cases one must resort to the more empirical model types for representing snow redistribution by wind drift.

In the following sections we will present two case studies. The aim of the case studies is to deterministically model spatially distributed snow processes in small catchments. The case studies differ greatly in terms of their hydrological and climatological settings, and in terms of the processes giving rise to spatial snow variations. The Kühtai study is set in steep alpine terrain with high annual precipitation, deep snowpacks and an extended ablation period. The Reynolds Creek study is set in undulating rangelands where precipitation is low and most of the snowmelt occurs during a relatively short period. Radiation along with sloughing, avalanching and wind redistribution are important at Kühtai, while wind drift is the most important factor causing spatial snow variations at Reynolds Creek. The catchments also differ in scale, the catchment of the Kühtai study (9.4 km^2) being about 40 times the size of the Reynolds Creek study catchment (0.26 km^2). Climate, terrain and scale have implications for the type of data of snow variability used in the two studies. In the case of Kühtai, remotely sensed snow cover patterns are used (binary values of snow/ no snow at about 15,000 pixels) while in the case of Reynolds Creek ground data of snow water equivalent (sampled on a regular grid of about 300 points) are used.

7.4 KÜHTAI CASE STUDY

The Längental catchment is located in the Kühtai region, Tyrol, in the Austrian Alps. The catchment is 9.4 km^2 in size and elevations range from 1900 to 3050 m above sea level. Geomorphologically, the basin consists of two major units (Figure 7.1). The lower part comprises east and west facing slopes including talus fans with typical slopes of 35 to 40°. The upper part in the south west is open to the east. The south-east edge of the basin is formed by three prominent cirques. Most of the catchment lies above the timber line and there are only a few scattered larches and cembra-pines. The flat areas are covered by alpine meadows and the steep areas are rock and debris. Average annual precipitation is about 1200 mm, 50 % of which falls as snow. Temperatures average 10 °C in summer and −5 °C in winter (Figure 7.2, dotted line). In the lower parts of the catchment the snow cover period typically lasts from November to May, reaching maximum snow depths of about 1.5 m in April. The upper parts of the basin are bare only for a few weeks in August or September and maximum annual snow depths are on the order of 4 m. Snowmelt occurs in several episodes during the period from March to late June. Redistribution caused by wind drift, avalanching and sloughing substantially affects the spatial distribution of snow.

In the mid 1980s the Kühtai snow monitoring station was established next to the catchment outlet and was the place of detailed snow hydrological studies (Kirnbauer and Blöschl, 1990). When we considered extending the point-scale studies to the entire Längental catchment, we soon realised that the key to successfully representing the spatial patterns of snow processes would be to get data on the spatial variability of snow depth. There were a number of problems specific to the Längental catchment not usually encountered in research catchments. First, the catchment is inaccessible for weeks at a time due to avalanche hazard. Some of the steep slopes and cliffs sometimes do have snow accumulation but are time consuming to access. Also, the size of 9.4 km^2 with the given logistic constraints made exhaustive sampling of snow depth not an option. We considered a number of possibilities to work around these problems. The first idea was to place a large number of snow stakes in the catchment during the

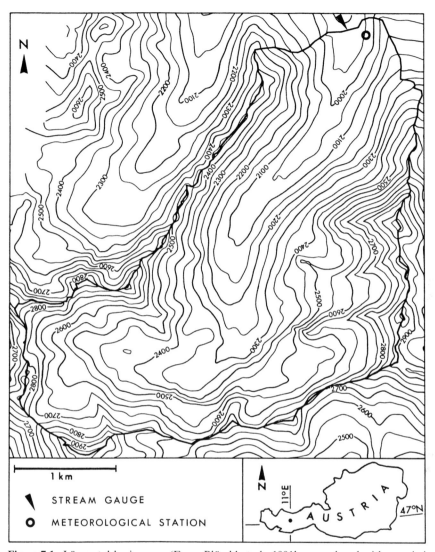

Figure 7.1. Längental basin map. (From Blöschl et al., 1991b; reproduced with permission.)

summer and read snow depth using binoculars. We were hoping to get snow depth to an accuracy of 0.2 m at 100 locations within the catchment. However, it soon became clear that this was not a feasible option because of the rocky subsurface, snow creep and potential problems with conservationists. As an alternative we considered using aerial stereo photographs to estimate the elevation of the snow surface and then calculating snow depth by taking the difference between snow and terrain elevations. With the scale of the photos envisaged we were hoping to get snow depth to an accuracy of 0.5 m exhaustively in the catchment. However, initial tests showed that there is not enough visual contrast on the snow surface for accurate stereo photo interpretation and we therefore abandoned this option. Finally, we decided to use the spatial patterns of snow cover only. Nine aerial surveys were undertaken during the 1989 ablation period,

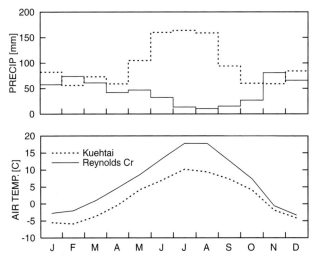

Figure 7.2. Mean monthly precipitation and air temperature at Kühtai, Längental (dotted line), and at Upper Sheep Creek (solid line). The averages are over 1981–1990 and 1983–1994 respectively.

and oblique visual photos of the catchment were acquired. We marked the snow boundary lines manually on the prints. The snow lines were subsequently digitised as vector data, rectified by digital mono-plotting methods (Hochstöger, 1989), and rasterised. As a first step, a 5×5 m grid was used which was then generalised to a 25×25 m grid based on the majority of snow-covered or snow-free 5×5 m pixels in any one 25×25 m pixel (Blöschl and Kirnbauer, 1992). Each pixel value therefore represents the average over the pixel area rather than a grid point value. Although this methodology provided only binary information of snow-covered and snow-free pixels, comparisons indicated that this information was extremely accurate. We chose to have a large number of points with simple information (i.e. binary values from photo interpretation) rather than fewer points with detailed information such as is possible with snow courses (see discussion on the trade-off between accuracy and spatial detail in Chapter 2, pp. 24–5).

However, we did also get some snow course data to complement the aerial survey. A field program was undertaken in late April to assess the distribution of water equivalent in the basin. As we could only sample a small number of sites, the selection of sites was based on typical terrain types as outlined by Woo et al. (1983) and Yang and Woo (1999). These sites included different elevations, slopes and aspects. Measurements were designed to be representative of an area of roughly 50×50 m each, accomplished by numerous snow-depth measurements over that area and a few density profiles.

A snowmelt model (the Vienna University of Technology Snow – VUTS model) was set up for the Längental catchment based on a 25 m grid. For each grid element the energy balance components were simulated and the coupled heat and mass flow within the snowpack was simulated by a multilayer model (Blöschl and Kirnbauer, 1991). Atmospheric data used to drive the model included incoming shortwave radiation, air temperature, humidity, wind speed and precipitation on an hourly basis. These variables were observed at the Kühtai station (1930 m elevation) near the basin outlet. Cloudiness was determined from visual observations. Additional air temperatures at Finstertal (2330 m elevation, 700 m east of the catchment boundary) were also used. Inputs of air temperature were assumed to decrease linearly with elevation based on the readings at the two stations. Wind speed and relative humidity were taken as invariant across the catchment. Horizon shading, and aspect and slope dependence of solar radiation input were accounted for by using equation (7.1). One of the essential assumptions was that terrain attributes could be used to represent the effects of wind drift and sliding as discussed below. Also, as we were running the VUTS model only for the ablation period, initial conditions for the spatial distribution of snow water equivalent within the catchment had to be stated. For both snowfall and initial snow water equivalent we postulated a wind drift factor F of the form

$$F = (a + b \cdot H) \cdot (1 - f(S)) \cdot (1 + e \cdot C) \qquad \geq 0 \qquad (7.3a)$$

$$f(S) = \begin{cases} 0 & \dots \quad S < c \\ \dfrac{S - c}{d - c} & \dots \quad \text{otherwise} \end{cases} \qquad (7.3b)$$

where H is elevation, S is slope and C is terrain curvature at the grid scale of the digital elevation model. For the case of solid precipitation a and b were chosen so as to give a 30 % increase of precipitation with elevation from the lowest to the highest part of the catchment. For the case of initial snow water equivalent a and b were estimated from snow course data. c, d and e were derived from an interpretation of the aerial photos of the snow cover as $c = 10°$, $d = 60°$, and $e = 50$ m (Blöschl and Kirnbauer, 1991). A discussion of this approach is given in Moore et al. (1996).

Figure 7.3 (top) shows the initial snow cover pattern on April 24, 1989 as used for the model initialisation. The other patterns are observations and simulations for May 22, June 14, and June 26, 1989. There is a good agreement of percent snow-covered area. Observed and simulated snow-covered areas, respectively, are 64 % and 70 % for May 22; 46 % and 46 % for June 14; and 31 % and 33 % for June 26. Observed and simulated snow patterns are, overall, also quite similar but there are some differences. We will use these differences to infer potential misrepresentations of snow cover processes in the model.

The simulations for May 22, 1989 in Figure 7.3 indicate that in the northern part of the catchment near the catchment outlet (particularly in the valley floor) snow cover is slightly overestimated, and the simulated snow cover is spatially more coherent than in the observations. This suggests that the VUTS model also overestimated snow water equivalent in this part of the catchment. Conversely, on June 14, 1989 the model tends to underestimate snow cover (and consequently probably snow water equivalent) in the same part of the catchment, which must be related to too fast a depletion of the snowpack from May 22 to June 14. These inconsistencies are believed to be due to two reasons. (a) There was fair weather with substantial melting from May 22–31, snowfalls in the entire catchment from June 1–9, and again fair weather from June 10–14. While the model does simulate

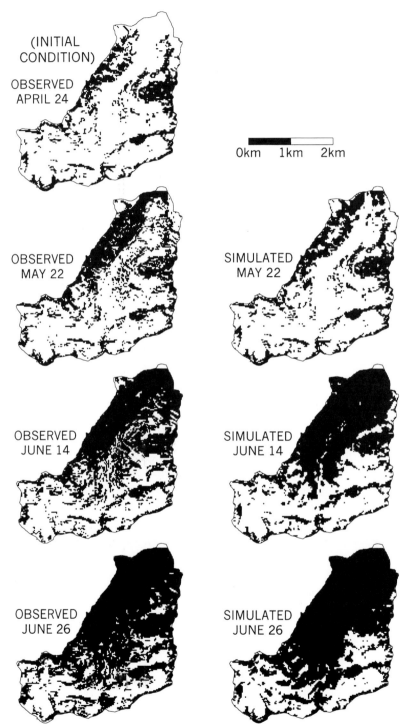

Figure 7.3. Observed and simulated snow cover patterns at Kühtai. Top left: observed snow cover April 24, 1989 (initial condition). Left column below: observed snow cover on May 22, June 14, and June 26, 1989. Right column: model simulations for the same dates. Dark areas denote bare ground and light areas denote snow cover.

albedo as a function of time after snowfall, it is likely that the parameter value for new snow albedo was set too low which caused an overestimation of melt, particularly after the snowfalls. (b) The inconsistencies may also be due partly to the effect of wind blown snow which, on May 22, accumulated in the valley floor more strongly and, on June 14, depleted the north-west facing slopes in the centre of the catchment more rapidly than predicted by equation (7.3). It is important to note that equation (7.3) uses terrain parameters (including curvature) at the grid scale of the terrain model (i.e. 25 m) while the scale of wind drift processes and the scale at which terrain affects wind drift patterns range from smaller to much larger scales than 25 m. One potential remedy would be to use a wind drift factor F that also uses terrain information at larger scales that can be derived from a lower resolution terrain model. This approach has shown potential in another Austrian catchment (Kraus and Blöschl, 1998). The simulations for June 26, 1989 in Figure 7.3 indicate an overprediction of snow cover in the south-eastern part of the basin which is formed by three prominent cirques. The rockwalls of the cirques are bare during most of the ablation period and hence may substantially enhance energy input to the snow cover in the cirque (Olyphant, 1986). We therefore believe that part of this overprediction derives from neglecting longwave radiation emissions from bare surfaces and their interaction with the snow cover.

To better visualise the effects of terrain on model results, the simulated snow pattern for June 26, 1989 in Figure 7.3 was plotted as a perspective view and compared to the oblique photo (Figures 7.4 and 7.5). One apparent inconsistency of observed and simulated snow patterns is an underestimation of snow cover at the base of the steep cliff in the centre of the photo. It is clear that a massive snow deposit had formed there due to sloughing and wind drift from the upslope area. Although equation (7.3) does account for wind drift, it does so in a simplified way and does not explicitly route blown snow and avalanches. Although the average conditions are captured well, situations such as the base of a steep cliff are not represented so well. One potential improvement over equation (7.3) would be a model that deterministically routes snow as a function of terrain and wind conditions.

Another minor discrepancy in Figures 7.4 and 7.5 is that the simulations tend to exhibit fewer small patches of snow. Clearly, they are related to small-scale (subgrid) variability not explicitly accounted for in the model. Although it is not clear how this subgrid variability affects the mean catchment simulations it does highlight the limitations of using point measurements for representing spatial averages of snow water equivalent and snowmelt.

Figure 7.6 shows an evaluation of simulation errors on an element-by-element basis for June 26. The elements are subdivided into classes according to slope and aspect separately for the upper part (> 2400 m, dashed lines) and lower part (⩽ 2400 m, solid lines) of the basin. The labels on the vertical axes relate to the disappearance of the snow cover as simulated by the model. The percentage denoted by "too late" refers to elements with snow cover simulated and bare ground observed, i.e. an overestimation of snow cover, and the percentage

Figure 7.4. Air photo of the upper part of the Längental catchment on June 26, 1989, showing grid elements 25 × 25 m. (By permission of Bundesministerium für Landesverteidigung. From Blöschl et al., 1991b; reproduced with permission.)

denoted by "too early" (negative frequencies in Figure 7.6) refers to elements with bare ground simulated and snow cover observed. For example, on south-facing slopes with slopes around 30° and elevations > 2400 m (i.e. the upper part of the catchment) Figure 7.6 indicates that for 25 % of the pixels in this class, snow cover was simulated but bare ground observed, and for 10 % of the pixels in this class, bare ground was simulated but snow cover observed. The rest of the pixels in this class (i.e. 65 %) were correctly simulated as either snow covered or bare. For most terrain classes, the simulation errors are less than 10 % which indicates good model performance. Figure 7.6 also indicates that there is a certain symmetry about west and east facing slopes, whereas the graph for north- and south-facing slopes is nearly antisymmetric, with a tendency for south-facing

Figure 7.5. Simulated snow cover on June 26, 1989. Dark areas denote bare ground and light areas denote snow cover. (From Blöschl et al., 1991b; reproduced with permission.)

slopes to have too much snow in the model. This tendency suggests that errors are related to solar radiation and specifically to albedo. There are two possible reasons for this. (a) Albedo tends to decrease with the increasing grain size associated with metamorphism (Colbeck, 1988). On south-facing slopes more energy is available for metamorphism and hence albedo will decrease more rapidly with time than on north-facing slopes. This aspect dependence of albedo has not been accounted for in the model. (b) An alternative explanation is a general overestimation of albedo along with an overestimation of sensible and latent heat fluxes or longwave radiation inputs. Although on average over the catchment these two potential errors may compensate, their aspect dependence does not, as there may be too little net solar radiation input on south-facing slopes as compared to north-facing slopes.

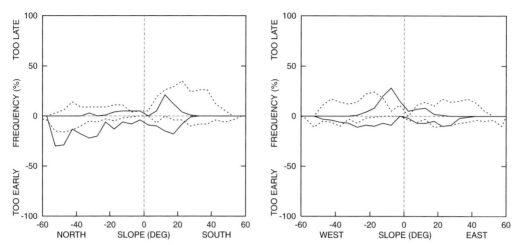

Figure 7.6. Percent errors in snow cover for various slope and aspect classes, June 26, 1989 (too late: snow cover simulated, bare observed; too early: bare simulated, snow cover observed). The dashed lines refer to the upper part of the catchment (above 2400 m) and the solid lines refer to the lower part.

Overall, the comparison of observed and simulated snow cover patterns indicates that the basic model assumptions are realistic. However, there are subtle differences that are very useful in diagnosing model inadequacies. In a parameter sensitivity study (Blöschl et al., 1991a) it was found, not surprisingly, that catchment runoff volume was very sensitive to a parameter controlling the average snow water equivalent in the catchment while it was much less sensitive to the spatial variability of snowmelt due to differential melting. On the other hand, percent error in snow cover (as in Figure 7.6) was highly sensitive to a number of model parameters associated with differential melting including albedo and parameters of equation (7.3). It can therefore be expected that catchment runoff and snow cover patterns are complementary in identifying an appropriate model structure, but the snow cover patterns allow a better identification of individual processes. This complementary information underscores the value of spatial pattern measurements and comparisons in model validation if one is interested in a model of snow cover processes that is close to reality.

7.5 REYNOLDS CREEK CASE STUDY

The Upper Sheep Creek sub-basin within the Reynolds Creek Experimental Watershed has been the location of a detailed study on snowpack variability (Cooley, 1988). Upper Sheep Creek (Figure 7.7) is a 26 ha sub-basin located on the east side of the Reynolds Creek Experimental Watershed (Robins et al., 1965) in the western U.S. rangelands, Idaho. Elevations range from 1840 to 2040 m. The terrain is undulating with maximum slopes of 25°. The vegetation is mostly low sagebrush and mountain sagebrush. Aspen grow in a strip along the north-east facing slope where snow drifts form. Severe winter weather and winds keep the aspen dwarfed to a height of about 4 m. Average annual precipitation is

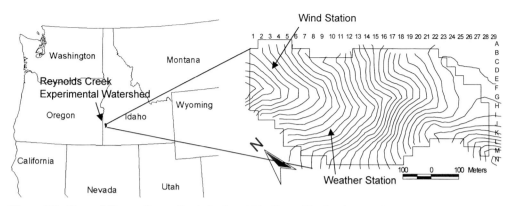

Figure 7.7. Map of Upper Sheep Creek basin within Reynolds Creek Experimental Watershed. Contour interval is 5 m. The catchment outlet of Upper Sheep Creek is on the left.

about 500 mm, with ephemeral runoff usually between February and July when it is generated by snowmelt from deep drifts on the north-east facing slopes. Temperatures average 17 °C in summer and −3 °C in winter (Figure 7.2, solid lines). Various instruments that continuously monitor precipitation, incoming solar radiation, wind direction and speed, air temperature, relative humidity, snowmelt (snowpack outflow) and soil moisture and temperature were operated from 1984 to 1996. In addition, snow depth and snow water equivalent measurements at 30 m grid spacing were obtained on a number of occasions during this interval using standard snow sampling techniques and the Rosen type snow sampler (Jones, 1983). Each snow sample consisted of inserting the snow tube into the snowpack to the soil surface, recording the depth of the snowpack, removing the tube and recording the snow water equivalent as the residual of the weight of the tube and snow sample minus the weight of the empty tube. Manpower limitations were such that it required two storm-free days to fully sample the complete 30 m grid (i.e. about 300 sampling points in space). As a result, typically from four up to nine surveys were done in each year of sampling, attempting to measure the build up and peak snow accumulation followed by ablation. Density was determined for each sample at each grid point by dividing measured snow water equivalent by measured depth. The advantages of this type of snow sampling procedure are the amount of information obtained, i.e. snow depth, snow water equivalent and snow density at each sample point as opposed to more common methods of taking numerous snow depth measurements but only very few snow density samples. When the snow cover exhibits considerable variability in depth as is the case at Reynolds Creek, the density of the snowpack also exhibits considerable variability, and this variability cannot be described by only a few measurements. The disadvantage of this type of sampling procedure is the amount of manpower required to get enough samples to define a pattern. Also, there are snowpack conditions that limit its applicability, such as shallow very dry snow where the snow sample will not stay in the tube and therefore a snow water equivalent cannot be determined, although a depth can still be recorded. Ice lenses in the deeper snowpack can also make it difficult or impossible to collect samples of the snow water equivalent, but depth can usually be obtained by repeated insertions of the snow tube in the same hole until the soil surface is reached.

The patterns of snow accumulation and melt at Upper Sheep Creek are dominated by drifting (Figure 7.8). Snow accumulation usually begins in November and first appears on north-facing slopes and in brushy pockets. As snowfall increases, the upper edges of drifts start to build on the leeward side of the ridges and a general snow cover forms over most of the remaining catchment area. Ridges and south-facing exposures usually experience several periods of snow accumulation and melt during the winter due to strong winds and solar radiation. The general snow cover and drifts normally continue to increase in depth (and width in the case of drifts), often absorbing rain which occurs during occasional warm periods, until maximum accumulation is reached, typically near the beginning of April. After maximum accumulation occurs and melt begins, the ridges and south-facing slopes are generally depleted of snow in a matter of hours. The general snow cover melts next and most of the snow is melted within a few warm days, leaving only the isolated drifts. These drifts persist, sometimes into June or July, sustaining streamflow into late spring and summer.

A typical snow depth distribution is shown in Figure 7.9 for April 4, 1984 which represents conditions near the time of maximum accumulation and shortly after a snowfall event (also see Cooley, 1988). This illustrates the pattern of accumulation as it has been influenced by wind redistribution and variable radiant energy. Depths varied from 0 to 3.8 m. The spatial variations in density for April 4, 1984 are shown in Figure 7.10. Densities on this day were noted to vary from less than 0.15 g/cm^3 to over 0.50 g/cm^3, and appeared to be related mainly to depth, with density larger where the snow is deeper. This figure illustrates the obvious limitations of uniform density assumptions, particularly when wind drift is important.

The model used to simulate the spatial patterns of snow accumulation and melt at Upper Sheep Creek was the Utah Energy Balance (UEB) model which was applied at each 30 m grid point at the same locations as the snow sampling grid. The UEB model is a single layer physically based point energy and mass balance model for snow accumulation and melt (Jackson, 1994; Tarboton et al., 1995; Luce et al., 1997, 1999). The snowpack is characterised using two primary state variables, snow water equivalent, W [m], and the internal energy of the snowpack and top layer of soil, U [kJ m^{-2}]. U is defined as zero when the snowpack is at 0 °C and contains no liquid water. These two state variables are evolved according to energy and mass balance equations accounting for all terms in the energy and mass balance, namely: net solar radiation, incoming longwave radiation, outgoing longwave radiation, heat from precipitation, ground heat flux, sensible heat flux, latent heat flux, heat removed with melt water, precipitation, melt rate and sublimation rate. The model is driven by inputs of precipitation, air temperature, humidity, wind speed and incoming solar radiation. Physically based representations for the energy and mass fluxes are used. Snow surface temperature, a key variable in calculating latent and sensible heat fluxes and

(a)

(b)

Figure 7.8. Photographs of snowdrifts (a) at Upper Sheep Creek; and (b) in Reynolds Creek Experimental Watershed.

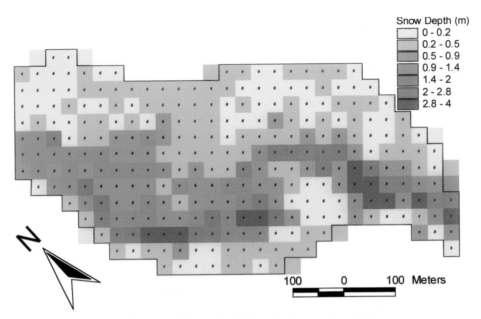

Figure 7.9. Snow depths at Upper Sheep Creek Watershed on April 4, 1984.

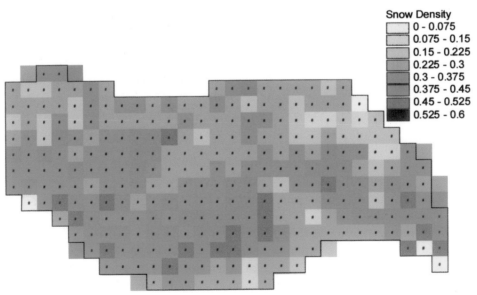

Figure 7.10. Snow density at Upper Sheep Creek Watershed on April 4, 1984. Units are [g/cm^3].

outgoing longwave radiation, is calculated from the energy balance at the surface of the snowpack where incoming and outgoing fluxes must match. This allows the snow surface skin temperature to be different from the average temperature of the snowpack as reflected by the energy content. This reflects the insulating effect of snow and facilitates good modelling of the surface energy balance without needing to introduce multiple layers and detail of within-snow energy transfers. The model was run on a six-hourly time step.

Inputs of precipitation, temperature, relative humidity and incoming radiation were measured at a weather station located centrally within the catchment (Figure 7.7). This location is sheltered and below the drift, so it is subject to minimal wind deposition and transport effects. Wind speed was measured at a more exposed location (Figure 7.7) in order to be more representative of general wind flow. With the exception of solar radiation, the climate variables were assumed to be spatially uniform. Distributed solar radiation was calculated in two steps. Pyranometer (incoming solar radiation) data at the weather station was used to calculate an effective atmospheric transmission factor. Local horizons, slope and azimuth were used to find local sunrise and sunset times and integrate solar radiation received on the slope for each time step. The calculated atmospheric transmission factor characterised cloudiness for incoming longwave radiation calculations.

The UEB model does not represent the physics of snow drifting. Since observations show this to be important at Upper Sheep Creek, we accommodated this in the modelling through the use of a snow drift factor (Jackson, 1994; Tarboton et al., 1995). The fraction of precipitation (measured at a gauge) falling as rain or snow is modelled as a function of temperature. The fraction falling as snow is assumed to be susceptible to drifting. Snow accumulates in some areas (mainly the lee of ridges) and is scoured from other areas (mainly ridges and windward slopes). In the model this redistribution process, which really occurs after snowfall, is lumped together in time with the occurrence of snowfall. Snow accumulation in a grid is modelled as snowfall multiplied by a drift factor, F, which is a spatial field of distinct factors for each grid location. F does not change in time. F is greater than 1 where accumulation is enhanced by drifting and less than 1 where scour occurs. In the application to Upper Sheep Creek, F was estimated by calibrating the snow water equivalent obtained from the snow model at each cell, W_m, against the observed values, W_o. The discrepancy between observations and predictions over an interval between measurements is attributed to drifting and F is adjusted until W_m equals W_o at the end of the interval. The calibration of F assumes that the snowmelt model correctly accounts for all other processes (melt, sublimation, condensation, etc.) affecting the accumulation and ablation of snow water equivalent. Figure 7.11 gives drift factors F calibrated to match the snow water equivalent on February 25 and March 26, 1986 (Jackson, 1994; Luce et al., 1998). Values of F ranged from 0.2 to 6.8 with an average of 0.975.

The UEB model was used with drift factors calibrated from February–March 1986 to predict snow-cover patterns and surface water inputs for the 1993 water year. This is a genuine split sample test (see Chapters 3 and 13), as the calibration

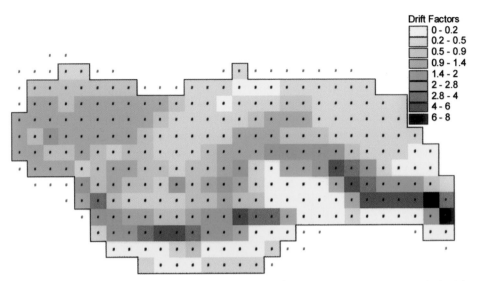

Figure 7.11. Upper Sheep Creek drift factors calibrated from the 1986 snow cover period. (After Jackson, 1994.)

and verification periods do not overlap. The results comprising maps of observed and simulated snow water equivalent over Upper Sheep Creek are shown in Figure 7.12a and b, respectively. From a visual comparison, the observed and simulated patterns are quite similar but there are a few subtle differences. First, the simulated drift is more sharply defined than the observed drift, and snow water equivalent is overestimated in the north-west of the catchment. This suggests that there was less snow drift in 1993 than in 1986. One potential remedy would be to use a deterministic wind drift model (e.g. Liston and Sturm, 1998) to better represent the variability of drifting from year to year. This is an option we are pursuing in current research. Another difference between observed and simulated patterns is that the model has a tendency to melt snow too rapidly, as evidenced by the disappearance from the model of general snow cover on and around pixel J10 (Figure 7.7) in early April. In the observations this snow cover persists about two weeks longer. This rapid melting tendency is also noted in Table 7.1, where during the ablation period simulated basin average totals are less than observed. Plotting observed against simulated grid snow water equivalent for each date (Figure 7.13) shows that the model generally overestimates snow water equivalent for locations with moderate to high snow water equivalents, but underestimates snow water equivalent where there is little snow, with systematic overestimation most apparent in the early melt season. This is consistent with the interpretation of a more uniform snow pattern in 1993 than in 1986, made above.

A sensitivity study was performed in order to assess the relative importance of the various sources of spatial snow variability (Luce et al., 1997, 1998). Results from one of the scenarios are shown in Figure 7.12c and Table 7.1 where the model was run without the effect of drifting, i.e. the drift factor was set to 1 everywhere. In this scenario, spatial variability in snow water equivalent is

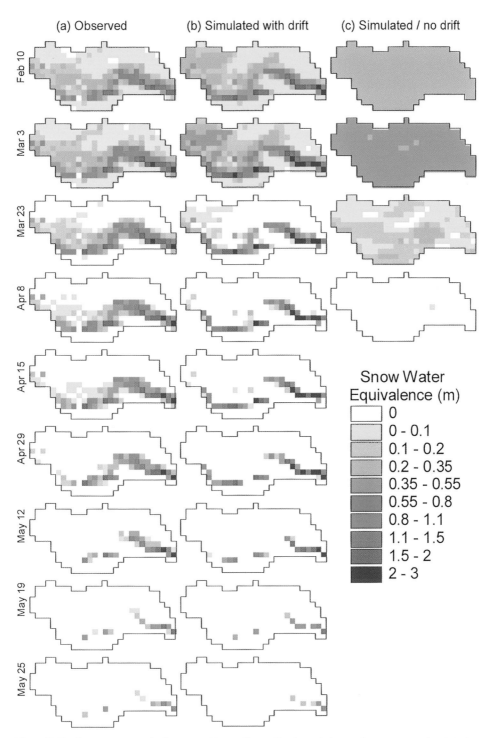

Figure 7.12. Snow water equivalent over Upper Sheep Creek on 9 dates of snow survey in 1993 for: (a) observed; (b) simulated with drift; and (c) simulated without drift. (From Luce et al., 1998. Copyright John Wiley and Sons Ltd. Reproduced with permission.)

Table 7.1. Basin-averaged snow water equivalent (m) from observations and models

Date	Observed	Model with drift	Model no drift
Feb 10, 1993	0.22	0.28	0.28
Mar 3, 1993	0.28	0.38	0.39
Mar 23, 1993	0.23	0.23	0.10
Apr 8, 1993	0.18	0.16	0.00
Apr 15, 1993	0.17	0.16	0.00
Apr 29, 1993	0.13	0.13	0.00
May 12, 1993	0.09	0.07	0.00
May 19, 1993	0.04	0.03	0.00
May 25, 1993	0.02	0.01	0.00

mainly due to topographically induced variation of radiation inputs. Figure 7.12c shows that in the no-drift case the spatial variability in snow water equivalent is much smaller than in the original case where snowdrift is included. This smaller variability highlights that, at Upper Sheep Creek, variation due to wind drift is vastly more important than any other source of variation including radiation. For a model to approach reality it is essential to properly represent wind drift processes. The other difference between the no-drift case and the original case with snowdrift is that in the no-drift case, the snow cover disappears much earlier, and the average basin snow water equivalent is significantly lower (Table 7.1). This bias is clearly due to the nonlinear nature of snowmelt processes, where the spatial average of spatially distributed simulations may be very different from simulations based on spatial averages (Blöschl, 1999). This sheds some light on the limitations of the use of effective parameters. In environments where the snow cover is as heterogeneous as in this case study, effective parameter values of snow models are likely to be greatly in error. More extensive sensitivity analyses (Luce et al., 1997, 1998) including comparisons with single or two-region models, corroborate these findings and show that the fully distributed model with distributed drift multiplier is the only model (of the ones tried) that predicts significant melt late in the season, coinciding with the observed rise of the streamflow hydrograph.

This underscores the value of a spatially distributed modelling approach incorporating spatial patterns describing the variability of the drift multiplier. The essential prerequisite for this type of modelling is the availability of spatial data, such as the observed patterns of snow water equivalent at Upper Sheep Creek. For applying this modelling approach to larger catchments, methods will be needed for predicting drift factors or alternative methods for representing small scale variability.

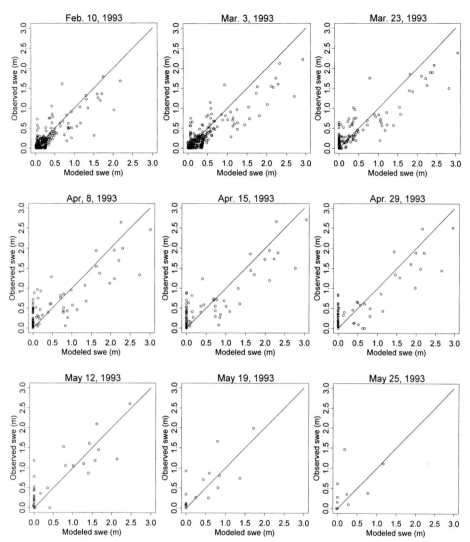

Figure 7.13. Comparison of observed and simulated snow water equivalent for each snow survey date. Each point represents one grid value within the catchment. (From Luce et al., 1998. Copyright John Wiley and Sons Ltd. Reproduced with permission.)

7.6   CONCLUSIONS

This chapter has examined the processes that lead to spatial variability in snow accumulation and snowmelt in the context of two case studies. Although the settings and methods of the two case studies were very different, the basic strategy of model evaluation was similar and consisted of process-based reasoning and analysis of both visual comparisons and pointwise statistical comparison of simulated and observed patterns. In both case studies, observed spatial patterns could be simulated only when particular processes were represented spatially.

At Kühtai the general snow patterns could be well represented by the spatial variability of radiation and a statistical representation of snow drifting but some aspects of the measured patterns could not be simulated. These aspects included enhanced snowmelt in cirques due to longwave emission from surrounding terrain, which was diagnosed from an overprediction of snow cover in the cirques; formation of a snow deposit at the base of a cliff due to avalanching which was diagnosed from an underprediction of snow at the base of the cliff; significant small-scale spatial variability of snow which was diagnosed from a visual comparison of patterns; and enhanced metamorphism and hence more rapid decrease of albedo on south-facing slopes as compared to north-facing slopes which was diagnosed from a slight tendency for overestimating snow on south-facing slopes. For improved simulations of the spatial variations of snow accumulation and melt these processes need to be modelled explicitly.

At Reynolds Creek the general snow patterns could be represented well by the spatial variability of snow drifting. However some aspects of the measured patterns could not be simulated. These included a slightly more uniform snow distribution due to less redistribution of snow in the later year relative to the earlier year. This discrepancy was diagnosed from a visual comparison of patterns as well as from a slight overestimation of snow water equivalent for locations with moderate to high snow water equivalents but a slight underestimation of snow water equivalent for locations with below average snow water equivalents as indicated by error statistics. This indicates that the drift factors computed from the 1986 snow data were more variable than the actual drift in 1993. More generally speaking, this means that the model calibrated on one data set does not necessarily perform as well on an independent data set. At Reynolds Creek the dominance of wind drift was illustrated by running a model which ignored drift but incorporated spatial variability in other processes. This resulted in an almost uniform pattern.

Both case studies have demonstrated the value of observed spatial patterns for diagnosing the performance of individual model components and for identifying model structural issues and parameterisation. It is clear that the patterns have provided more insight than a few point data or catchment average values from runoff would have provided. We believe that, in the future, comparisons of distributed model output with observed snow patterns will become part and parcel of any snow-modelling exercise.

ACKNOWLEDGEMENTS

The work of the first author on this was supported by the U.S. Environmental Protection Agency (agreement no: R824784) under the National Science Foundation/Environmental Protection Agency Water and Watersheds program. The second author wishes to thank the Fonds zur Förderung der Wissenschaftlichen Forschung, Vienna, project nos. P6387P, P7002PHY, and J0699-PHY for financial support.

8

Variable Source Areas, Soil Moisture and Active Microwave Observations at Zwalmbeek and Coët-Dan

Peter Troch, Niko Verhoest, Philippe Gineste,
Claudio Paniconi and Philippe Mérot

8.1 INTRODUCTION

The variable source area concept is now widely accepted to explain storm runoff production in humid regions. The concept was first introduced by Hewlett and Hibbert (1967): "The yielding proportion of the catchment expands and shrinks depending on rainfall amount and antecedent wetness of the soil". A major feature of variable source areas is that the area over which return flow and direct precipitation are generated vary seasonally and throughout a storm. The theory was developed because of inadequacies with the Hortonian runoff production mechanism (Horton, 1933) for describing storm runoff in humid catchments. In most humid regions, infiltration capacities are high because the vegetation cover protects the soil from rain packing, and because the supply of humus creates an open soil structure. Under such conditions, rainfall intensities generally do not exceed infiltration capacities. Therefore, Hortonian overland flow does not occur on large areas of the catchment.

Research in the 60s and 70s on the variable source area concept was supported by intensive field studies in small catchments (Hewlett and Hibbert, 1967; Dunne et al., 1975; Dunne, 1978). These authors mapped the spatial patterns of saturated areas and their seasonal fluctuations (e.g. Dunne et al, 1975, show seasonal variation of the saturated zone in a small catchment at Randboro, Quebec). Since then a number of modelling strategies, based on terrain analysis, have been developed to explain and predict these spatial patterns of saturated areas (Beven and Kirkby, 1979; Sivapalan et al., 1987; Barling et al., 1994). These modelling efforts recognise the control that catchment topography, soils, antecedent storage capacity and rainfall characteristics exert on the spatial extent of the contributing areas. To link these hydrologic and geomorphologic characteristics of the catchment to variable source areas, static and/or quasi-dynamic wetness indices were introduced. Wetness indices have been used to define, at a moment prior to storm rainfall, the readiness of a catchment to produce storm runoff through the saturation excess mechanism.

Rodger Grayson and Günter Blöschl, eds. *Spatial Patterns in Catchment Hydrology: Observations and Modelling* © 2000 Cambridge University Press. All rights reserved. Printed in the United Kingdom.

Routinely collected hydrologic data from catchments generally does not allow full validation of these models. Often validation is solely based on a comparison between observed and predicted streamflow records. This type of validation is insufficient to draw conclusions about the accuracy with which these models describe the temporal and spatial patterns of the contributing areas. More comprehensive validation procedures are required if we want to improve our understanding of these important hydrological processes. At first glance, it therefore seems that the difficulty in collecting information on saturated areas in larger catchments through field work hinders further progress.

Recently, however, new instruments have become available to the hydrologist, in the form of active microwave remote sensors. Active microwave instruments on board satellites offer tremendous opportunities to increase our observation capacities of large catchments. First of all, they are all-weather instruments, practically undisturbed by atmospheric conditions. Second, they are day and night instruments since they do not depend upon an additional energy source but produce their own electromagnetic energy to scan the Earth's surface. Third, when using a special technique called "synthetic aperture radar" (SAR), they produce images with the required spatial resolution to be of use for catchment modellers (pixel resolution typically on the order of 10 m). But the reason why these instruments are powerful for observing variable source areas is the sensitivity of the backscattered energy to soil moisture. Recent studies have demonstrated the potential of observing soil moisture by means of SAR instruments (Ulaby et al., 1982; Cognard et al., 1995). The main difficulty with SAR imagery is that, not only soil moisture, but also surface roughness, vegetation cover and topography have an important effect on radar backscatter. These interactions make retrieval of soil moisture difficult and only achievable under particular conditions, such as bare soil or surfaces with low vegetation cover (Altese et al., 1996). It should be possible to separate the vegetation, roughness, topography and soil moisture effects on radar response using multifrequency and/or multi-polarisation measurements (Ulaby et al., 1996), but currently operational satellites are not equipped with sensors that provide such data. It should also be noted, the frequencies at which SAR operates provide soil moisture estimates for only the top few cm.

In this chapter we present recent research on the use of multitemporal SAR imagery to map the seasonal extent of variable source areas (Gineste et al., 1998; Verhoest et al., 1998). The rationale of the proposed technique is based on the observation that the seasonal variability of surface soil moisture content is highly related to the occurrence or absence of contributing areas, as illustrated in Figure 8.1. At hillslope and catchment scales, soil moisture and its spatial and temporal variability are fingerprints of several hydrologic processes. During rainy periods, flow convergence results in relatively low temporal variability of surface soil moisture in the vicinity of the drainage network because the soil is saturated for much of the time. In contrast, areas located at or near hillslope tops will exhibit more pronounced soil moisture variation in time due to successive wetting during rainfall events and drying through evapotranspiration and redistribution

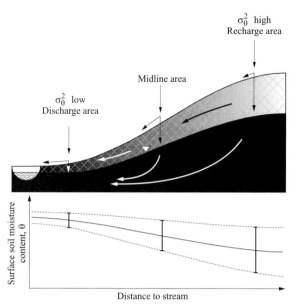

Figure 8.1. Redistribution of soil moisture along a hillslope during wet season conditions (winter) resulting in recharge areas with high temporal soil moisture variability and discharge zones with low temporal soil moisture variability (σ_θ^2: temporal variance of soil moisture). (From Verhoest et al., 1998; reproduced with permission.)

during interstorm periods. Therefore, by analysing the temporal variability of the observed radar signal during a winter season when the catchment is wet, particularly in the streamside zones, it should be possible to map the variable source areas at the catchment scale.

The chapter is organised as follows. In Section 8.2 we give a short description of the two experimental catchments used in this study, the Zwalmbeek in Belgium and the Coët-Dan in France. In Section 8.3 we describe the field survey data that is used to test the accuracy with which the proposed remote sensing techniques predict the spatial patterns of seasonally observed variable source areas. Section 8.4 gives an overview of the multitemporal SAR data collected over the experimental catchments and lists the different image processing procedures required to prepare the SAR data for multitemporal analysis. Section 8.5 starts with a review of earlier attempts to map variable source areas from SAR images. We then give a detailed discussion of two new techniques that appear promising for mapping the spatial extent of seasonal saturation-prone areas from multitemporal SAR images. The first method uses the temporal standard deviation of radar backscattered energy to map variable source areas in humid catchments with low relief. The second method, based on a principal component analysis, was developed to overcome the restriction in the first method of needing low-relief catchments. It is shown that principal component analysis allows separation of the different influencing factors (topography, land use, soil moisture) on the backscattering coefficient, and therefore results in a more robust method of mapping variable source areas. Finally, we summarise the main findings of this research in Section 8.6.

8.2 DESCRIPTION OF TEST SITES

8.2.1 The Zwalmbeek Catchment

The Zwalmbeek catchment (Figure 8.2a) is situated about 20 km south of Ghent in Belgium (50°45'N to 50°54'N and 3°40'E to 3°50'E). It is a fifth Strahler-order basin with a total drainage area of 114 km^2, and a drainage density of 1.55 km/km^2. Rolling hills and mild slopes, with a maximum elevation difference of 150 m, characterise the topography (Figure 8.3a). Land use is mainly arable crop farming and permanent pasture, but the south of the catchment is partly forested. The degree of urbanisation is 10%, and is mainly clustered in three small towns. The soil type in the catchment is predominantly sandy loam (Belgian soil classification), with minor isolated patches of sand and clay. The climatic regime is humid temperate with a mean annual rainfall of 775 mm, distributed almost uniformly over the year, and a mean annual pan evaporation of 450 mm. The catchment is described in detail in De Troch (1977).

8.2.2 The Coët-Dan Catchment

The Coët-Dan catchment (Figure 8.2b) is located near the town of Naizin, Brittany, France (48°N and 2°50'W). It is a second Strahler-order basin with a drainage area of 12 km^2. Gentle concave slopes (in general less than 5%, especially in the northern part) reflect the brioverian schists substratum with their top overlaid by dystric or aquic eutrochrepts (brown acidic, weakly leached soils) and their bottom by glossaquals (degraded hydromorphic soils) and fluvents (alluvial soils). Agriculture is intensive (Figure 8.3b), and in winter the vegetation cover is particularly low. A land use survey performed in the winter of 1992 revealed that about 22% of the catchment was covered with meadows and young winter cereals (crops 5 to 15 cm high), while about 44% was bare soil, sometimes covered with corn stubble. The mean annual rainfall is about 700 mm uniformly distributed over the year, and the mean annual runoff is estimated around 300 mm. For a full description of the hydrology of the catchment see Mérot et al. (1994).

8.3 FIELD SURVEY OF VARIABLE SOURCE AREAS

8.3.1 The Zwalmbeek Catchment

The field data used to investigate the spatial patterns and seasonal extension of the variable source areas are derived from the Belgian soil map (scale 1:20,000). The Belgian soil map was produced during the 1960s and early 1970s for the whole territory, and contains information on soil texture, natural drainage conditions, and profile development. These soil characteristics were derived from auger observations (using augers 1.25 m deep and 5–10 cm

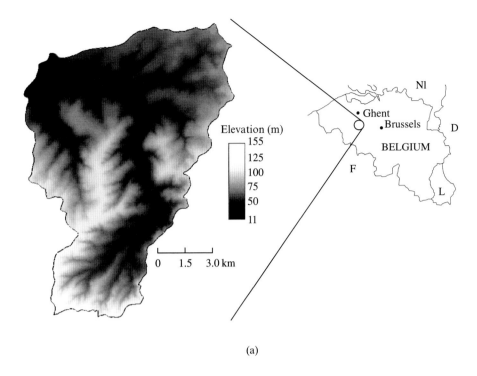

(a)

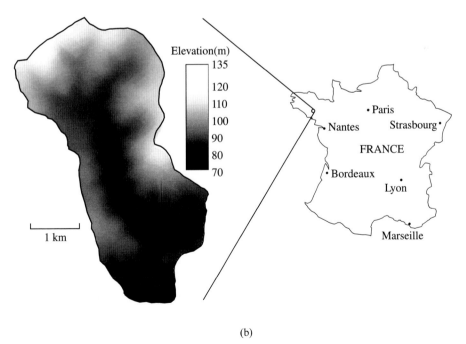

(b)

Figure 8.2. (a) Location map of the Zwalmbeek catchment; (b) location map of the Coët-Dan catchment.

(a)

(b)

Figure 8.3. (a) Photograph of part of the Zwalmbeek catchment; (b) photograph of part of the Coët-Dan catchment (© Christian Walter, ENSAR-INRA, Rennes).

diameter) taken in the field with an average density of two samples per ha. For our study site this means a total of about 23,000 samples. In addition, soil profile pits were dug with a 1 m^2 area, a depth of 1 to 2 m, and a density of 1 pit each 1.5 km^2. In this study we are mainly interested in the natural drainage conditions of the soils. The drainage map of the Belgian soil map classifies the different soils into classes ranging from well-drained to poorly-drained soils, according to the bore hole field observations (Table 8.1). These bore hole samples were used to measure the depth to gley and mottle. Gley can be described as a blue-grey waterlogged soil layer in which iron is reduced to the ferrous form. This layer can turn into a soil containing brownish mottles due to oxidation of iron during intermittent dry periods. The occurrence of these features therefore indicates the change in water table height between winter (mottle) and summer (gley).

Figure 8.4 shows daily rainfall for the winter period of 1995–1996 together with the mean backscattered signal for the whole catchment, calculated from the tandem pairs of the eight ERS-1/2 images (see Section 8.4 for more details on the SAR images used in this study.)

8.3.2 The Coët-Dan Catchment

The field data associated with the study performed in the Coët-Dan catchment were collected in one of the first-order subcatchments (drainage area: 1.2 km^2), located in the northwest of the catchment. This survey involved the mapping of the saturated areas during the winter of 1992 by visual inspection and auger hole sampling. The inset in Figure 8.5 shows the areal extent of the saturated areas as observed on February 15, 18 and 21, 1992 (Salahshour Dehchali, 1993). During the winter period of 1992, rain fell between February 8 and February 16 and was followed by a drydown period which lasted till March 3 (Figure 8.12).

Table 8.1. Natural drainage classes, Belgian nomenclature

Drainability index	Average winter water table depth (= depth to mottle) (cm)	Average summer water table depth (= depth to gley) (cm)
b	> 125	—
c	80–125	—
d	50–80	—
h	30–50	—
e	30–50	> 80
f	0–30	40–80
g	—	< 40
A (= b + c + d)	> 50	—
D (= c + d)	50–125	—

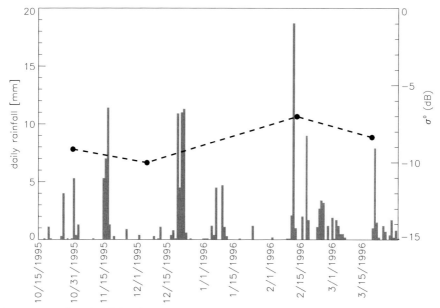

Figure 8.4. Daily rainfall for the 1995–1996 winter period observed at the Zwalmbeek catchment. Also indicated by the dashed line and right-hand scale is the average radar backscatter σ^0 for the catchment calculated for each tandem pair of ERS-1/2 images (small σ^0 corresponds to dry conditions). (From Verhoest et al., 1998; reproduced with permission.)

8.4 SAR DATA COLLECTION AND PRELIMINARY PROCESSING

8.4.1 The ERS-Satellite SAR System

The SAR images used in this study originate from the ERS-satellite system. The first ERS satellite (ERS-1) was launched in 1991. This satellite carries several advanced Earth observation instruments, such as the Active Microwave Instrument (AMI) which combines the functions of a synthetic aperture radar and a wind scatterometer. The SAR instrument is a C-band (5.3 GHz) radar operating in VV polarisation (Attema, 1991). One of the products generated by the Processing and Archiving Facilities (PAFs) are precision images with a spatial resolution of 30 by 30 m and a pixel size of 12.5 by 12.5 m. During the winter of 1992, the satellite was put in the so-called "ice-phase", allowing Earth observations at a limited number of locations with a repeat cycle of three days. The Coët-Dan catchment is located in one of these areas with three-day repeat coverage. Between January 28 and March 28, fifteen precision images (PRI) were collected over the catchment (Table 8.2). In 1995, a second satellite (ERS-2) was launched and put in the same sun-synchronous orbit as ERS-1, such that the time difference between overpasses is exactly 24 hours (the so-called "Tandem-phase"). From October 1995 to April 1996, four tandem pairs (8 PRI images) were collected over the

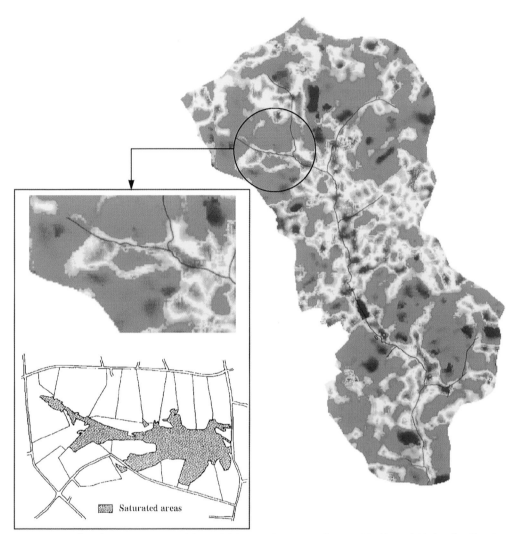

Figure 8.5. Saturated areas observed from field campaigns on February 15, 18, and 21 for the Coët-Dan catchment. The right image shows the saturation potential index (SPI) for the Coët-Dan basin, calculated on the sequence of ERS-1 images taken during the winter period of 1992. The SPI varies from low (red) to high (blue). Low (red) values indicate saturation prone areas.

Zwalmbeek catchment (Table 8.2). In both cases, winter-time images were selected in order to minimise changes in soil roughness, due to agricultural activities, and vegetation characteristics, thereby minimising their effect on the total radar backscatter.

8.4.2 SAR Image Calibration

After georeferencing the twenty-three images (eight for Zwalmbeek and fifteen for Coët-Dan), the data had to be calibrated in order to be useful for multi-temporal analysis. As stated before, in this study we have used SAR PRI (pre-

Table 8.2. Identification of satellite images used in the analysis of Coët-Dan and Zwalmbeek

Date	Mission	Orbit	Track	Frame	Desc./Asc.
Images for the Coët-Dan Catchment					
01/28/1992	ERS–1	2807	1	963	Ascending
02/06/1992	ERS-1	2936	1	963	Ascending
02/09/1992	ERS-1	2979	1	963	Ascending
02/12/1992	ERS-1	3022	1	963	Ascending
02/15/1992	ERS-1	3065	1	963	Ascending
02/21/1992	ERS-1	3151	1	963	Ascending
02/24/1992	ERS-1	3194	1	963	Ascending
02/27/1992	ERS-1	3237	1	963	Ascending
03/01/1992	ERS-1	3280	1	963	Ascending
03/04/1992	ERS-1	3323	1	963	Ascending
03/10/1992	ERS-1	3409	1	963	Ascending
03/13/1992	ERS-1	3452	1	963	Ascending
03/16/1992	ERS-1	3495	1	963	Ascending
03/22/1992	ERS-1	3581	1	963	Ascending
03/28/1992	ERS-1	3667	1	963	Ascending
Images for the Zwalmbeek Catchment					
10/31/1995	ERS-1	22455	423	2583	Descending
11/01/1995	ERS-2	2782	423	2583	Descending
12/05/1995	ERS-1	22956	423	2583	Descending
12/06/1995	ERS-2	3283	423	2583	Descending
02/13/1996	ERS-1	23958	423	2583	Descending
02/14/1996	ERS-2	4285	423	2583	Descending
03/19/1996	ERS-1	24459	423	2583	Descending
03/20/1996	ERS-2	4786	423	2583	Descending

cision image) data. SAR PRIs are subjected to engineering corrections and relative calibration to compensate for well-understood sources of system variability. Absolute calibration of the precision images, on the other hand, has to be performed by the user. The calibration procedure used here is described in Laur et al. (1997). The digital numbers in the PRIs are related to the backscattering coefficient through the following formulas:

$$A_{ij}^2 = DN_{ij}^2 \cdot \frac{1}{K} \cdot \frac{\sin \alpha}{\sin \alpha_{ref}} \cdot C \cdot \frac{PRP}{RRP} \cdot PL \qquad (8.1)$$

and

$$\sigma^0 = \frac{1}{N} \sum_{i,j}^{N} A_{ij}^2 b \qquad (8.2)$$

where N is the number of pixels within the area of interest (AOI), i.e., a group of pixels corresponding to a distributed target in the image (e.g. bare soil field); i, j are the range and azimuth locations of the pixels within the distributed target containing N pixels; DN_{ij} is the digital number corresponding to the pixel at location (i, j); A_{ij} is the amplitude corresponding to the pixel at location (i, j);

σ^0 is the backscattering coefficient corresponding to all N pixels within the AOI; K is a calibration constant; α is the average incidence angle corresponding to the AOI, which is calculated from the image geometry, the earth surface being represented by a reference ellipsoid (Goddard Earth Model, GEM 6); α_{ref} is the ERS reference incidence angle, i.e., $23.0°$; C is a factor that accounts for updating the gain due to the elevation antenna pattern implemented in the processing of ERS SAR PRI products; PRP is the power of the replica pulse used to generate the PRI product and is given in the header file of the PRI image; RRP is the replica pulse power of a reference image taken over Flevoland (The Netherlands); and PL is the analogue to digital convertor (ADC) power loss. For more details on the calibration we refer to Gineste et al. (1998) and Verhoest et al. (1998).

8.4.3   Speckle Filtering

Radar images of homogeneous rough surfaces always show a granular pattern called speckle. This noise-like phenomenon is the result of changes in the distances between elementary scatterers and the receiver caused by surface roughness, so that the received waves, although coherent in frequency, are no longer coherent in phase. SAR systems rely upon the coherence properties of the scattered signals, making these systems susceptible to speckle to a much greater extent than noncoherent systems, such as side-looking airborne radars (Porcello et al., 1976). The presence of speckle noise in an imaging system reduces resolution, and thus the detectability of a target, and also degrades the quality and interpretability of the scene.

In SAR practice, speckle is suppressed by creating n-look images. These multi-look images are obtained by averaging n independent samples (looks) of the same scene. This reduces the variance of speckle by a factor n, but deteriorates the spatial resolution by that same factor (see Chapter 2, p. 23). During the last decade techniques that do not deteriorate resolution have been proposed for speckle reduction. In homogeneous areas a box or lowpass filter (i.e. a moving window averaging filter) is efficient. However, this kind of filter blurs edges, strong point targets or high frequency texture variations (Nezry et al., 1991). Therefore speckle filters which are adaptive to the local texture information are recommended to apply to radar images. These filters smooth speckle in homogeneous areas and preserve texture and high frequency information in heterogeneous areas (Shi and Fung, 1994). Numerous types of adaptive filters have been proposed and they generally weigh the central pixel value with the neighbouring values (e.g. local mean value of the applied kernel as a function of the local coefficient of variation). Nevertheless these filters suppress the noise insufficiently along edges, roads or point targets (Nezry et al., 1991). In order to get better speckle reduction results, Lopes et al. (1990) proposed to divide an image into three classes. The first class corresponds to the homogeneous areas within the image. In these areas the speckle can easily be reduced using a box filter. The second class represents the heterogeneous areas in which texture has to be preserved, while reducing speckle. In this area the pixel value is replaced by a weighted average of the original central pixel value and the mean value obtained for the applied kernel. The third class corresponds to areas which contain isolated point targets. In this case, the filter preserves the observed value of the central pixel. The criteria for assigning pixels to one of these three classes and the weighting function applied for the pixels in the second class, define the different filters. In this study an adapted Lee sigma filter (Gineste et al., 1998) and the Gamma Map filter (Lopez et al., 1990) have been used to reduce speckle in the 23 SAR images. For more information on these filters, we refer to Verhoest et al. (1998) and Gineste et al. (1998).

8.5   SPATIAL PATTERNS OF VARIABLE SOURCE AREAS THROUGH MULTITEMPORAL SAR ANALYSIS

8.5.1   Introduction

During the last decade, several SAR data analysis techniques have been proposed to map the saturation-prone areas in catchments. Brun et al. (1990), based on helicopter-borne C-band scatterometer data, proposed to map variable source areas by applying a threshold on the backscattering coefficient. The reasoning behind this method is that, when ponding occurs, the radar signal drops due to specular reflection. They found that a threshold of −7 dB allowed estimation of the spatial patterns in the variable source areas. If the method could be applied successfully it would allow mapping of the extent of saturated areas on each day of radar observations, thereby providing a sequence of variable source area maps. This technique was tested by Gineste et al. (1998) based on the 15 ERS-1 images given in Table 8.2, but was rejected as an accurate way to map variable source areas. The main difficulty with an absolute threshold is that other surface characteristics that influence radar backscatter, such as vegetation cover and surface roughness when the terrain is not completely inundated, are not taken into account.

Another method, proposed by Rignot and van Zyl (1993) and Gineste et al. (1998), uses difference images to overcome the problems occurring with absolute thresholding. These researchers found that a two-date difference image yields more valuable information than the threshold method, but is still limited because the analysed images should reflect extreme hydrologic conditions (inundated versus dry) before the method becomes reliable. The method of differencing in itself is further susceptible to other problems: changes will not be detected in the same fashion in high intensity regions compared to low intensity areas, which renders the method less reliable. Moreover, the differencing method is not very robust since the radiometric errors introduced in the imagery during SAR processing are multiplicative factors to the total radar intensity, which will not be eliminated during the differencing.

This problem can be overcome by dividing the intensity values of the two dates instead of subtracting them. This ratio method is shown to be better adapted to the statistical characteristics of the radar data, but only works well when the number of looks is very high, since the method is very sensitive to speckle noise. Rignot and Van Zyl (1993) found that, in order to detect changes in radar intensity less than 2 dB with a confidence level better than 90 %, the number of looks must be greater than 64.

It is apparent from these trials, using change detection techniques, that accurate and reliable mapping of saturated areas from one image or from a pair of images is restricted to atypical situations. As an alternative, one can try to analyse a sequence of images taken during a complete season. In the following sections we present two recently developed techniques that appear promising for mapping the spatial patterns in variable source areas.

8.5.2   Saturation Potential Index

Given the strong differentiation in temporal variability of soil moisture as a function of the position along a hillslope (Figure 8.1), Gineste et al. (1998) developed a technique based on the backscatter temporal standard deviation to indicate the local saturation likelihood during the period of observations. This standard deviation is termed the saturation potential index (SPI). It directly reflects the fact that the more the backscattering coefficient from a point in the landscape varies in time, the greater is the soil moisture variation at that point, whereas saturation is expected to develop on parts of the catchment that are usually wetter and thus subject to less soil moisture variation in response to the hydrologic forcing conditions. The method uses the logarithmic transform of the backscattering coefficient given by (8.2). Therefore the speckle noise becomes additive and consequently exhibits the same strength regardless of the absolute backscatter level. Moreover, it has been shown that the possible range of backscatter variation with soil moisture remains about the same (on the order of 5 dB for soil conditions varying from dry to wet) independent of the roughness of the surface (Altese et al., 1996). A measure of the local backscatter temporal variation should thus allow an assessment of the extent to which the soils in the catchments have departed from wet conditions.

Results of this method applied to the Coët-Dan data are discussed here. Figure 8.5 shows spatial patterns of backscatter temporal variation (red areas indicate low temporal standard deviation and thus saturation prone areas). The pattern generally shows that the drainage lines as indicated by the stream network have low temporal standard deviation indicating consistently wet areas. The hillslopes have high temporal standard deviation indicating wetting and drying. Zooming in on the subcatchment where detailed field observation of the extent of saturated areas has been performed (inset in Figure 8.5) allows us to compare the SPI with in situ observed saturated areas. Saturation around the main streamline is generally well defined although the SPI appears to overpredict the saturated area on the downstream end of the image. The small saturated branch on the south side in the middle of the image is well defined by the SPI. The SPI underestimates the extent of the saturation up the main streamline. Gineste et al. (1998) also showed that there was good agreement between spatial patterns of saturation potential indices (SPI) and wetness index across the Coët-Dan catchment. However, the backscatter temporal standard deviation cannot be used directly over the whole catchment as a measure of saturation, as other areas where little variability is to be expected (e.g., areas of dense vegetation such as forests where microwave penetration is impeded) are not discriminated (e.g., the mid-east ridge area in Figure 8.5). Some of the areas with low temporal standard deviation are not in forested or urban areas and so are likely to actually be saturated areas but these are not well connected to the stream network. These may be due to saturation caused by human activities in agricultural catchments such as compaction of top soil by cattle, harvesting, ploughing, etc. but will not contribute to runoff since they are not connected to the stream network.

The same method was applied to the data for the Zwalmbeek catchment. Figure 8.6 shows the SPI with a reversed colour scale compared to Figure 8.5, i.e. blue relates to small temporal standard deviation associated with wet areas and red areas indicate high temporal standard deviation associated with areas that drain freely. Here we can compare the SPI to the drainage network, and to the drainage classes derived from the soils map. In Figure 8.6 the predicted saturated areas broadly follow the drainage network. However, there is a shift of the predicted saturated areas with respect to the drainage network, indicating that there may be some topographic effects on computation of the SPI. We can also compare Figure 8.6 with the right panel of Figure 8.8 which represents the soil drainage classes. The poorly drained soils (class $h+e+f+g$) in the main drainage line are reasonably well predicted but the spatial shift is highlighted in the north west. Given the apparent confounding effects of topography, we sought a more robust technique which can separate the topographic and land-use effects from soil moisture influences on the total backscattering described in the next section.

8.5.3 Principal Component Analysis

The principal components transformation is a standard tool in image enhancement, image compression, and classification (Richards, 1986; Singh, 1989; Lee and Hoppel, 1992). It linearly transforms multispectral or multidimensional data into a new coordinate system in which the data can be represented without correlation. The new coordinate axes are orthogonal to each other and point in the direction of decreasing order of the variances, so that the first principal component contains the largest percentage of the total variance (hence the maximum or dominant information), the second component the second largest percentage, and so on. Images transformed by principle component analysis (PCA) may make evident features that are not

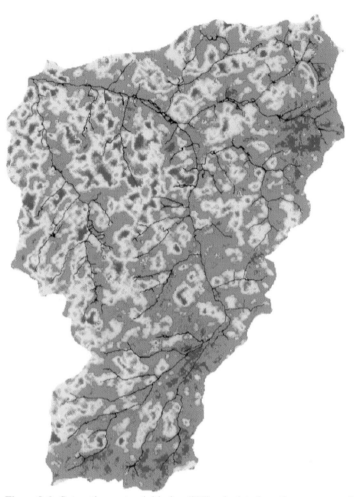

Figure 8.6. Saturation potential index (SPI) calculated on the sequence of eight ERS images for the Zwalmbeek catchment, ranging from low (blue) to high (red). (Note: colour map is reversed with respect to the one of Figure 8.5 so blue areas are now saturation prone). The stream network (black) is given in overlay. Notice the shift of the low SPI with respect to the river network.

discernable in the original data – local details in multispectral images, changes and trends in multitemporal data – that typically show up in the intermediate principal components.

PCA is widely used in optical remote sensing but less so in the more recent area of SAR image processing. One example is provided by Lee and Hoppel (1992), who used a modified principal component transformation on multi-frequency polarimetric SAR imagery for reducing speckle and for information compression. Another example is given by Henebry (1997) who used PCA on a temporal series of twelve images for the production of a high spatial resolution/low spatial noise image that served as a template for georeferencing. One of the principal components obtained could then be used for land cover segmentation.

Figures 8.7–8.9 show the images constructed for the first three principal components computed from the eight Zwalmbeek images. Applying PCA to these eight images leads to the separation of the information into several components that can be attributed to different factors influencing the backscatter. The first principal component (PC) accounts for 76.6 % of the total variance, the second component for 6.6 %, the third for 5.9 %, and each of the remaining PCs for less than 4 % (Verhoest et al., 1998).

Figure 8.7 compares the first component (left image) with a local incidence angle image computed from the digital elevation model of the catchment. The similarity between these two images suggests that topographic effects are responsible for the largest contribution to the total variance in the sequence of SAR images and dominate the backscattering signal. A sequence of images taken with the same radar configuration and footprint (frame and track) will show a very high correlation: slopes facing the satellite will consistently return more energy than slopes turned away from the sensor. The principal component analysis has brought out these highly correlated features in the first PC.

The left image in Figure 8.8 represents the second principal component. This image displays a strong spatial organisation, with the highest values grouped along the drainage network of the catchment. To test the hypotheses that the information contained in this image is related to the drainage conditions of the

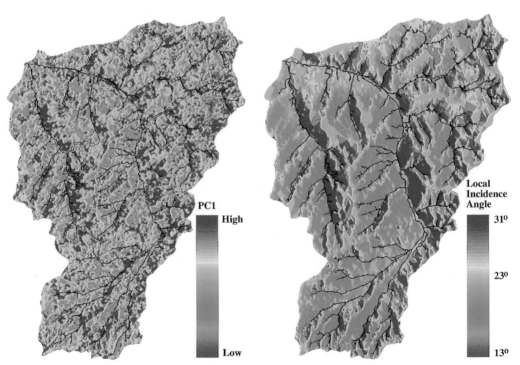

Figure 8.7. First principal component image calculated for the Zwalmbeek catchment (left) and an image of the local incidence angles calculated from the digital elevation model of the catchment (right). The stream network (black) is shown in overlay in both images. (From Verhoest et al., 1998; reproduced with permission.)

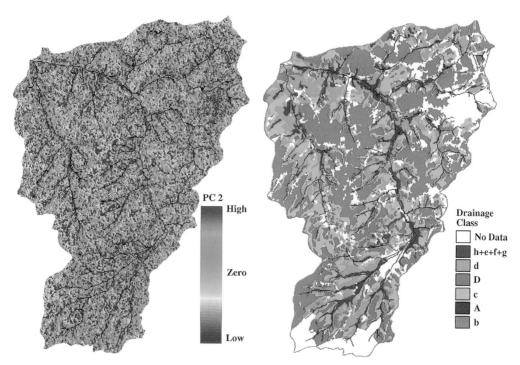

Figure 8.8. Second principal component image calculated for the Zwalmbeek catchment (left) and the drainage map for the catchment (right). The drainage classification scheme is explained in Table 8.1. (From Verhoest et al., 1998; reproduced with permission.)

catchment, a drainage map for the Zwalmbeek was generated from the Belgian soil map, and is shown in the right image of Figure 8.8. As can be noticed from this figure, the poorly-drained soils (types e–h) tend to occur in the valley regions of the catchment and correspond well with the areas with high second PC values. This suggests a radar response, brought out in the second principal component, to the soil moisture patterns that result from the drainage characteristics of the basin. These patterns are not attributable to any single event, but reflect the overall response of the soil to the rainfall and interstorm periods spanned by the images, as illustrated by Figure 8.1. The pattern of the second PC in Figure 8.8 is similar to the SPI pattern in Figure 8.6, but the shifting due to terrain is slightly less as is apparent in the drainage line in the north west of the catchment.

The third principal component (left image in Figure 8.9) shows the influence of land cover and land use, as evidenced by its strong correlation with the Landsat-derived map (right image in Figure 8.9) that highlights the forested areas in the south of the catchment and the few towns on the basin. The land use map is the result of a supervised classification performed on a Landsat TM image of October 12, 1994. The classification resulted in ten classes, such as woods, urbanised areas and several agricultural uses which are grouped together as shown in Figure 8.9. In SAR images urban areas typically appear as bright objects, and in a sequence of images such areas, with their relatively static features, will produce a consistent backscattering signal. This is apparent if Figures

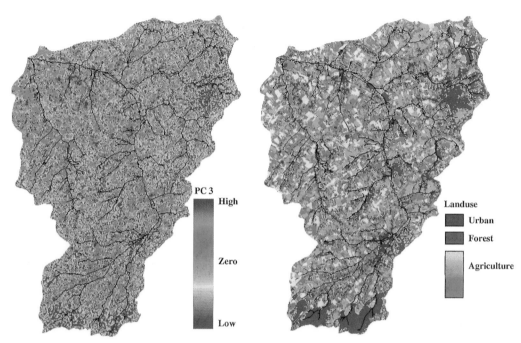

Figure 8.9. Third principal component image for the Zwalmbeek (left) and a classified Landsat image with forested and urbanised regions in the catchment highlighted (right). (From Verhoest et al., 1998; reproduced with permission.)

8.6 and 8.9 are compared with low values in Figure 8.6 corresponding to woods and urban areas. If there are few changes in major vegetation features over the same sequence of images, each canopy type will also produce a typical and temporally consistent radar response.

The fourth and subsequent principal components account for a very small fraction of the total variance in the sequence of SAR images, and they do not seem to reveal significant features. These PCs are characterised mostly by noise (including speckle).

As was already mentioned, the second component shown in Figure 8.8 is strongly related to the soil moisture response expected from rainfall and drainage/redistribution episodes and reflects the drainage characteristics of the soil. This can be further illustrated by investigating the signal's behaviour for the negative and positive values of the second principal component. Pixels in each image were assigned to two groups, one for positive PC2 and the other for negative PC2. The average backscatter was then calculated for each group in each image. Figure 8.10 plots, for each SAR image, the average backscatter value of these two classes in the second PC. The negative class generally corresponds to the well-drained soils, which are found upslope, while the positive class mainly coincides with the poorly-drained areas. As was mentioned before, the discharge areas exhibit a lower temporal variability in soil moisture content than the upslope areas, and this is reflected in the lower variability of the radar backscattering for the positive PC2 areas during the

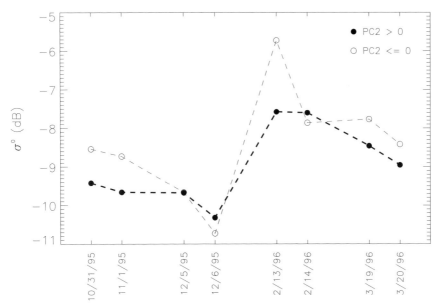

Figure 8.10. Average radar backscatter values for the negative and positive PC2 values for the Zwalmbeek catchment. (From Verhoest et al., 1998; reproduced with permission.)

winter period. During a drydown period, upslope areas show a larger decline in soil moisture content than near-stream areas, which corresponds to a larger decline in backscattered signal of the negative PC2 areas for the first four images. The large rain event of February 13 is reflected in the large increase in signal for both positive and negative PC2 areas. Over the course of the following 24 hours, soil moisture was redistributed over the basin, leading to a decrease of soil wetness in the recharge zones but little change in the already saturated near-stream zones, as evidenced in Figure 8.10. Shrinkage of the variable source areas over the following weeks accounts for the decrease of the backscattered signal of the positive PC2 class.

Applying PCA on the sequence of ERS-1 images taken over the Coët-Dan catchment, omitting the frost dates (January 28 and February 21), leads to similar results as obtained for the Zwalmbeek catchment. As the Coët-Dan basin is quite flat (slopes are generally less than 5 %), the backscattered signal is less influenced by the topography. Therefore, the first PC does not show the topographic effects on the backscattered signal but rather shows variations in land use. The second PC, as shown in Figure 8.11, generally corresponds well to the SPI computed over the catchment (as in Figure 8.5). In the area covered by the inset in Figure 8.5, again the general patterns are similar although there is a difference in the upper section of the inset where the PC2 predicts a wet area extending to the north of the stream, which is not apparent in either the SPI image or the observed saturated area. Nevertheless, it can be concluded that generally the saturation prone areas can be mapped using the principal

Figure 8.11. Second principal component calculated for the Coët-Dan catchment. The stream network (black) is given in overlay. (Red: negative values (saturation prone), blue: positive values.)

component technique on a sequence of SAR images. Again, slicing the histogram of the second PC into positive and negative classes leads to a similar behaviour in the radar backscatter as was observed in the Zwalmbeek catchment. In this analysis, the negative PC2 areas correspond to the saturation prone areas observed during the field campaign. Figure 8.12 shows the large temporal variability exhibited by the positive PC2 pixels which reflects wetting from the first rainstorm in February and the drydown of the soil in those areas thereafter, while the negative areas almost remain at the same level due to their high moisture content.

8.6 CONCLUSIONS

Remote sensing offers great potential for mapping saturation prone zones within a catchment. For two humid catchments in Western Europe, several methods based on change detection techniques have been tested for this pur-

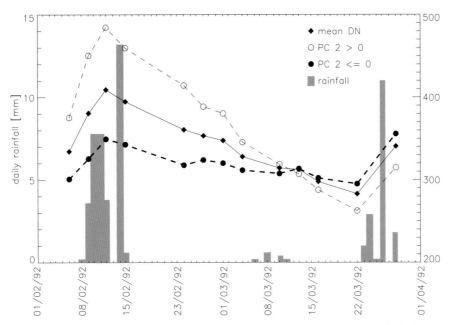

Figure 8.12. Daily rainfall for the Coët-Dan catchment during the winter period of 1992. Also indicated by the solid line is the average backscatter over the basin (expressed in digital numbers) and the average radar backscatter values for the negative and positive PC2 values (dashed lines).

pose. Applying a threshold on the backscattering coefficient is not succesful in delineating saturated areas since the choice of an absolute threshold cannot take account of the several surface characteristics that influence the backscatter. Other simple methods consist of differencing or using the ratio of SAR images. The differencing technique yields more valuable information than the threshold method, but still is limited due to statistical problems related to the speckle in SAR images.

Based on the observation that soil moisture variability is a function of its position along a hillslope, the saturation potential index was introduced. This index is based on the temporal standard deviation of the backscattering coefficient at a certain location which is directly related to the variation of soil moisture at that spot where small temporal variation indicates saturated areas. This index compares well to field observed saturated areas for the Coët-Dan catchment. However, for the Zwalmbeek catchment, which has a more pronounced topography, influences of the change in local incidence angle were introduced in the SPI. This problem is addressed by performing a principal component analysis on the sequences of images. This technique can separate topography and land use effects from the soil moisture influence on the total backscattering. In particular, it was possible to detect changes between scenes in the second principal component that could be linked to soil moisture variations. The soil moisture patterns observed are consistent with the rainfall-runoff dynamics of a catchment and coincide with the saturation prone areas derived from information on the natural drainage condition of the soils in the Zwalmbeek catchment. It should be noted that the presented methods have been tested during the wet season only, as it can be assumed that they will only be applicable under these conditions.

Full quantitative validation of these techniques would require extensive observations of saturated zones over large areas. Nevertheless, the results of these ERS1/2 SAR analyses indicate that a quantitative representation of saturation prone areas is now possible.

9

Soil Moisture and Runoff Processes at Tarrawarra

Andrew Western and Rodger Grayson

9.1 INTRODUCTION

It has been recognised for at least two to three decades that saturation excess is the dominant surface runoff process operating in most landscapes with a humid climate (Betson, 1964; Dunne et al., 1975). This mechanism is associated with soil moisture patterns characterised by high moisture zones in depressions and drainage lines (Anderson and Burt, 1978a,b,c; Anderson and Kneale, 1980, 1982; Burt and Butcher, 1985, 1986; Dunne and Black, 1970a,b; Moore et al., 1988a). It has also been known that event based hydrologic models are sensitive to initial conditions (Stephenson and Freeze, 1974). Blöschl et al. (1993) and Grayson et al. (1995) showed that event hydrographs simulated by spatially distributed hydrologic models were sensitive to the way in which the antecedent moisture was arranged spatially (see Chapter 1, pp. 12–13 for more detail). However, until recently there has been little spatial soil moisture data available that could be used to determine the characteristics of soil moisture patterns in natural catchments. Most of the data that were available were not sufficiently detailed to provide spatial soil moisture patterns without significant ad hoc interpretation of the data. It is important to recognise that catchment runoff does not give us much insight into internal catchment processes (Grayson et al., 1992a,b). Given that we want to understand runoff processes in catchments, it is necessary to measure internal information, i.e. soil moisture patterns in this case.

The apparent importance of spatial soil moisture patterns, combined with the limitations of existing data, motivated us to perform a series of experiments in which we measured actual soil moisture patterns and used these patterns for testing and developing distributed models. In addition to the soil moisture patterns, we measured other key variables that would enable us to interpret the observed patterns in terms of the controlling hydrologic processes. Our aims centred on understanding the spatial variability of soil moisture, its representation in geostatistical and hydrological models and its importance from the

Rodger Grayson and Günter Blöschl, eds. *Spatial Patterns in Catchment Hydrology: Observations and Modelling* © 2000 Cambridge University Press. All rights reserved. Printed in the United Kingdom.

perspective of the hydrological response of the landscape. We were particularly interested in the issue of spatial organisation of soil moisture patterns. In this chapter we discuss the above aims, while keeping a particular focus on the interaction between the field experiments and modelling and on using patterns in model development and testing. Initially some background material is presented, and the design of the experiment and the justification for making specific measurements is discussed in some detail. Next the field site and the observed soil moisture patterns are described. Then we discuss the Thales model structure and parameterisation and the role of the data and our qualitative field observations in choosing them. Some model simulations are then presented. The chapter concludes with a discussion of the model results, two problems with the model structure that contribute to simulation errors, the importance of different sources of spatial variability, and some general modelling insights.

9.2   MODELLING BACKGROUND

This section is intended to provide a brief overview of the types of models that can be used for predicting soil moisture and runoff in small catchments. There are essentially three different approaches that can be used: lumped models; distribution models; and distributed models. Lumped models are of no interest here since they do not represent spatial variability. Distribution models represent the spatial variability of soil moisture (or some related state variable such as saturation deficit) using a distribution function. This distribution function can be derived from the catchment topography, as is the case with Topmodel (Beven and Kirkby, 1979; see also Chapter 3), or it can be a theoretical distribution function, as is the case with VIC (Wood et al., 1992). From a water balance perspective, these models operate as lumped models (i.e. the whole catchment is represented using one store). However, where the distribution function is based on the catchment topography, it is possible to map simulated soil moisture back into the catchment to produce a simulated pattern (Quinn et al., 1995). To our knowledge there has not been any detailed testing of soil moisture patterns actually simulated by models such as Topmodel. However, a number of studies have compared various terrain index patterns (on which Topmodel is based) with soil moisture patterns (e.g. Anderson and Burt, 1978a; Anderson and Kneale, 1980, 1982; Barling et al., 1994; Burt and Butcher, 1985, 1986; Moore et al., 1988a; Western et al., 1999a). Models incorporating distribution functions that are not related to the topography do not allow the soil moisture patterns to be mapped back into space; however, the statistical distribution functions used in these models can be compared to the equivalent distributions derived from measurements (see e.g. Western et al., 1999b).

Fully distributed, physically based hydrologic models explicitly predict the spatial pattern of soil moisture by simulating the water balance at many points in the landscape. These models are usually based on combinations of differential equations that describe the storage and fluxes of water within the catchment. There have been some qualitative comparisons of soil moisture patterns with these models, including that by Barling et al. (1994); however, more detailed comparisons are rare. This chapter presents detailed comparisons between simulated and observed soil moisture patterns for the Tarrawarra catchment and describes the utility of high resolution spatial patterns for providing insights into the modelling of hydrologic response.

9.3 THE TARRAWARRA EXPERIMENTS

9.3.1 Key Experimental Requirements

The aims of the Tarrawarra experiment can be broadly stated as being: to understand the characteristics of the spatial variability of soil moisture at the small catchment scale; to determine how well various techniques can represent that variability; and to understand how that variability impacts on the hydrologic response of the landscape. Of specific interest was the issue of how organised the spatial variation of the soil moisture is in general and specifically how any organisation might reflect the organisation of the topography. These objectives could not be met without detailed patterns of soil moisture, collected at a scale that was appropriate and collected using a measurement methodology that could be reliably interpreted. Because suitable data did not exist, it was necessary to collect reliable soil moisture pattern data. The objectives above were of primary importance when developing our field methodology.

The methods chosen for collecting the soil moisture data had to meet four criteria and followed the general approach to sampling design presented in Chapter 2 (pp. 45–9). First, measurements had to be at sufficiently high resolution (small spacing) to resolve the important details of the pattern. Second, the soil moisture measurement technique had to have an accurate calibration to soil moisture. Third, the depth over which the soil moisture was measured needed to be sufficiently large to be of hydrologic significance. Fourth, the extent of the study area had to encompass at least a complete catchment, not just one hillslope. Furthermore it was desirable to maximise the rate at which measurements could be made and to minimise disturbance while taking the measurements since they were to be repeated several times.

High-resolution measurement was essential to this project because we wanted to examine the degree of organisation in the spatial soil moisture pattern. Low-resolution data has a strong tendency to make spatial variation appear disorganised. This is because details of the spatial pattern are not resolved (Williams, 1988). This problem is illustrated in Figure 9.1, which shows time domain reflectometry measurements of soil moisture in the top 30 cm of the soil profile collected at Tarrawarra on: (a) 10×20 m; and (b) 30×60 m grids. The 30×60 m resolution pattern was obtained by subsampling from the 10×20 m resolution pattern. Tarrawarra has two main drainage lines, which contained narrow bands

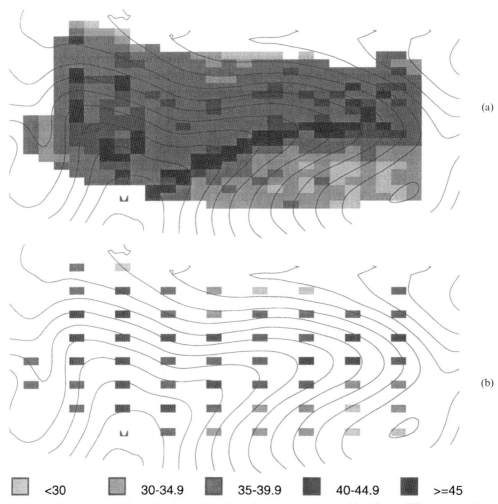

Figure 9.1. Comparison of volumetric soil moisture patterns resulting from different sampling resolutions. The pattern on the top is point soil moisture sampled on a 10×20 m grid. The pattern on the bottom is the same data sub-sampled to obtain a pattern of point measurements on a 30×60 m grid. The top pattern appears organised while the bottom pattern appears random, even though the underlying pattern is the same.

of high soil moisture at the time of these measurements. These bands are evident in the 10×20 m resolution pattern. The important point is that the top pattern looks organised while the pattern at the bottom looks random, even though the latter has been simply subsampled from the former. The difference in apparent soil moisture pattern demonstrates the need for high-resolution sampling.

The reason for measurements that can be accurately calibrated to soil moisture is, in one sense, obvious: it is always desirable to minimise noise in the data. In another it is less obvious. We were interested in the spatial pattern of soil moisture. If we had used a sensor that was significantly influenced by other factors, such as surface roughness or vegetation, as well as by soil moisture, the patterns we obtained would reflect both the soil moisture pattern and the patterns of the confounding variables. With current technology, this excluded remote sensing due to significant data interpretation problems with current space-borne systems, which are all influenced by factors other than soil moisture (see e.g. Jackson and Le Vine, 1996; and Chapter 8).

In landscapes where saturation excess runoff is important, it is the saturation deficit over the whole soil profile that is of most interest. This means that it is desirable to measure soil moisture over the entire vadose zone. This is a key shortcoming of current soil moisture remote sensing systems, which are only influenced by the surface soil layer (usually < 5 cm depth) and require assumptions about moisture profiles to estimate deficits.

The requirement for a complete catchment was primarily due to our interest in the role of topography in controlling the soil moisture pattern and our interest in the response of the landscape in general rather than individual hillslopes. By sampling a whole catchment, a diverse range of topography was sampled. When we were selecting the study site we looked for catchments that contained both convergent and divergent slopes and a wide range of aspect. By doing so we hoped to sample an area that would be fairly representative of the general landscape. A number of other criteria were also used when selecting the study site. For experimental reasons it was desirable to have a catchment in which: saturation excess was the dominant runoff mechanism; there was no artificial drainage, roads, dams or irrigation; and the vegetation was permanent pasture. Permanent pasture was desirable, as it would minimise the impact of spatial variation in land management on the results. For logistical reasons the catchment had to be trafficable and the soils had to enable TDR probes to be easily inserted (i.e. not too many rocks).

9.3.2   Catchment Description

Tarrawarra is a 10.5 ha catchment in southern Victoria, Australia (37°39′S, 145°26′E) (Figure 9.2). The climate is temperate, the mean annual rainfall is 820 mm and the areal potential evapotranspiration is 830 mm. Compared with evapotranspiration, there is a significant rainfall deficit in summer and excess in winter. The terrain consists of smoothly undulating hills (Figure 9.3). There are no perennial streams and no channels within the catchment. The land use is cattle grazing and the vegetation consists of perennial improved pastures. At the catchment scale the vegetation cover was relatively uniform. On a few occasions the vegetation was variable at a patch scale of 1–10 m due to preferential grazing by the cattle. There was also variability at the individual plant scale; however, the percentage of plant cover was generally greater than 90%. This plant and patch-scale variability may have contributed some small-scale variability to the soil moisture.

The bedrock is lower Devonian siltstone with interbedded thin sandstone and local bedded limestone (the Humevale formation) (Garratt and Spencer-Jones, 1981) at a depth of 0.5–1.5 m on the hillslopes and deeper in the

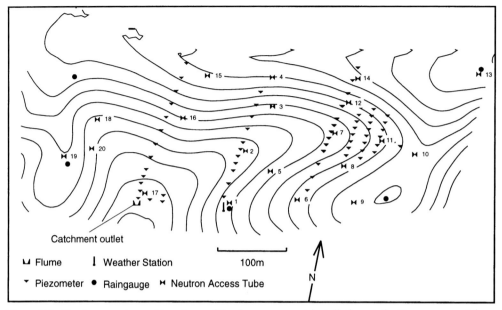

Figure 9.2. The topography and locations of fixed measurement installation at the Tarrawarra catchment. The contour interval is 2 m.

Figure 9.3. A view of the Tarrawarra catchment looking north east from the catchment outlet.

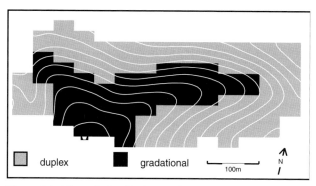

Figure 9.4. The soil distribution at Tarrawarra.

drainage lines. The soils in the Tarrawarra catchment have been studied in detail by Gomendy et al. (in preparation) and consist of two units (Figure 9.4). The soil on the upper slopes is a pedal mottled-yellow Duplex soil (Dy3.4.1, Northcote, 1979), which is a texture contrast soil. The Duplex soil has a 10–30 (typically 20) cm thick silty loam or silty clay loam A1 horizon, a silty clay loam A2 horizon up to 45 (typically 20) cm thick and a medium to heavy clay B horizon. Ironstone gravel often occurs, mainly in the A2 horizon. The A1 and B horizons are moderately to strongly structured, while the A2 horizon shows massive to weak structure. The soil on the lower slopes and in the depressions is a gradational grey massive earth (Gn3.9.1 or 2.8.1, Northcote, 1979) soil with a silty loam A horizon 10–35 (typically 20) cm thick which gradually transitions into a silty clay loam B horizon. The A horizon is strongly structured and the B horizon is massive to weakly structured. The solum depth varies from 40 cm at some points on the ridge-tops to over 2 m in the depressions. Water tables form in the A horizon during the wetter months of the year. There is little physical difference between the surface soils of each of these two groups. The surface soils crack during dry periods.

9.3.3   Experimental Methods

The system adopted to measure the soil moisture patterns was the Terrain Data Acquisition System (TDAS) developed at the Centre for Environmental Applied Hydrology, University of Melbourne (Tyndale-Biscoe et al., 1998). TDAS is a ground-based system that includes Time Domain Reflectometry (TDR) equipment for measuring soil moisture, a real time Differential Global Positioning System (DGPS) for measuring spatial location and a computer which displays real time spatial location, guides the user to the desired sampling location and records the measurements. The real time DGPS system allows accurate relocation of sampling sites. All the instrumentation is mounted on an all-terrain vehicle, which is fitted with hydraulics for inserting the TDR probes. A range of other instrumentation and soil sampling equipment can also be mounted and recorded (Tyndale-Biscoe et al., 1998). TDAS is shown in

Figure 9.5. TDAS allows 50–100 soil moisture measurements to be collected per hour, which greatly reduces the logistical problems of collecting high-resolution soil moisture maps. This would have been impossible with manual sensor insertion. Another advantage of this system is that using hydraulics to insert the TDR probes ensures that the probes are inserted much more smoothly, and with minimal sideways force, compared to manual insertion. This minimises disturbance to the soil and air gaps, which reduces the accuracy of TDR measurements.

The actual soil moisture patterns collected at Tarrawarra generally consisted of approximately 500 measurements on a 10×20 m grid. This grid spacing was chosen as a compromise between time constraints, maximising the detail obtained up and down the hillslopes, which was the direction in which we expected the greatest variability to occur, and the advantages of maintaining a similar spatial resolution in both grid directions (see Chapter 2, p. 48–9). Thirty centimetre TDR probes were used to make the soil moisture measurements. This means that the measurements represent the average volumetric soil moisture over a depth of 30 cm at the measurement point. This depth was chosen to minimise problems with bending TDR probes when they were inserted while obtaining a good measure of the moisture in the unsaturated zone (at least during the wet season). In space (i.e. as a map view) the TDR measurements represent the average soil moisture over a very small area (i.e. support), which is of the order of 10 cm × 10 cm (Ferré et al., 1998).

Figure 9.5. The terrain data acquisition system (TDAS).

The accuracy of the TDR measurements was assessed by comparing them with gravimetric measurements collected in the field. The variance of the difference between the gravimetric and TDR soil moisture measurements was 6.6 $(\%V/V)^2$. An analysis of the magnitude of different error sources indicates that during normal operating conditions, approximately half of this variance is due to errors in the gravimetric measurements and half due to errors in the TDR measurements. This is also consistent with variogram nuggets found by Western et al. (1998a) and with average variances for repeated TDR measurements in small (0.25 m^2) patches.

Neutron moisture meter measurements were also made at twenty points across the catchment. This was done for two reasons. First, the TDR measurements were not deep enough to extend over the whole unsaturated zone during dry periods and it was important to obtain some information about soil moisture behaviour at depth. Second, we wanted some information about the vertical soil moisture dynamics. Subsequent analysis of the neutron moisture meter data showed that between 40 and 60 % of the active soil moisture storage was measured by the TDR and that during wet periods most of the variation in soil moisture occurred in the top 30 cm of the soil profile. Neutron data were calibrated to soil moisture for the Tarrawarra soils.

A range of other data was also collected to complement the soil moisture measurements. Further details of these data can be obtained from Western and Grayson (1998). A series of piezometers measuring shallow water table elevations were manually read during the wet period to obtain information about surface saturation directly. Detailed meteorologic and surface runoff measurements were also made to enable the meteorological forcing and the catchment scale runoff response to be characterised. A range of soil properties was also measured.

When designing the measurement network we aimed to target our measurements on the basis of our interests and on the scales at which variability might be important. Because they were our main focus, the soil moisture patterns were measured in as much detail as possible, at a scale that would allow us to characterise the influence of topographic and terrain-related soil parameters. However, logistical constraints limited the number of surveys to about twelve per year. This means that we obtained twelve soil moisture patterns in a year, although more frequent surveys would have been desirable. To some extent this was overcome by making more frequent measurements with the neutron moisture meter. Most meteorological measurements were conducted at only one site since they were not expected to vary significantly in space. However, they were measured frequently in time due to their high temporal variability. Rainfall was measured at five locations around the field site. This was done to measure the spatial variability of rainfall at the event and longer timescales. These measurements illustrated that spatial variability of rainfall was not a significant influence on the spatial pattern of soil moisture measured with the TDR.

9.4 OBSERVED VARIABILITY

Field observations are important for generating and testing hypotheses about the processes controlling the behaviour of the catchment and hence are an important aid in determining an appropriate model structure. Field observations may be either quantitative measurements or qualitative observations of behaviour. Both are useful when developing a model and are used here to progressively develop the model structure. This approach allows us to maintain a relatively simple model structure and a small number of parameters.

Figure 9.6 shows soil moisture patterns measured at Tarrawarra on twelve occasions during 1996. Table 9.1 summarises the statistics of each pattern. These observations can be used to generate hypotheses about how the catchment behaves, particularly about the factors that control the observed spatial patterns. Looking at the wet patterns (e.g. May 2, 1996), it can be seen that there is a strong topographic influence on the soil moisture pattern. During the dry periods (e.g. February 23) there is little topographic influence. This behaviour suggests that lateral movement of water in the shallow subsurface is important during wet periods but not during dry periods. This observation, together with the observation that the temporal behaviour of the soil moisture is dominated by persistent wet periods and persistent dry periods, with rapid changes between the two, led us to develop a hypothesis of preferred wet and dry states (Grayson et al., 1997). The wet and dry states are characterised by spatial soil moisture patterns that are topographically controlled and random (unrelated to topography) respectively. This behaviour means that a model that can simulate the moisture dependent lateral movement of water is required to simulate the patterns.

Most of the soil moisture patterns collected at Tarrawarra show some aspect-related differences in soil moisture. The northerly facing slopes are drier because they receive more solar radiation (being in the Southern Hemisphere). This suggests that there may be some spatial differences in evapotranspiration related to radiation input. Hence, including this effect in the model is also important. However, another potential explanation of this aspect-related effect is a difference in soil properties since, compared to the gradational soils, the duplex soils are relatively more important (extend further downslope) on the north-facing hillslopes. Thus this difference in soil type and associated characteristics also needs to be incorporated in the model.

Several important qualitative observations were made at Tarrawarra that can be used either in selecting the model structure or in testing model performance. On one occasion (April 11, 1996) a rainfall event with quite high rainfall intensities (up to 50 mm/hr) was observed at Tarrawarra and there was no visual evidence of infiltration excess runoff being produced and no runoff response at the catchment outlet. Even if infiltration excess were produced under such conditions, the cracking nature of the soils means that the water would soon flow down a crack. On several other occasions, high rainfall intensities have been measured without any runoff being produced at the flume. These observations

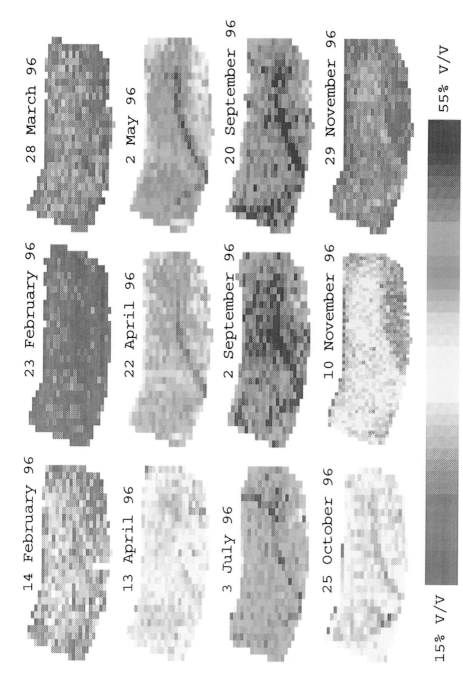

Figure 9.6. Observed soil moisture patterns at Tarrawarra for twelve occasions during 1996. Note that each rectangle represents one point measurement of the average volumetric soil moisture in the top 30 cm of the soil profile made using time domain reflectometry. The measurement error standard deviation is 1.7% V/V.

28 March 96

23 February 96

14 February 96

2 May 96

22 April 96

13 April 96

20 September 96

2 September 96

3 July 96

29 November 96

10 November 96

25 October 96

55% V/V

15% V/V

219

Table 9.1. Soil moisture statistics for the twelve moisture patterns used here

	Soil moisture			Antecedent precipitation	
	Mean (%V/V)	Variance (%V/V)^2	Coefficient of variation	10 days (mm)	40 days (mm)
14-Feb-96	26.2	10.6	0.12	58	98
23-Feb-96	20.8	5.3	0.11	0	97
28-Mar-96	23.9	7.1	0.11	7	89
13-Apr-96	35.2	12.3	0.10	65	145
22-Apr-96	40.5	14.6	0.09	71	215
2-May-96	41.4	19.4	0.11	6	172
3-Jul-96	45.0	14.0	0.08	20	25
2-Sep-96	48.5	13.9	0.08	22	108
20-Sep-96	47.3	15.2	0.08	36	117
25-Oct-96	35.0	19.2	0.13	15	84
10-Nov-96	29.3	10.8	0.11	35	71
29-Nov-96	23.9	6.28	0.11	12	61

suggest that infiltration excess runoff is not a significant process at Tarrawarra and that detailed infiltration algorithms are not required in the model.

Another qualitative observation is that when the catchment "wets-up" during the autumn period, the highly convergent areas around neutron access tubes 7, 11 and 18 are the first to become saturated. The gullies below these areas then gradually saturate due to runon infiltration of the runoff from upslope and any lateral subsurface flow that may be occurring. These initial wet areas can be seen on April 13, 1996 (Figure 9.6) and the runon infiltration process was observed visually and saturated areas were mapped (Figure 9.7) following further rain on April 14, 1996. These observations suggest that runon infiltration should be included in the model.

The above observations, together with the recognition that a long-term soil moisture simulation requires allowances for evapotranspiration and possibly

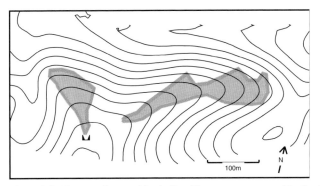

Figure 9.7. Saturated areas (shaded) at Tarrawarra mapped in the field on April 14, 1996. These had developed since April 13 and were expanding down the drainage lines via a process of overland flow and runon infiltration.

deep seepage, provide the basis for selecting an initial model structure which is described in the next section.

9.5 THE THALES MODEL

The Thales modelling framework (Grayson et al., 1995) is used here. Several significant modifications to the model described in Grayson et al. (1995) were required. A general description of the model and a detailed description of the modifications are provided here. Thales uses a computation network based on a mesh of streamlines and elevation isolines. Pairs of streamlines form stream tubes down which the lateral movement of water is simulated. Figure 9.8 shows the stream tube network for Tarrawarra, as well as a schematic of the fluxes through a computational element. As applied at Tarrawarra, rainfall and potential evapotranspiration force the model. The model incorporates the following processes.

- Saturated subsurface lateral flow (kinematic wave).
- Saturation excess overland flow (quasi steady-state).
- Exfiltration of soil water.
- Runon infiltration of overland flow.
- Deep seepage.
- Evapotranspiration.

Except for deep seepage, these processes were included because our field observations suggested that they might have an important impact on the spatially distributed water balance and hence on the soil moisture patterns. Deep seepage was incorporated because our water balance calculations and initial model results suggested that evapotranspiration and surface runoff could not account for all the outflows from the catchment.

Thales uses a water balance to simulate soil moisture for each element (Figure 9.8). Inputs of water to an element are rainfall, subsurface flow from upslope and surface flow (with runon infiltration) from upslope. Outputs of water are evapotranspiration, subsurface flow to downslope, and surface flow to downslope. Exfiltration is possible. Surface flow is generated when an element is saturated. Infiltration excess overland flow is not simulated due to low (six minute) rainfall intensities and the lack of surface runoff during periods without saturated conditions (Western and Grayson, 1998).

In this version of Thales, a single store is used to represent the soil moisture for each element and the element-average soil moisture, θ, varies between the permanent wilting point, θ_{pwp}, and saturation (i.e. porosity), θ_{sat}. When θ is less than the field capacity, θ_{fcp}, the vertical soil moisture profile is assumed to be uniform. When $\theta > \theta_{fcp}$, the vertical soil moisture profile is assumed to consist of an unsaturated zone with a moisture content of θ_{fcp} above a saturated zone with a moisture content of θ_{sat}. Under these moisture conditions, the depth to the water table, d_{wt}, can be calculated as:

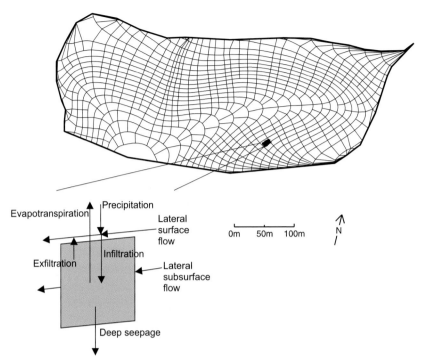

Figure 9.8. The flow net for Tarrawarra used in Thales. A schematic of a typical computational element is also shown.

$$d_{wt} = \frac{(\theta_{sat} - \theta)}{(\theta_{sat} - \theta_{fcp})} \cdot d_{soil} \qquad \text{given } \theta > \theta_{fcp} \tag{9.1}$$

where d_{soil} is the soil depth. These soil moisture profile approximations are appropriate for the shallow soils at Tarrawarra but are likely to be invalid for deep soils where the dynamics of the unsaturated zone are important. For calculating lateral subsurface flow, the soil profile is assumed to consist of two layers. These layers can loosely be thought of as soil horizons. This flow is routed using a kinematic wave description and Darcy flow. Lateral subsurface flow, Q_{sub}, is allowed when part or all of the upper soil layer is saturated. It is calculated as:

$$Q_{sub} = W \cdot k_{sat} \cdot \tan(\beta) \cdot (d_{upper} - d_{wt}) \qquad d_{wt} < d_{upper} \tag{9.2a}$$

$$Q_{sub} = 0 \qquad d_{wt} \geq d_{upper} \tag{9.2b}$$

where d_{upper} is the depth of the upper layer, W is the width of the stream tube (computational element), k_{sat} is the saturated hydraulic conductivity, and β is the surface slope. The numerical scheme is similar to that used by Grayson et al. (1995); however, the state variable for the subsurface has been changed from water table depth to soil moisture for convenience. The depth of the upper layer is assumed to coincide with the relatively high transmissivity surface soil horizons (often the A horizon). This representation of the soil moisture store and lateral flow was used because the A horizon is significantly more permeable than the B horizon, and is thus the dominant horizon for lateral flow, while the B horizon contributes to active storage during dry periods.

Exfiltration occurs under saturated conditions where the capacity of the hillslope to transmit subsurface lateral flow is decreasing downslope. This may occur due to topographic convergence, a downslope decrease in slope, transmissive layer depth, or hydraulic conductivity or a combination of these factors. Overland flow from upslope becomes runon infiltration if the element is unsaturated. Deep seepage has been added to the long-term model. The rate of deep seepage per unit plan area, q_{deep}, decreases linearly from the effective vertical saturated hydraulic conductivity of the soil profile, k_{deep}, when $\theta = \theta_{sat}$ to zero when $\theta = \theta_{fcp}$.

Overland flow discharge is calculated as the total inflow to the cell, minus the sum of subsurface outflow, deep seepage and evapotranspiration. Any water required to bring the cell up to saturation is also subtracted and the overland flow is assumed to be at steady-state during the time-step implying that the water was immediately routed to the downslope element. This is different to Grayson et al. (1995) where the surface flow is routed using a kinematic wave. The quasi steady-state approach to simulating the surface flow taken here allowed daily time-steps to be used, which resulted in large computational savings compared to the short time-steps required for a kinematic wave solution of the overland flow. Given that our interest was in soil moisture patterns and water balance, rather than event runoff, daily time steps were quite adequate.

Evapotranspiration is assumed to occur at the potential evapotranspiration rate (PET) when the soil moisture exceeds a stress threshold, θ_{stress}, and below θ_{stress} it decreases linearly to zero at the permanent wilting point, θ_{pwp}. Evapotranspiration, ET, is calculated as follows.

$$\text{ET} = P_i \cdot \gamma \cdot \text{PET} + (1 - \gamma) \cdot \text{PET} \qquad \theta \geq \theta_{stress} \qquad (9.3a)$$

$$\text{ET} = P_i \cdot \gamma \cdot \text{PET} + (1 - \gamma) \cdot \text{PET} \cdot \left(\frac{\theta - \theta_{wp}}{\theta_{stress} - \theta_{wp}} \right) \qquad \theta < \theta_{stress} \qquad (9.3b)$$

P_i is the potential solar radiation index (Moore et al., 1991) and it accounts for slope and aspect effects. γ is a weighting factor which is applied to P_i and varies between 0 (spatially uniform) and 1 (spatial pattern of potential evapotranspiration fully weighted by direct solar radiation). The potential solar radiation index varies seasonally as the solar declination varies.

9.6 DATA ISSUES

The key steps in this work really revolved around the design and execution of the field experiments, the thought that went into developing the model structure and the interpretation of the data and simulations. Most of the data handling steps in this study were straightforward. The input data required by Thales and the methods used for model testing are described below.

9.6.1 Input Data

There are three categories of input data used by Thales as applied at Tarrawarra: terrain data, soil/vegetation data and meteorological data. An initial moisture pattern is also required. The terrain data consists of the geometric properties of each element (excluding soil depths) and information about the connectivity between elements. The derivation of this data from a vector (contour) digital elevation model is discussed in detail by Moore and Grayson (1991) and Grayson et al. (1995). The digital elevation model used at Tarrawarra is based on a detailed ground survey (Western and Grayson, 1998) and is significantly more accurate than many standard large (say 1:25,000) scale topographic maps. TAPES-C (Moore et al., 1988b) was used to derive the stream tube network for Tarrawarra (Figure 9.8).

The soils at Tarrawarra have been studied in detail by Gomendy et al. (in preparation) who have identified two morphologically distinct soil profiles, a Duplex (texture contrast) soil and a Gradational soil, which were described earlier. In the model of Tarrawarra, the properties and depths within each of these two soil units were assumed to be uniform and model elements were assigned a soil type on the basis of the mapping by Gomendy et al. (in preparation).

Three time-varying quantities are used to force the model. Daily rainfall depths were obtained from the weather station data and were applied to the catchment in a spatially uniform manner. Hourly potential evapotranspiration depths were calculated using net radiation, wet and dry bulb temperature, wind and soil temperature data from the automatic weather station and the Penman–Monteith model (Smith et al., 1992). These were aggregated to daily values and applied to the catchment in a spatially uniform manner. The potential solar radiation index was calculated daily for each model element and was used as a modifier to the potential evapotranspiration as specified in Section 9.3.

We set the initial soil moisture pattern on the basis of the soil moisture pattern measured on September 27, 1995. The first results analysed were for February 1996, which was long enough after initiation to make the results independent of initial conditions. It therefore would have been possible to have used an estimated soil moisture pattern. It is worth noting that the period required to remove the effect of initial conditions from a simulation (i.e. the spin-up time) depends on the response time of the system. For soil moisture in a small catchment, this period is likely to be several months. For this reason, event models require some other approach that better reflects the true antecedent conditions at the time (see, e.g., Chapter 10). For systems that take a long time to respond (e.g. most groundwater systems) a much longer period would be required to spin-up the model from estimated initial conditions.

9.6.2 Testing data and methods

Three different data types were used to compare to model simulations for this exercise. The most important are the soil moisture patterns measured with TDR. These were measured on a regular grid, not at the computation nodes used in Thales, hence some interpolation was required to allow quantitative pattern comparison. The model results were interpolated to the sampling grid for the following two reasons. First, the model results are very smooth in space and hence can be interpolated with minimal smoothing error. Second, using a regular grid implies a uniform spatial weighting in any comparison, whereas the computational network tends to be more heavily weighted (smaller elements) to the hilltops. Other data that were compared with model simulations were the catchment outflow hydrograph and the saturation deficit data calculated from the NMM data.

The simulated soil moisture was compared with the measured soil moisture using several different approaches. Simulated and observed patterns were compared visually and error (difference) patterns were calculated and examined. Summary statistics including the mean, standard deviation and root-mean-square of the soil moisture simulation errors were calculated for each measured soil moisture pattern. Time series of saturation deficits calculated from the NMM data and simulated by Thales were plotted and compared for both individual tubes and for all tubes averaged together.

Plots of the simulated and observed hydrographs and the difference between the two were also examined, and the mean error and prediction efficiency for the hydrographs were also calculated. The hydrographs were regarded as the least important of the data sets for comparison since they only contain information at the catchment scale about the percentage of saturated area well connected to the catchment outlet and about baseflow processes integrated up to the catchment scale.

9.7 MODEL PARAMETERS

There are eight parameters that need to be set for each element in this version of Thales and one global parameter, γ. For each element, the soil parameters are θ_{sat}, θ_{fcp}, k_{sat}, d_{soil}, d_{upper}, k_{deep}, and the soil/plant parameters are θ_{pwp} and θ_{stress}. With the exception of k_{deep}, all the parameter values used in the simulations here were initially set on the basis of field measurements. Here we describe how these parameters were initially set and then we discuss how the parameter values were changed for the different simulations that were conducted.

Given that we are comparing the simulations with moisture measurements over the top 30 cm of the soil profile, the water retention parameters were set to reflect this soil layer. Bulk density measurements across the catchment indicated that the mean porosity was approximately 50%. There were only small differences in mean bulk density between the different soil types and it was not clear whether this was related to porosity differences or particle density differences (which might arise due to the spatial distribution of iron stone and stone). Therefore θ_{sat} was set to 0.5. The driest set of TDR measurements (March 20, 1997 – not shown here) were used to set θ_{pwp}. θ_{pwp} was set equal to $\mu_\theta - \sigma_\theta = 0.09$, where $\mu_\theta = 0.116$ and $\sigma_\theta = 0.025$ are the mean moisture and the standard deviation of soil moisture observed on that occasion, respectively. The reason for selecting a moisture less than the mean to represent the wilting point is related to the possibility that the moisture at some sites might be slightly above wilting point. On this occasion, the neutron moisture meter measurements at each site, were the driest recorded, or very close to the driest recorded.

The soil moisture patterns were carefully examined to determine when lateral subsurface flow was evident and the mean moisture on these occasions was used to set θ_{fcp}. On April 13, 1996 ($\mu_\theta = 0.35$) lateral redistribution is becoming evident and it is also evident on October 25, 1996 ($\mu_\theta = 0.35$). On March 28, 1996 ($\mu_\theta = 0.24$) and August 3, 1997 ($\mu_\theta = 0.27$) lateral flow is not evident, while on November 10, 1996 ($\mu_\theta = 0.29$) little or no lateral flow is evident. Therefore θ_{fcp} was set to 0.3. θ_{stress} was assumed to be equal to θ_{fcp}.

The total active storage observed at each neutron access tube was estimated by integrating the difference between the wettest and driest profiles at each site over the depth of the access tube. This calculation assumes that the wettest profile is representative of saturated conditions. Qualitative field observations suggest that this is a reasonable assumption. The total storage used in the model [i.e. $d_{soil} \cdot (\theta_{sat} - \theta_{pwp})$] was then set equal to the mean total storage (210 mm) plus one standard deviation (38 mm). A larger value than the mean was used since the total storage at some sites may have been underestimated, due to either the wettest measured profile not being saturated or the driest measured profile not being representative of the driest possible conditions. The depth of the upper layer was set to 400 mm on the basis of observations of the depth of the A horizon in the duplex soil.

A series of saturated hydraulic conductivity measurements were conducted in the A horizon of the soil using the constant head well permeameter technique (Talsma, 1987; Talsma and Hallam, 1980). These were used to set the values of k_{sat} in the model. The mean measured k_{sat} was 19 mm/hr and the standard deviation of the measurements was 26 mm/hr. Clearly these measurements are highly uncertain. It is also important to note that the small-scale well permeameter measurements are not necessarily representative of lateral flow at larger scales, due to the influence of soil heterogeneity. Nevertheless a value of 20 mm/hr was initially used for k_{sat}. For most of the simulations k_{deep} was set to zero.

The parameter in Thales that represents the spatial variability of potential evapotranspiration is treated as a global parameter in the model. In these simulations it took on two different values, 0 and 1. A value of zero was used initially, which implies that potential evapotranspiration is spatially uniform. Subsequently a value of 1 was used, which implies that potential evapotranspiration varies spatially in direct proportion to the potential solar radiation index.

9.8 MODEL SIMULATIONS

A series of simulations was conducted with the Thales model. The first of these used the basic model parameterisation above. This provided a simulation in which the only source of spatial variability in soil moisture was lateral flow. Subsequent runs modified this initial parameterisation and introduced spatial variability in the potential evapotranspiration and soils. Table 9.2 provides a summary of the parameters used in each run. When the initial run (Run 1) was performed, it appeared that the drainage was too slow. Therefore two runs with higher k_{sat} values were performed. Run 3 resulted in drainage that appeared too rapid. Thus run 2 most closely represented the effects of lateral flow. It should be noted that the value of k_{sat} used in run 2 is within one standard deviation of the mean observed k_{sat} and is greater than the observed mean, as would be expected due to any effects of preferential flow. The only source of spatial variation in runs 1–3 is lateral flow routed by the terrain. Spatially variable potential evapotranspiration was added and runs 4–6 performed. These essentially repeat runs 1–3 with spatially variable potential evapotranspiration.

In run 7, soil variability is introduced via the total depth. A regression relationship between the wetness index and the total soil depth ($R^2 = 66\%$), measured at each neutron access tube, was used to scale d_{soil}. This resulted in shallower soils (less storage) on the ridge tops and deeper soils (more storage) in the gullies. In run 8 soil variability related to soil type is introduced by using a smaller depth for d_{upper} for the gradational soil. The depth of 200 mm was chosen to coincide with the typical A-horizon depth observed in this soil.

Run 9 illustrates the performance of a single soil layer model (i.e. the whole soil profile is laterally transmissive) with the same transmissivity as run 5. This allows us to examine the value of using the two-layer soil representation. Run

Table 9.2. Parameter values used in each simulation. Note that $\theta_{sat} = 0.50$, $\theta_{fcp} = 0.30$, $\theta_{stress} = 0.30$, and $\theta_{pwp} = 0.09$ were fixed for all simulations.

Run	k_{sat} (mm/h)	k_{deep} (mm/h)	d_{soil} (mm)	d_{upper} (mm)	γ	Comment
1	20	0	600	400	0	Parameters set from field measurements
2	40	0	600	400	0	k_{sat} doubled
3	60	0	600	400	0	k_{sat} trebled
4	20	0	600	400	1	Spatially variable ET
5	40	0	600	400	1	k_{sat} doubled
6	60	0	600	400	1	k_{sat} trebled
7	40	0	varies	400	1	$d_{soil} = 216 \cdot \ln(a/\tan(\beta)) - 1480$, $250 < d_{soil} < 760$
8	40	0	600	400[1] 200[2]	1	Variable upper layer depth related to soil type [1]duplex soil, [2]gradational soil
9	26.7	0	600	600	1	Single layer model, run with same transmissivity as run 5
10	40	0.013	600	400	1	Deep seepage calibrated to correct mean runoff

10 looks at how the simulations are affected by adding deep seepage to match the mean simulated and observed runoff volumes (i.e. to close the water balance).

9.9 RESULTS

The soil moisture pattern results of what we judge to be the most realistic simulation (run 5) are presented first. Run 5 has spatially uniform soil depths and parameters and includes potential evapotranspiration that is spatially weighted according to the potential solar radiation index. Then, the patterns from the other simulations are compared to this run. Finally, we return to run 5 and compare simulated and observed saturation deficit time series and catchment runoff. The runs are described in some detail to highlight how the measured patterns and other data were used to identify model problems and successes. Run 5 was judged to be the best by making visual comparisons of the simulated and observed soil moisture patterns and visual examinations of the spatial pattern of simulation errors (see Chapter 3, pp. 78–9). The winter and spring periods were emphasised in this comparison because strong spatial organisation was observed during these periods and because there was little difference between the simulations during the summer period. It should be noted that the differences between the runs were often subtle and that no one run was the best on every occasion. Some indication of the differences between the runs can be obtained from Table 9.3, which summarises the simulation error statistically. Three statistics are shown: the mean error; the root mean squared error; and the error standard deviation. The root mean squared error incorporates the effect of both bias and random error, while the error standard deviation is a measure of the random error.

9.9.1 Soil Moisture Patterns – Run 5

Figure 9.9 shows the simulated soil moisture patterns for run 5 for 1996. Some care in interpreting the small-scale variability is required when comparing Figures 9.6 and 9.9. It is clear that the simulated moisture patterns are smoother than the observed patterns. There are three key reasons for this. First, the soil parameters are assumed to be spatially uniform in the model, whereas some small-scale variability would be expected in reality. Second, the observed patterns contain some measurement error. Third, the model support scale is of the order of 20 m while the measurement support scale is only 0.1 m. If the observation support scale was 20 m, the small-scale (< 20 m) variability apparent in the observed patterns would be averaged out and the observed pattern would appear much smoother. Thus some of the small-scale variability in the observed data is not relevant to the model formulation and can be ignored when comparing the observations and simulations. However, there are some small-scale features that are critical to the runoff response of the catchment. These are the narrow bands

Table 9.3(a). The mean error in simulated soil moisture (%V/V) for each run and each soil moisture pattern

Run	1	2	3	4	5	6	7	8	9	10
14-Feb-96	0.9	0.8	0.7	1.1	1.0	0.9	0.5	1.5	1.2	0.6
23-Feb-96	0.7	0.6	0.5	0.9	0.9	0.8	0.5	1.4	1.0	0.6
28-Mar-96	−0.4	−0.5	−0.5	0.0	0.0	−0.1	−0.1	0.3	0.0	−0.1
13-Apr-96	−0.6	−0.6	−0.7	0.1	0.1	0.0	−0.1	0.5	0.1	0.0
22-Apr-96	3.7	3.7	3.7	4.6	4.6	4.6	4.1	4.9	4.7	4.2
3-Jul-96	4.8	4.0	3.1	4.8	4.2	3.4	3.3	3.1	3.5	3.5
2-Sep-96	0.1	−0.1	−0.9	0.2	0.0	−0.7	−0.7	−0.8	−0.1	−0.5
20-Sep-96	2.3	2.1	1.6	2.4	2.1	1.7	1.9	1.7	2.0	1.9
25-Oct-96	6.1	5.8	5.0	6.4	6.2	5.4	4.6	4.8	6.1	5.0
10-Nov-96	6.6	6.3	5.5	7.0	6.8	6.1	4.8	5.2	6.8	5.4
29-Nov-96	3.4	3.4	3.0	3.9	3.9	3.5	2.0	3.4	4.1	2.6
20-Mar-97	−0.1	−0.1	−0.1	0.0	0.0	0.0	0.0	0.1	0.0	0.0
3-Aug-97	−4.2	−4.2	−4.2	−3.0	−3.0	−3.1	−3.1	−3.1	−3.1	−3.1

Table 9.3(b). The standard deviation of soil moisture simulation errors (%V/V) for each run and each soil moisture pattern

Run	1	2	3	4	5	6	7	8	9	10
14-Feb-96	3.2	3.4	3.5	3.1	3.3	3.5	3.1	4.9	4.1	3.2
23-Feb-96	2.2	2.3	2.4	2.2	2.2	2.3	2.2	3.9	2.7	2.2
28-Mar-96	2.6	2.6	2.6	2.4	2.4	2.4	2.4	2.9	2.5	2.4
13-Apr-96	3.4	3.4	3.4	3.2	3.2	3.2	3.3	4.1	3.1	3.2
22-Apr-96	3.6	3.5	3.4	3.8	3.7	3.6	3.8	4.6	3.5	3.7
3-Jul-96	3.7	3.6	3.7	3.7	3.6	3.7	3.8	4.0	3.8	3.7
2-Sep-96	3.6	3.4	3.5	3.5	3.4	3.4	3.4	3.4	3.4	3.3
20-Sep-96	3.7	3.5	3.5	3.7	3.6	3.5	3.5	3.5	3.6	3.5
25-Oct-96	4.0	3.7	3.9	3.9	3.6	3.7	3.7	4.4	3.7	3.6
10-Nov-96	2.9	3.3	4.3	2.7	3.1	4.1	2.8	5.9	3.9	2.9
29-Nov-96	2.3	3.1	4.3	2.2	3.0	4.3	2.2	7.1	4.4	2.5
20-Mar-97	2.6	2.6	2.6	2.6	2.7	2.6	2.7	2.7	2.6	2.7
3-Aug-97	2.9	2.9	2.9	2.4	2.4	2.4	2.4	2.5	2.4	2.4

of high soil moisture along the drainage lines. The model should be able to capture these.

Figure 9.10 shows a smoothed version of the observed data. This smoothing was performed in an attempt to remove the measurement error and small-scale variability (i.e. increase the support scale) of the observed data, thereby making it more comparable to the simulations (see also Chapter 3, p. 79). There are a number of simple interpolation techniques available, such as block kriging, that are capable of removing the measurement error variance and increasing the support by some sort of smoothing procedure. However,

Table 9.3(c). The root mean square soil moisture simulation error (%V/V) for each run and each soil moisture pattern

Run	1	2	3	4	5	6	7	8	9	10
14-Feb-96	3.3	3.4	3.6	3.3	3.4	3.6	3.1	5.1	4.2	3.3
23-Feb-96	2.3	2.4	2.4	2.4	2.4	2.4	2.2	4.1	2.9	2.3
28-Mar-96	2.6	2.6	2.6	2.4	2.4	2.4	2.4	2.9	2.5	2.4
13-Apr-96	3.5	3.5	3.5	3.2	3.2	3.2	3.3	4.1	3.1	3.2
22-Apr-96	5.2	5.1	5.1	5.9	5.9	5.8	5.5	6.8	5.8	5.6
3-Jul-96	6.0	5.4	4.8	6.0	5.6	5.0	5.0	5.1	5.2	5.1
2-Sep-96	3.6	3.4	3.6	3.5	3.4	3.5	3.4	3.5	3.4	3.4
20-Sep-96	4.3	4.1	3.9	4.4	4.1	3.9	4.0	3.9	4.1	4.0
25-Oct-96	7.3	6.9	6.4	7.5	7.1	6.6	5.9	6.5	7.1	6.2
10-Nov-96	7.2	7.1	7.0	7.5	7.5	7.3	5.6	7.9	7.8	6.1
29-Nov-96	4.2	4.6	5.3	4.5	4.9	5.6	3.0	7.9	6.0	3.6
20-Mar-97	2.6	2.6	2.6	2.6	2.6	2.6	2.7	2.7	2.6	2.6
3-Aug-97	5.1	5.1	5.1	3.9	3.9	3.9	3.9	4.0	3.9	3.9

these techniques invariably remove organised small-scale features such as narrow bands of high soil moisture. Therefore, a more sophisticated approach was adopted here. The smoothing was done in three steps. Firstly, a regression relationship was fitted between the soil moisture and a linear combination of the wetness index and potential radiation index (see Western et al., 1999a for a discussion of these terrain indices). The residuals from this regression exhibited minimal spatial organisation. Secondly, residuals from this regression were smoothed using a thin plate spline (Hutchinson and Gessler, 1994) so that their spatial variance was reduced by an amount equal to the nugget of the soil moisture variogram for that pattern (Western et al., 1998a). Note that the nugget variance is assumed to be the sum of the measurement error variance and the variance of the small-scale ($<10\,\text{m}$) variability (see Chapter 2, p. 32). Thirdly, the smoothed residuals were added back to the estimated soil moisture obtained from the regression. By doing this we were able to smooth the observed data without unduly distorting the moisture pattern observed in the drainage lines. This was possible because the residuals from the regression did not exhibit strong discontinuities around the drainage lines, unlike the soil moisture itself.

Comparing Figures 9.10 and 9.9 indicates that the model generally captures the seasonal trends in soil moisture. It also generally captures the existence of topographic control. That is, when the observed pattern shows strong topographic control, so does the model, and when the observed pattern is random (i.e. uniform with some random variation), the model predicts a uniform pattern. Figure 9.11 shows maps of the errors (simulated minus smoothed observed moisture) in the simulation. The error maps are a useful way of examining the spatial characteristics of the simulation errors. For example, they can show whether the errors are consistently related to terrain features, as is the case on

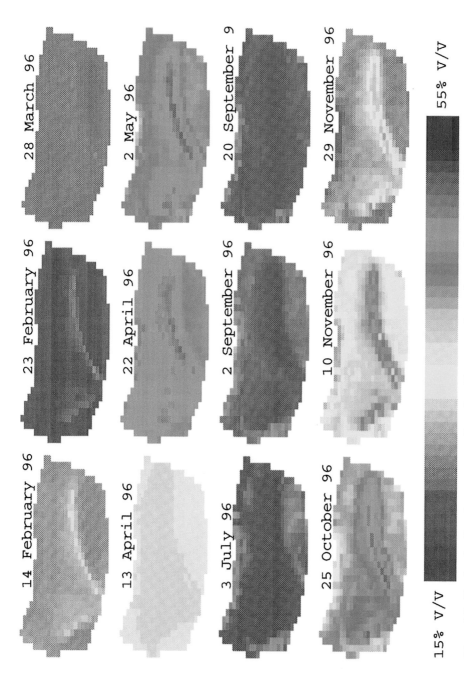

Figure 9.9. Simulated soil moisture patterns (Run 5) at Tarrawarra for twelve occasions during 1996.

231

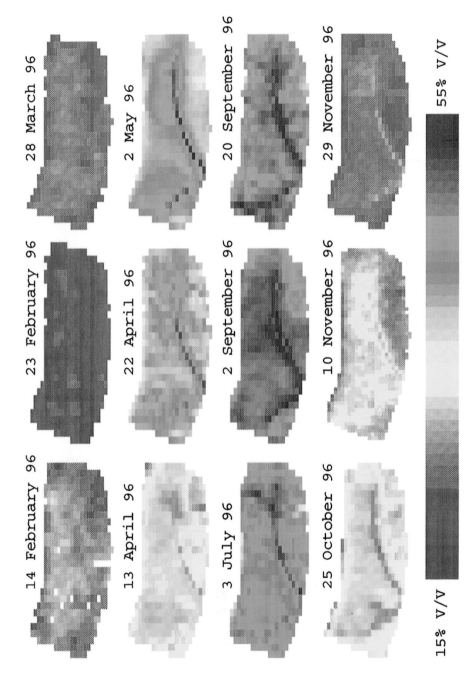

Figure 9.10. Smoothed patterns of observed soil moisture at Tarrawarra obtained from the patterns in Figure 9.6 by removing the small-scale variability and measurement error (see text).

28 March 96

2 May 96

20 September 96

29 November 96

23 February 96

22 April 96

2 September 96

10 November 96

14 February 96

13 April 96

3 July 96

25 October 96

15% V/V

55% V/V

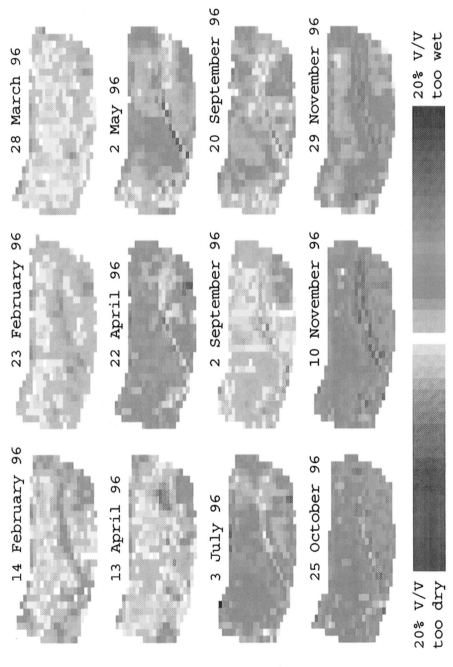

Figure 9.11. The difference between simulated (Run 5) and observed (smoothed) soil moisture patterns at Tarrawarra for twelve occasions during 1996, i.e. Figure 9.9 minus 9.10.

233

November 10, when there was a strong relationship between aspect and simulation error.

A detailed examination of the simulation results indicates that some of the observed patterns are simulated very well, while for others there are some inconsistencies between the model and the observations. The three summer patterns (February 14, February 23, March 28) are well simulated in terms of the average moisture and the spatial pattern. The only inconsistencies are that the model predicted some slight lateral redistribution during the rainfall events immediately before February 14 and that there is a slight aspect bias in the simulation errors for these patterns.

During April a large amount of rain fell and the catchment became significantly wetter (Table 9.1). The model predicted the average soil moisture well on April 13, but overestimated it on April 22. The observed patterns illustrate the effect of lateral redistribution on the development of the wet bands in the drainage lines. On April 13 the highly convergent areas at the upper end of the drainage lines were becoming wetter and on April 22, fully connected wet bands were present in the drainage lines (Figures 9.6 and 9.10). Neither of the simulated patterns showed evidence of significant lateral redistribution (Figure 9.9) and the error maps (Figure 9.11) show connected bands in the drainage lines where the model has underpredicted the soil moisture. The effect of lateral redistribution was also underestimated on May 2. This points to a problem with the model formulation, which is discussed later. There is no obvious aspect bias in the simulation errors for the April and May patterns.

During winter the catchment soil moisture is very high and extensive areas are saturated. The model predicts the average soil moisture correctly on September 2 and slightly overpredicts it on July 3 and September 20. Generally, the soil moisture pattern is well predicted during this period. It should be noted that the high (55 %V/V) measurements observed in the drainage lines on September 2 and 20 are likely to be slightly too high due to surface water ponding introducing some measurement errors (by filling any gaps introduced by the TDR probes).

At the end of September, the amount of rainfall reduced significantly and the catchment began to dry. The patterns indicate that the model performs relatively poorly during this period. The average soil moisture levels are overpredicted, the effect of lateral redistribution is overpredicted, and there is a strong aspect bias in the soil moisture errors (Figures 9.6, 9.9, 9.10 and 9.11).

In summary, when compared to the observed spatial patterns of soil moisture, the model performs well during dry and wet periods. During the transition periods in autumn (dry to wet) and spring (wet to dry) there are some differences between the simulated and observed patterns. Lateral redistribution is too slow during the autumn and it persists for too long during the spring. We will return to the implications of this for the model parameterisation later.

9.9.2 Soil Moisture Patterns – Other Runs

Here we compare the effects of different model parameter sets on the simulated soil moisture patterns. These comparisons fall into four groups: the effect of spatially variable PET; the effect of the k_{sat} parameter; the effect of adding deep seepage; and the effect of using different soil profile representations. Figure 9.12 shows observed and simulated soil moisture patterns for April 13, October 25 and November 10 for several model runs.

The first row of Figure 9.12 shows smoothed observed patterns and the second row shows a simulated pattern from run 5, which is the base case. The third row shows patterns from run 2. These patterns differ from run 5 in that they were simulated with spatially uniform potential evapotranspiration ($\gamma = 0$, run 2). On these dates, a strong aspect effect was apparent in the observed data (Figures 9.6 and 9.10). Comparing these patterns to the corresponding patterns from run 5, we see that the differences between the south and north facing slopes are smaller in run 2 than in run 5 on all three occasions. Introducing spatially variable PET leads to the south facing slopes being up to 2 %V/V wetter on April 13 and up to 1 %V/V wetter on October 25 and November 10. The north facing slopes are up to 1 %V/V drier on April 13 and up to 0.7 %V/V drier on October 25 and November 10. For run 5, the differences between the steepest south facing and steepest north facing slopes is 3 %V/V on April 13 and 2 %V/V on October 25 and November 10. This compares well with the observations on April 13. On October 25 and November 10, the observed difference between the steepest south facing and steepest north facing slopes is approximately 7 %V/V, significantly greater than for the simulations on these dates. This will be discussed further below.

In the fourth row of Figure 9.12, simulated patterns are shown for run 6. Comparing these patterns to those from run 5 illustrates the effect of increasing k_{sat} from 40 mm/h to 60 mm/h. In early autumn the increase in k_{sat} makes little difference to the simulated soil moisture pattern (less than 1 %V/V on April 22), while on May 2 the higher k_{sat} leads to the gully being approximately 4 %V/V wetter, compared with run 5. The increasing effect of k_{sat} during autumn is due both to increased moisture content and increased drainage time. In winter (July 3, September 2 and 20 – not shown on Figure 9.12) the increased k_{sat} leads to a decrease of approximately 2 %V/V in the soil moisture on the upper parts of the hillslopes. In spring (October 25 and November 10) the difference is greater, with the upper parts of the hillslope being 3–4 %V/V drier than run 5 and the drainage lines being 3–8 %V/V wetter. The soil moisture pattern is most sensitive to k_{sat} during the spring when the hillslopes have been draining for a significant period. In winter the hillslopes are regularly re-saturated by rainfall and the intervening drainage times are relatively short.

Three alternative methods (to runs 1–6) for representing the soil in the model were explored. Results from simulations 7, 8 and 9 are shown in rows 5, 6 and 7 of Figure 9.12. These included spatially variable soil depths (run 7), a thinner transmissive layer for the gradational soil (run 8), and a single layer soil (run

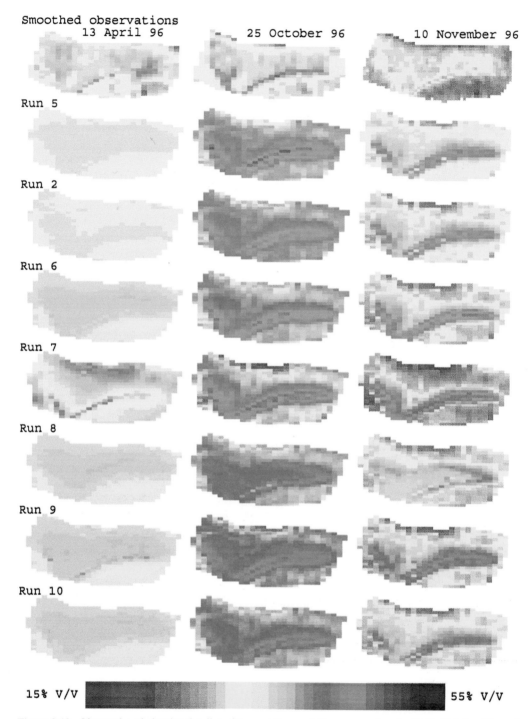

Figure 9.12. Observed and simulated soil moisture patterns at Tarrawarra for April 13, October 25, and November 10, 1996. First row: smoothed observations. Second row: run 5. Third row: run 2 ($\gamma = 0$ i.e. spatially uniform PET). Fourth row: run 6 ($k_{\mathrm{sat}} = 60\,\mathrm{mm/h}$). Fifth row: run 7 (variable soil depth related to wetness index). Sixth row: run 8 (variable soil depth related to soil type). Seventh row: run 9 (single soil layer). Eighth row: run 10 (as for run 5 but k_{deep} adjusted to match catchment annual runoff).

9 – entire soil profile is transmissive). Runs 7 and 8 include spatially variable soils and the spatial variability is consistent with that suggested by field observations. Run 9 includes a spatially uniform single layer soil. Introducing spatially variable soil depths (run 7) has a marked effect on the simulated soil moisture patterns due to the local changes in storage capacity. The topographic effects were enhanced on many occasions due to the relationship between topographic location (in terms of wetness index space) and storage capacity. Generally this relationship caused the hilltops to respond more quickly and the drainage lines to respond more slowly, compared with run 5. The simulations with spatially variable soil depth generally led to similar or poorer simulations of the patterns. Relative to the average soil moisture, in autumn the hilltops were too wet, in winter the hilltops were too dry, and the wet gully persisted for too long in the spring.

The changes to the soil representation in runs 8 and 9 both affect the profile average moisture content above which lateral redistribution occurs. Introducing the thinner transmissive layer for the gradational (lower slopes and drainage lines) soil reduces the amount of lateral redistribution in the gradational soil by a factor of two for fully saturated conditions and by more for unsaturated conditions. Due to the discontinuity in lateral flow introduced at the boundary between the duplex and gradational soils, the soil moisture patterns from run 8 tended to have a band of high soil moisture at the soil unit boundary. This band is unrealistic. However, compared to run 5, some other features of the patterns were simulated more realistically, such as the more rapid drying of the drainage lines in spring.

Treating the entire soil profile as being laterally transmissive (run 9) led to an increase in lateral flow for all moisture contents greater than field capacity, except for saturated conditions when the lateral flow was the same as for run 5. This increase led to lateral flow influencing the pattern more under drier conditions. The topographically controlled pattern developed slightly more quickly during the autumn but persisted for too long during the spring. There was also slightly more lateral redistribution predicted during summer, which is inconsistent with the observations.

Deep seepage was introduced in run 10 by calibrating k_{deep} to obtain the correct annual runoff (this led to a decrease in simulated surface runoff from 0.51 mm/d to 0.39 mm/d for 1996, compared with run 5). As would be expected, introducing deep seepage reduced the simulated average soil moisture. This reduction was greater in the drainage lines during autumn and spring and on the ridge tops during winter (when the drainage lines remained saturated). The reductions in soil moisture rarely exceeded 1 %V/V. The exception was in the drainage lines during spring where reductions were up to 4 %V/V (on November 29), compared with run 5, and the degree of topographic control was slightly less for the simulated pattern on November 29. These results represent some improvement over run 5, but some calibration against runoff was required to achieve them.

In summary, comparing the different simulation runs leads to the following conclusions. To get the best representation of the soil moisture pattern at Tarrawarra with Thales it is necessary to include spatially variable potential evapotranspiration. The saturated hydraulic conductivity is only important during wet periods following a long (compared to the subsurface flow response time) period of drainage. Introducing spatially variable soil depths provided little overall improvement and caused problems near soil type boundary interfaces where abrupt changes in soil properties caused anomalies. Deep seepage was needed to close the water balance and the deep seepage parameter (k_{deep}) required calibration against runoff.

9.9.3 Saturation Deficits

Figure 9.13a shows time series of the simulated and observed saturation deficit averaged over all twenty neutron access tubes. Note that simulated saturation deficits are plotted only on days when observations are available to aid comparison of the time series. The timing of the fluctuations in average saturation deficit are modelled correctly. The magnitude of the fluctuations is also reasonably well simulated; however, during the summer of 1996 (Jan–Mar), the saturation deficit is systematically overestimated by approximately 50 mm. During the transition periods when the catchment is wetting up (autumn) or drying down (spring), the rate of change of simulated saturation deficit tends to be too quick, particularly in spring. While this is consistent with the comparisons between the TDR data and the simulations during autumn (particularly April 13–22), the simulated soil moisture reduces more slowly than the soil moisture in the upper 30 cm (TDR data) during the spring (September 20 – November 29).

Simulated and observed time series of saturation deficit are shown in Figure 9.13b for neutron access tubes 7 and 8 (see Figure 9.2). From the perspective of the timing of saturation deficit fluctuations, tube 7 is typical of tubes in locally convergent areas (i.e. tubes 2, 7, 11, and 18) and tube 8 is typical of all the other tubes. We exclude tube 17 from the convergent group because it is not very convergent from a local perspective and the saturation deficit observations suggest that it fits better in the other group. The observations show that there is little difference between the timing of the wetting between sites. The simulations are consistent with this. During the spring, the drydown at tube 7 lags tube 8 slightly, probably as a result of lateral redistribution. There is a much greater lag between the drying of these two sites in the simulations.

9.9.4 Surface Runoff

The annual surface runoff during 1996 is overestimated by 31 % in run 5. This is equivalent to a 32 % underestimation of evapotranspiration during the main runoff-producing period (June 23 to October 9). This error can be easily cor-

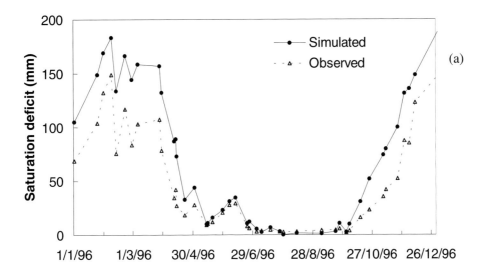

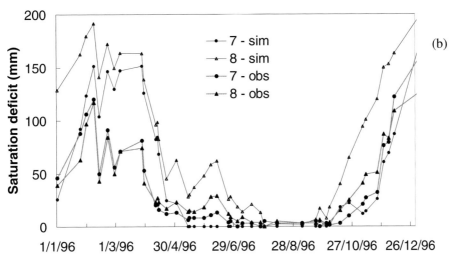

Figure 9.13. (a) Time series of simulated and observed saturation deficits at Tarrawarra during 1996 for run 5. The saturation deficit was calculated as the mean saturation deficit at the twenty neutron access tubes (Figure 9.2). (b) Time series of simulated and observed saturation deficits at neutron access tubes 7 and 8 during 1996 for run 5. Note that simulated values are only shown for dates on which observations were made.

rected by adding a small amount of deep seepage to the model. Generally the model simulates the occurrence of runoff events well. That is, events are simulated on days when significant runoff occurred but not on days when no runoff events were recorded. The notable exceptions to this are the first significant recorded runoff events (Figure 9.14 – "A"), when no runoff is simulated. This is because simulated saturation deficits are too large in April (Figure 9.13a). It should be noted that there is some uncertainty as to the exact magnitude of these events as some flow bypassed the flume and the hydrographs had to be estimated. Also, during a period beginning with the large event on July 30 (Figure

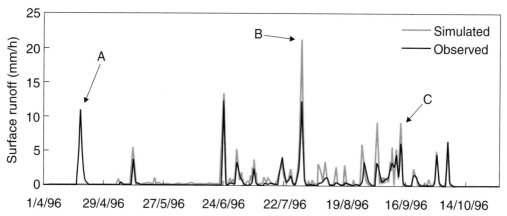

Figure 9.14. Simulated and observed runoff hydrographs at Tarrawarra during 1996 for run 5. "A", "B" and "C" identify events referred to in the text.

9.14 – "B") and ending on September 12 (Figure 9.14 – "C"), the model system-atically overestimates the magnitude of the runoff events.

It should be noted that it is not possible to correct the error in mean runoff by calibrating the evapotranspiration parameters, as these control the spatial dis-tribution of evapotranspiration rather than its absolute value during high soil moisture periods when the runoff occurs. Also it is unlikely that the potential evapotranspiration is underestimated during this period, since this was a wetter than average winter and the estimates used are already slightly greater than the available average monthly regional estimates of potential evapotranspiration (Wang et al., 1998). Obviously, it is possible to simulate the annual runoff cor-rectly by adjusting the total storage in the model, but this results in a poorer simulation when the timing of runoff events is considered.

The quality of the runoff predictions can be assessed using the prediction efficiency (Nash and Sutcliffe, 1970); however, this is very sensitive to any bias in the runoff predictions. For run 5 the predictive efficiency is only 41%. For run 10, when the deep seepage was calibrated such that the model correctly simulated the annual runoff, the predictive efficiency is 63%. The events in April that the model failed to simulate (Figure 9.14 – "A") account for half of the remaining errors. If the runoff events on 17–19 April are ignored, and the deep seepage is calibrated such that the model correctly predicts the total runoff in the remaining events, the predictive efficiency can be increased to 82%.

9.10 DISCUSSION AND CONCLUSIONS

In this chapter, the Thales model framework has been applied to simulate the hydrologic behaviour of the Tarrawarra catchment for a period of one year. Detailed spatial soil moisture patterns, saturation deficit, soils, meteorological and runoff data are available for this catchment. The simulations have been compared to spatial patterns of soil moisture in the top 30 cm of the soil profile, to saturation deficit measurements at individual neutron moisture meter tubes and at the catchment scale, to catchment scale runoff. The simulations were conducted using a daily time step. The model parameters were set by a careful analysis of the available field data, with some subsequent minor adjustment of the saturated hydraulic conductivity parameters.

Given the limited degree of calibration, the model performed very well. The seasonal changes in the soil moisture patterns were accurately simulated and the seasonal changes in saturation deficit were also reasonably well simulated at the catchment scale, although there was some tendency for the model to overestimate the saturation deficit during summer. Of the three sets of data that the simulations were compared to, the poorest performance was for the surface runoff. This relates, at least in part, to the difficult challenge of simulating threshold processes such as saturation excess runoff. The predicted runoff can be greatly improved by calibrating the deep seepage component of the model so that the annual runoff is correctly simulated.

The difference in the model's ability to predict the soil moisture patterns and the runoff has important implications. If one were to calibrate the model against the soil moisture patterns, and then use the model to predict runoff (i.e. take run 5 as being good), then one would overestimate surface runoff by 31% and explain only 41% of the observed variability in the runoff. Such results would be considered poor in most modelling studies and would be "calibrated out" by adjusting parameters. However, if a model is calibrated on catchment runoff only, one can expect substantial errors in the representation of the internal hydrological processes, which may cause predictions of erosion and transport to be grossly in error. Thus, it is important to test models against appropriate data, that is, data that directly tests the prediction in which you are interested. Obviously, the type of data that is appropriate depends on the purpose to which the model is to be put. Conceptually similar problems have been documented by several other modellers (e.g. Grayson et al., 1992b; Chapters 1, 3, 13).

While the results reported here were obtained with little calibration effort, it is important to recognise that the highly detailed data set and the preliminary analysis of the data were critical. For example, the storage values used in the model were calculated from comprehensive neutron moisture meter data. Due to the extreme conditions sampled by these measurements (we were lucky to have had a very wet winter followed by one of the driest 18 month periods in the last one hundred years), it was possible to calculate the active storage from observations. Similarly, we had soil moisture patterns at key points in time that allowed us to estimate the field capacity and wilting point, and bulk density data that allowed us to estimate the porosity. Due to the relatively simple, yet physical, basis of the model structure and our detailed understanding of the behaviour of the catchment (gained from the data), it was possible to obtain good estimates of the parameter values a priori, rather than via detailed calibration, as is often required. How well this approach would work in other catchments is an open question. The key point is that, without such data, confidence that the model is performing for the right reasons will be lacking.

One aspect of this approach that is critical is the design of the field experiments. It is important that the measurements made can be related to the processes controlling the hydrologic response. For example, we would not have been able to set the soil depth in the model without measurements of profile storage. It is also important to have measurements that can be related to the limits between which soil moisture can vary. Therefore, some knowledge of the (type of) key controlling processes in the landscape of interest is highly valuable when designing a field experiment. Attention should also be given to the degree of climatic variability expected during a field experiment. It is important that the experiments do not rely on the occurrence of a rare event, relative to the life of the experiment.

While the modelling results reported here are satisfying, there are two problems with the model that have been identified by comparison with the measured patterns of soil moisture. In autumn, the topographically controlled soil moisture pattern is too slow to develop in the model and in spring the simulated soil moisture and saturation deficits do not compare particularly well with the observations. There are likely to be separate causes for these two problems.

In autumn the key problem is insufficient lateral redistribution around April 13 and 22. One potential explanation of this behaviour relates to the fact that the soils at Tarrawarra crack during dry periods. It is possible to simulate the observed patterns on April 13 and 22 quite well using $k_{sat} = 400\,mm/h$ and a total soil depth of 500 mm. The key variable here is the high k_{sat} value (the shallower soil depth was used to correct the overestimated saturation deficit in run 5 during this period). Given that the soils at Tarrawarra have a relatively heavy (clayey) texture, such a high effective value of k_{sat} is indicative of some form of preferential flow. Using such a high k_{sat} value during late autumn (May 2) and winter, results in a gross overestimation of the amount of lateral flow and high base flow contributions. These results suggest that preferential flow paths may be much more effective in autumn (as the catchment switches from dry to wet) than in winter, when the catchment is very wet. The reduction in preferential flow appears to be quite rapid, with little evidence for high effective k_{sat} values on May 2. Soil cracks would appear to be a likely candidate for explaining this behaviour, given its temporal characteristics. Cracks persist during long dry periods but close quickly at the end of such periods. The under-prediction of lateral flow during the autumn period is the explanation for the failure of the model to predict the onset of surface saturation in the drainage lines and the first runoff events of the autumn (events "A" on Figure 9.14). This leads, in turn, to the overestimation of soil moisture on April 22.

In spring, some systematic problems with simulating the drying of the catchment exist. These are somewhat complicated in that comparison of the simulations and the TDR soil moisture patterns indicates that the model dries too slowly, while comparison of the simulated and observed saturation deficits indicates that the model dries too quickly. In terms of the overall soil water balance, the saturation deficit data is the more relevant data. However, the TDR data provides more spatial detail. It is important to recall that the TDR data represents the top 30 cm of the soil profile while the saturation deficit data represents the complete profile.

When the soil profile dries there is a tendency for the upper soil layers to dry more quickly due mainly to greater root activity in this zone. Thus the zone measured by the TDR is drying more quickly than the zone measured by the NMM and the zone simulated by Thales. This means that the poor comparison with the TDR data is likely to be explained by the fact that the model does not properly simulate the vertical dynamics of the soil moisture. This is also likely to be a major reason why the model predicts a much weaker aspect effect than is observed during the spring drying period. If all the water were lost from the upper 30 cm of the soil profile, the simulated aspect effect would effectively be doubled (i.e. a 4 %V/V difference between north and south facing slopes) and would compare much more favourably to the observations ($\sim$ 7 %V/V difference). To enable completely consistent comparisons between the model and the TDR data, it would be necessary to model the vertical soil moisture dynamics. This would involve using at least two soil moisture stores, with the associated cost of additional complexity and parameterisation issues. However, the pattern data indicate that this will be necessary to further improve the model performance.

The rapid drying of the model compared to observed saturation deficits during spring occurs on the hillslopes rather than in the drainage lines (Figure 9.13). No surface runoff was observed or predicted during this period and the problem remained when the deep seepage was set to zero. Therefore, we can deduce that the evapotranspiration was overestimated during the spring period. Two potential explanations exist. These explanations may also be relevant to the overestimation of the saturation deficit during the summer period (January–March). Firstly, the evapotranspiration could be overestimated as a consequence of the soil moisture in the root zone (upper 30 cm) being overestimated, while in reality low root zone soil moisture was exerting a control on the evapotranspiration rate. The problem of overestimating the drying rate was evident on October 25 when the observed catchment average (upper 30 cm) soil moisture was 35 %V/V, and the soil moisture exceeded 30 %V/V at most sites. While this may provide part of the explanation, these soil moisture levels are not likely to exert a sufficiently strong control on the evapotranspiration rate to fully explain the error. Secondly, the evapotranspiration could be overestimated due to invalid assumptions about the vegetation. For the purpose of estimating potential evapotranspiration, it was assumed that the pasture could always be represented by the "reference crop" (Smith et al., 1992). The reference crop is an extensive surface of green grass of uniform height that is actively growing, completely shades the ground and is not short of water. At the end of winter in 1996, the pasture at Tarrawarra was very short, and had been badly damaged by cattle "pugging" the soil, which had been structurally weak due to water logging. The relatively poor health of the pasture at that time may have led to a reduction in the amount of evapotranspiration compared with that expected for the reference crop.

While there are some problems with the model's ability to simulate transitions between wet and dry seasons, the comparison of measured and simulated patterns does provide substantial insight into the importance of different sources of spatial variability at Tarrawarra. The only source of spatial variability in runs 1–3 is the effect of terrain on the routing of lateral flow. Runs 4–6 also include the effect of terrain on incident radiation and, hence, potential evapotranspiration. Of these two terrain-related sources of spatial variation, the lateral flow contributes most to the spatial variation of the simulated soil moisture. On most occasions, introducing radiation effects leads to a small improvement in the simulated pattern. Introducing spatially variable soils (runs 7 and 8) does significantly change the simulated soil moisture patterns, which implies that correctly specifying the spatial variability of soil parameters is likely to improve the model simulations. However, we have not been able to make consistent improvements in the model performance at Tarrawarra using the currently available soils data. Reducing the transmissive layer depth for the gradational soil did lead to better simulations of the soil moisture in the drainage lines during summer and spring, but at the cost of poorer simulations on the hillslopes.

Two problems currently hinder our progress in incorporating better soils information in the model and are indicative of problems with soils data generally. Firstly, there does not appear to be a correspondence between features of the soil moisture patterns and the spatial distribution of soil types that have been mapped in the catchment. It may be that there is limited correlation between the soil type and relevant soil hydrologic parameters. In this case, treating the soils as distinct types is of limited value (see also Chapters 6 and 10). Nevertheless a close examination of the residual plots does indicate some consistency in the pattern of simulation errors (Figure 9.11) that may be related to soil variation. For example, the simulated soil moisture on the hilltop in the south-eastern corner of the catchment (Figure 9.2, near neutron tube 9) is consistently too wet (Figure 9.11). A second problem is obtaining useful spatial soil measurements. It is difficult to convert observations of soil profile characteristics (i.e. soil type, horizon depths) into values of hydrologic parameters such as field capacity, wilting point and storage. Also, the combination of small-scale variability and the time-consuming nature of soil property measurement often compromise quantitative measurement approaches. Spatial characterisation of soil properties for modelling remains a difficult challenge.

One conclusion that we can draw about the representation of soils in the model of the Tarrawarra catchment, is that characterising the soil as having a transmissive layer underlain by a layer that simply acts as a store is an improvement over using a single transmissive layer. While we have not explored other options for describing the changes in hydraulic conductivity with depth, we feel confident that this is a sound representation, at least for the duplex soil.

This modelling exercise has made extensive use of soil moisture pattern data. One of the issues faced when using data of this sort is the problem of comparing observed and simulated patterns. Here we did this qualitatively using two complementary approaches. The first was to compare simulated and observed patterns of soil moisture visually and the second was to calculate an error map and examine that map visually. The comparison of simulated and observed soil moisture patterns has given us greater confidence in the predictive ability of the model. It was also valuable for identifying problems with the model structure. For example, the poor simulation of the autumn patterns suggests that preferential flow paths due to soil cracks might be important. While this led to problems with the simulation of the runoff hydrograph during this period, the hydrograph by itself would not have allowed us to understand the problem. In all likelihood, the failure to simulate runoff during the initial events in autumn would have been corrected by reducing the model storage, if the model had simply been calibrated against runoff. Similarly, the soil moisture patterns and the saturation deficit data allowed us to identify specific problems in simulating the soil water balance during the spring period. The soil moisture patterns have proved to be much more informative as to the integrity of process representations in the model than has the catchment runoff. The only exception to this is that runoff proved useful to test the bulk water balance. This could not be tested with soil moisture alone.

The detailed spatial data have been extremely valuable for other analyses. The seasonal evolution of geostatistical characteristics of soil moisture patterns in this landscape were quantified and the scaling properties of the soil moisture patterns were characterised (Western and Blöschl, 1999; Western et al., 1998a,b). It was found that geostatistical regularisation techniques were able to predict the effects of changes in scale (in terms of spacing, extent and support) on the variance and correlation length (Western and Blöschl, 1999). Results of these analyses also have implications for sampling strategies in terms of the number of measurements required (Western et al., 1998a) and the measurement scale limitations (Western and Blöschl, 1999) for obtaining geostatistically representative samples of soil moisture. We have also made progress in the identification of representative locations for soil moisture measurement by demonstrating the existence of sites that always have soil moisture close to the catchment mean moisture (Grayson and Western, 1998). The soil moisture patterns have also been valuable for characterising and predicting the spatial organisation of soil moisture. Soil moisture patterns at Tarrawarra typically exhibit both random and topographically organised characteristics (Grayson et al., 1997; Western et al., 1999a). The degree of spatial organisation changes seasonally and the organised component of the variation can be predicted using terrain indices (Western et al., 1999a). The existence of spatial organisation is related to the processes controlling the spatial pattern. Spatial organisation is strongest when there is lateral flow occurring or when the soil moisture is influenced significantly by up-slope processes (non-local control). Little organisation is present when the soil moisture is locally controlled and the main fluxes of water are vertical (Grayson et al., 1997). Detailed event simulations indicate that spatial organisation has a significant effect on the rainfall-runoff behaviour at Tarrawarra (Western et al., 2000). Spatial organisation can also be analysed within a geostatistical framework. Spatial connectivity is a spatial organisation feature that is not captured by standard geostatistical techniques (variography). We were able to show that indicator geostatistics (indicator variograms) are also unsuitable for characterising connectivity, despite suggestions to the contrary in the literature (Western et al., 1998b). However, connectivity statistics (Allard, 1994, 1993) provide an appropriate statistical tool for characterising spatial connectivity (Western et al., 2000).

The use of patterns in model testing is valuable but has some limitations. Many of the simulated patterns were quite similar and it was difficult to assess visually which was the better simulation, especially when patterns for all twelve occasions were considered. There is a need for quantitative pattern comparison techniques that account for a range of different scales including points, hillslopes and catchments. The statistics used for these comparisons need to be chosen carefully so that hydrologically important aspects of the patterns are compared. This might require the comparisons to focus on specific components of the landscape, for example, drainage lines rather than ridge tops, or on correctly simulating pattern features such as connectivity. It is also important that any quantitative approach be able to deal with several different types of data. Here the combined use of soil moisture pattern, saturation deficit time series and runoff time series data was extremely valuable.

ACKNOWLEDGEMENTS

The Tarrawarra catchment is owned by the Cistercian Monks (Tarrawarra) who have provided free access to their land and willing cooperation throughout the project. Funding for the above work was provided by the Australian Research Council (project A39531077 and S3913076), the Cooperative Research Centre for Catchment Hydrology, the Oesterreichische Nationalbank, Vienna (project 5309), and the Australian Department of Industry, Science and Tourism (International Science and Technology Program).

10

Storm Runoff Generation at La Cuenca

Robert Vertessy, Helmut Elsenbeer, Yves Bessard and
Andreas Lack

10.1   INTRODUCTION

Dynamic, spatially-explicit models of storm runoff generation are needed to
underpin the prediction of particulate and solute movement through catchments.
When such runoff models are applied to these problems, the spatial pattern,
frequency and magnitude of overland flow must be simulated faithfully.
Remarkably, there are few studies reported in the literature that compare
model predictions and field observations of overland flow patterns at the catchment scale. This seems to be partly due to the fact that overland flow within
catchments is notoriously variable in time and space, and is thus problematic to
measure. However, it is also true that model predictions of overland flow patterns
are rarely flattering when compared to reality, and this partly explains why few
studies have reported such results. In this chapter we illustrate that part of the
key to getting good spatial predictions of overland flow is properly representing
spatial variability in soil hydraulic properties.

There is considerable debate in the literature regarding the need and manner
in which to represent spatial variability of soil properties in runoff models. Smith
and Hebbert (1979) compared storm runoff generation on a simple plane in
which saturated hydraulic conductivity (K_s) values were either held uniform in
space, systematically distributed or randomly distributed. The various parameter
sets they compared produced very different results, though they noted some
situations where a uniform K_s value produced similar runoff results as a randomly allocated log-normal distribution of K_s values. Binley et al. (1989) also
found that runoff predictions were affected by soil property representation in
their model, though they emphasised that the differences were only significant for
low-permeability soils, dominated by surface runoff processes. Grayson et al.
(1992a) detected only minor differences in discharge hydrographs on a catchment
dominated by saturated source area runoff when uniform and randomly distrib-

Rodger Grayson and Günter Blöschl, eds. *Spatial Patterns in Catchment Hydrology: Observations and Modelling* © 2000 Cambridge University Press. All rights reserved. Printed in the United Kingdom.

uted K_s values were compared. Merz and Plate (1997) and Grayson et al. (1995) illustrated that the spatial pattern of soil properties had different effects on predicted hydrographs depending on the intensity and amount of rainfall relative to the soil hydraulic properties and antecedent condition (see also Figure 1.6). Smith and Hebbert (1979) and Grayson et al. (1992a) argued that spatial variation in K_s should be, at least in part, deterministic. They recommended against allocating randomly variable K_s values in space without being sure there was no deterministic component to the K_s patterns.

What seems to be lacking in the literature is the coupling of deterministic and stochastic variation in soil properties and the joint representation of these in distributed hydrologic models. Many catchment surveys reveal considerable spatial variability in soil hydraulic properties (see for instance Loague and Kyriakidis, 1997), though the median and standard deviation of property values may still be distinguishable, statistically speaking, between different parts of these catchments. This is well illustrated in the field data discussed by Elsenbeer et al. (1992) for a tropical rainforest catchment in western Amazonia, where four different soil types are distinguished and each has its own distinctive internal variability. The superimposition of stochastic variation on top of deterministic patterns of soil hydraulic properties, or any other system property for that matter, appears to be an approach rarely employed in distributed hydrologic modelling. One of the few exceptions is the KINEROS model described in Chapter 6, where deterministic patterns of soil properties are imposed across the modelling domain, but grid-scale hydraulic conductivity varies stochastically within an element around its geometric mean.

Another key determinant in the success of any storm runoff modelling exercise can be the manner in which initial moisture conditions are set in the model. Stephenson and Freeze (1974) argued that the initial moisture state of the catchment is the factor most likely to determine the outcome of an event prediction, while Merz and Bárdossy (1998) have argued that initial conditions are less critical, particularly in the case of large events. These and other studies have demonstrated that the importance of initial conditions depends on the dominant runoff mechanisms. For saturated source area runoff, correct specification of the initial saturation deficit is critical to accurate storm modelling. In the case of infiltration excess runoff, the importance of initial conditions depends on the storm intensity relative to the infiltration characteristics of the soil. If the storm is very much larger or smaller than the infiltration rates of the soil, initial conditions are not critical. When these are of similar magnitude, predicted runoff becomes highly sensitive to initial conditions (see also discussion in Chapter 6, p. 155). A survey of the literature reveals significant variation in the manner in which modellers deal with initial moisture conditions in event-based simulations. For instance, Coles et al. (1997) treated initial moisture conditions as a "black box" parameter, freely adjusting it between events as a calibration parameter. Grayson et al. (1992a) used the $\ln(a/\tan \beta)$ index to spatially modulate a single known or estimated moisture value. Merz and Plate (1997) used actual soil moisture observations, interpolated in association with the $\ln(a/\tan \beta)$ index, to initiate their simulations.

In this chapter, we review our attempts to measure and simulate storm runoff generation patterns across a small tropical rainforest catchment in western Amazonia, La Cuenca. After many years of effort we have gained a thorough understanding of storm runoff generation mechanisms within this catchment, and have recognised the importance of spatial variability in soil hydraulic properties (Elsenbeer et al., 1992, 1995; Elsenbeer and Lack, 1996; Vertessy and Elsenbeer, 1999; Elsenbeer and Vertessy, 2000). We have used our field knowledge to develop and evaluate a simple, dynamic and spatially explicit storm runoff model called Topog_SBM (Vertessy and Elsenbeer, 1999). In our quest to predict the spatial distribution of overland flow, we have come to appreciate the importance of spatial variability in soil hydraulic properties and the manner in which this should be represented in distributed hydrologic models. We have also compared different ways of initialising soil moisture conditions for our event simulations and have shown that assuming steady-state drainage is an objective and adequate approach to simulating initial soil moisture patterns.

Below, we describe the La Cuenca catchment, focussing on the measured spatial variability in soil hydraulic conductivity across the catchment. We then discuss observed mechanisms of storm runoff generation within the catchment, yielded from a combination of hydrometric and hydrochemical studies. After describing the Topog_SBM model briefly, we discuss four different ways in which such a model can be parameterised in terms of soil hydraulic property representation. Running the model on La Cuenca with each of these four different parameter sets, we compare model performance in terms of outflow hydrographs and spatiotemporal patterns of overland flow occurrence. It is shown that the manner in which soil hydraulic properties are represented has modest consequences for outflow hydrographs, but a very significant impact on simulated spatial patterns of overland flow. We demonstrate that the best results are obtained when deterministic and stochastic variations in soil properties are represented in the modelling process.

10.2 THE LA CUENCA CATCHMENT

La Cuenca is located in the Rio Pichis Valley in the Selva Central of Peru ($75°5'$W, $10°13'$S) at about 300 m above mean sea level (Figure 10.1). It is a small first-order basin, covering an area of 0.75 ha and spanning a relative relief of 28 m. La Cuenca is characterised by short, steep convexo-linear sideslopes (up to 40°), narrow valley floors, and deeply incised gullies near stream heads (Elsenbeer et al., 1992). The catchment is covered by an undisturbed multi-storied primary rainforest (Figure 10.2). Despite its very small size, the catchment includes at least 57 different plant species, belonging to 25 families (Elsenbeer et al., 1994).

Mean annual temperature for the study site is 25.5 °C and mean annual rainfall is 3300 mm. Monthly rainfall is highest during December–March (up to 900 mm) and lowest during June–September (below 110 mm). Daily rainfall rarely exceeds 100 mm. During our intensive study period, June 1987 to April 1989, the

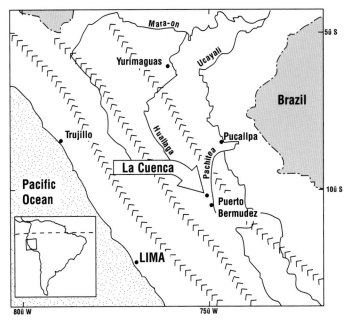

Figure 10.1(a). Location of the La Cuenca catchment. (From Elsenbeer et al., 1992; reproduced with permission.)

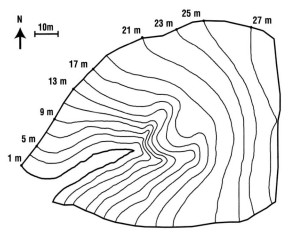

Figure 10.1(b). Topographic contours for the La Cuenca catchment (contour interval is 2 m).

maximum daily rainfall amount was 70.3 mm and the maximum five-minute rainfall intensity was 96.0 mm h^{-1} (Table 10.1).

Elsenbeer et al. (1992) defined three main land units in the catchment, differentiated by topography and soil properties. These were the steep lower sideslope (unit B), intermediate terrace (unit C), and gentle upper sideslope (unit D) (Figure 10.3). Ultisols, observed over extensive areas of western Amazonia (Buol et al., 1989), are the main soil type within the catchment, though Inceptisols are present in land unit B. Soil depth across the catchment averages

Figure 10.2(a). Photograph of the La Cuenca catchment.

Figure 10.2(b). Photograph of the overland flow detectors.

Table 10.1. Descriptive statistics pertaining to 214 individual rainfall events sampled at La Cuenca, June 1, 1987 – April 18, 1989. I_5 denotes maximum 5-minute rainfall intensity.

Variable	Units	Median	Maximum	Minimum
Rainfall amount	mm	8.7	70.3	0.3
Rainfall duration	min	170	960	10
I_5	mm h^{-1}	19.2	96.0	10

about 1.0 m, with only modest spatial variation. The soils tend to be slightly (< 30%) deeper in the valley bottom than on the sideslopes. Elsenbeer et al. (1992) measured saturated hydraulic conductivity (K_s) across the catchment at various depths in the soil, involving 740 undisturbed small cores. On the basis of a detailed statistical analysis, they were able to demonstrate statistically significant differences between the K_s value distributions measured in the various land units, despite the fact that each land unit contained huge variability within it. Hence, they observed random variation imposed on top of a deterministic spatial pattern of soil properties.

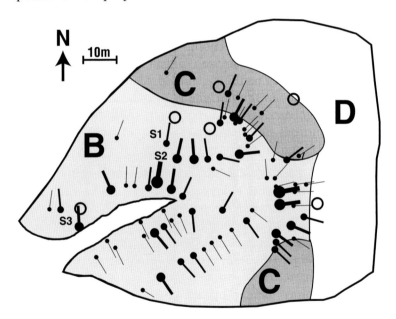

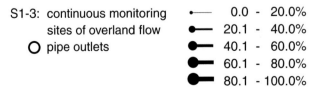

Figure 10.3. Position of land units B, C and D in the La Cuenca catchment, showing the location of major sub-surface pipes and the frequency of overland flow occurrence at 72 detector sites for 187 separate events, after Vertessy and Elsenbeer (1999).

Table 10.2. Median K_s values (m d^{-1}) for various soil layers in land units B, C and D, after Elsenbeer et al. (1992). Note that 1 m/d = 41.7 mm/hr

Unit	0–0.1 m	0.1–0.2 m	0.2–0.3 m	0.3–0.4 m	0.4–1.0 m
B	1.09	9.50×10^{-3}	2.16×10^{-3}	3.46×10^{-3}	4.58×10^{-3}
C	11.09	8.98×10^{-2}	1.99×10^{-2}	2.50×10^{-3}	2.40×10^{-4}
D	8.02	2.59×10^{-2}	1.30×10^{-2}	3.46×10^{-4}	2.40×10^{-4}

Median values of K_s at the surface varied by an order of magnitude between land units, with the highest values present in unit C and the lowest in unit B (Table 10.2). K_s was found to decrease sharply with depth in all three land units, though most sharply in land unit B. This has been attributed to higher clay content at depth in the soils of this unit. In Figure 10.4, we show the cumulative frequency distribution of surface K_s values for all three land units and for the catchment as a whole. Each of the four cumulative frequency distributions shown is approximately log-normally distributed. These distributions are used later in the modelling analysis reported in this chapter.

The spatial and temporal frequency of overland flow occurrence was measured at La Cuenca using 72 overland flow detectors, similar to those described

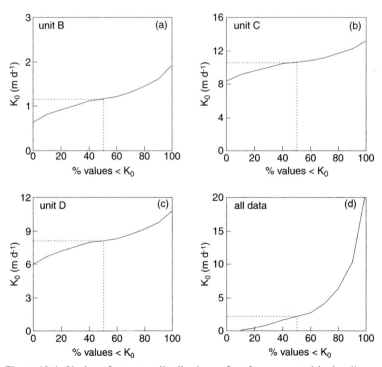

Figure 10.4. Various frequency distributions of surface saturated hydraulic conductivity, K_0, measured in the La Cuenca catchment; these pertain to land units B, C and D, and the catchment as a whole. The dashed lines highlight the median values. (From Vertessy and Elsenbeer, 1999; reproduced with permission.)

by Kirkby et al. (1976). The positions of the 72 detectors are shown in Figure 10.3, which also shows the observed frequency of overland flow occurrence at each of the installation sites. Each detector consisted of a 25 cm long, 5 cm diameter PVC tube, with one end sealed with a lid, and the other attached to a Y-junction (Figure 10.2b). The bottom end of this junction was sealed with a lid, serving as a collecting unit, the top end covered with a can. One third of the PVC tube's circumference was perforated with some 200 1-mm diameter holes. The detectors were installed in such a way that the perforated portion was in good contact with the soil, and that any intercepted overland flow would drain towards the collecting unit. Overland flow was judged to have occurred if the bottom of the collecting unit was completely covered by standing water, not just by a thin film.

Overland flow was also monitored continuously at three sites. The monitoring system at sites S1 and S2 (see Figure 10.3) was designed according to Riley et al. (1981). A triangular layer of topsoil, about 5 cm thick and two metres wide at the upslope base parallel to the contour lines, was carefully removed. The resulting cavity was filled with concrete, and the upslope contact moulded in such a way as to fit the microtopography of the soil surface. Strips of sheet metal were attached to both sides so as to route any intercepted overland flow towards the downslope apex equipped with a pipe, and further on to a series of connected 55 gallon drums. One of these at each site was equipped with a float-operated water level recorder. At site S3 (see Figure 10.3), a concentrated-flow line was intercepted with a simple device fabricated out of sheet metal.

Three sets of observed rainfall events are used in this chapter. The first is a set of 187 events for which data from the overland flow detectors is available and frequency of occurrence of overland flow can be computed. The second is a set of 34 events that were used for runoff simulations. Thirdly, the 10 events from the second set that overlapped with the first set are used to compare both the simulated runoff and the simulated spatial patterns of runoff occurrence with the respective observations.

10.3   STORM RUNOFF PROCESSES OPERATING AT LA CUENCA

Storm runoff in steep, humid, forested landscapes has traditionally been viewed to occur primarily via subsurface pathways (Dunne et al., 1975). Overland flow in such environments is often presumed to occur only as saturation excess in preferred topographic locations, namely valley bottoms and hillslope hollows. The study of Bonell and Gilmour (1978) was one of the first to demonstrate that overland flow could be widespread in a tropical rainforest setting, and display patterns of occurrence not necessarily dictated by topography. Our findings at La Cuenca concur with their observations.

For the La Cuenca catchment, previous studies have concluded that:

1. overland flow is generated frequently, both in the spatial and temporal sense (Elsenbeer and Lack, 1996);

2. changes in the K/SiO_2 ratio in streamflow during storms indicate a significant volumetric contribution of overland flow to storm runoff (Elsenbeer and Lack, 1996);

3. overland flow is generated by infiltration excess (Hortonian), saturation excess, and return flow mechanisms (Elsenbeer and Vertessy, 2000); though the relative proportions of these are not known it has been assumed that Hortonian runoff is infrequent as surficial hydraulic conductivities almost always exceed the maximum five-minute rainfall intensities;

4. topography exerts only a mild control on overland flow generation in this catchment (Elsenbeer and Vertessy, 2000);

5. there is significant storm runoff generated through a shallow subsurface pipe network, which emerges at the surface as return flow (Elsenbeer and Vertessy, 2000).

The widespread occurrence of both overland and subsurface flow at La Cuenca, and the manner in which they are generated, can be explained by the interaction of catchment soil hydraulic properties and local rainfall characteristics. The K_s of the soil decreases so abruptly with depth (Elsenbeer et al., 1992) that even low-intensity rainfall is likely to generate shallow subsurface flow, if rain persists for long enough (Figure 10.5). Low-intensity rainfall, however, is the exception rather than the rule in this environment, which, together with a high-rainfall frequency, causes a perched water table to reach the soil surface in many places, thus producing extensive saturation excess overland flow (Elsenbeer and Vertessy, 2000).

By comparing surface K_s values with the observed maximum and median 5-minute duration rainfall intensities (96 and 25 mm h^{-1}, respectively), it is evident that infiltration excess overland flow may also occur in this catchment (dashed lines in Figure 10.5). However, this tends to operate in small 'partial areas' only, usually

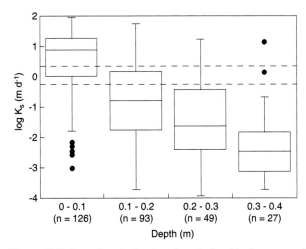

Figure 10.5. Box plots indicating the relationship between soil hydraulic conductivity at four different depth intervals and rainfall intensity at La Cuenca, after Elsenbeer and Vertessy (2000). The dashed horizontal lines denote the maximum and median 5-minute duration rainfall intensities recorded during the study period (96 and 25 mm h^{-1}, respectively).

confined to the steep lower sideslopes (unit B) where surficial K_s values are lowest (Elsenbeer et al., 1992).

For several places within the catchment, overland flow can be clearly traced back to the outlets of soil pipes (Figure 10.3). Elsenbeer and Lack (1996) argued that this return-flow mechanism is at least as prevalent as saturation overland flow in the catchment, and a strong determinant of the observed overland flow pattern (Figure 10.3). As noted earlier, they measured overland flow continuously at the three sites marked S1, S2 and S3 in Figure 10.3, across six events of varying magnitude and duration. The total event overland flow volumes at sites S1 and S2 ranged between 0 and 33 litres, whereas the total event overland flow volumes from S3 (the only one of the three sites associated with a subsurface pipe) ranged between 103 and 500 litres. At least six pipes of similar dimensions have been detected across the catchments (Figure 10.3), though we have no knowledge of the volume of flow emerging from these. On the basis of the flow volumes emerging from the single pipe we have instrumented, it is conceivable that the total volumes of runoff emerging from pipes account for a large proportion of total runoff during storm events. It is also worth noting that because of its point-source origin, return flow from pipes tends to occur more in concentrated flow lines, although this is also often a consequence of the rough micro-topography of the forest floor. This has consequences for the ability of such flows to re-infiltrate further downslope.

Hydrochemical measurements provide further insights into how storm runoff is generated within the La Cuenca catchment. Elsenbeer et al. (1995) showed that the chemical "fingerprints" of saturation overland flow, return flow, and subsurface flow at La Cuenca were each distinctive with respect to certain elements, most notably potassium (K) and Silica (SiO_2). Elsenbeer and Lack (1996) showed how the fingerprints of stream discharge varied systematically throughout storms in response to varying inputs of water from different hydrologic compartments in the catchment. Invariably, the K/SiO_2 ratio of the stream water rose and fell in association with discharge, reflecting the importance of overland flow as a major contributor to storm runoff (Figure 10.6).

Several studies have used hydrochemical information to "separate" the discharge hydrograph into time-varying fractions of storm runoff generated via different hydrologic pathways in catchments (Turner and Macpherson, 1987; McDonnell et al., 1990). In conducting such separations, it is commonly assumed that the signatures of the various hydrologic compartments do not vary in space. Our field experience at La Cuenca tells us that such assumptions are invalid and result in erroneous separations. Figure 10.7 shows the result of a chemical hydrograph separation into groundwater and overland flow components for a single event at La Cuenca. By treating overland flow chemistry as a constant in space, an apparent dominant volumetric contribution of overland flow results in the separation for this event. However, by accounting for the measured spatial variability in overland flow chemistry, confidence limits may be attached to the separation. The shaded area in Figure 10.7 shows that, even in the generous case of a 90% confidence limit, uncertainty regarding the relative contribution of overland

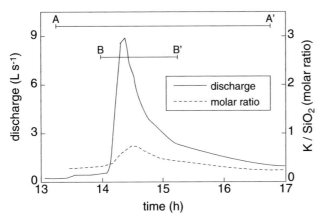

Figure 10.6. Catchment discharge and K/SiO₂ chemograph for the November 27, 1988 event at La Cuenca, after Elsenbeer and Lack (1996). A-A′ denotes duration of net rainfall (throughfall), and B-B′ denotes duration of overland flow.

flow, especially near peak flow, is considerable. In the case of a 95% confidence level (not shown), the lower limit for the overland flow contribution at peak flow overlapped with the groundwater contribution. Such data highlight the importance of spatial variability in overland flow and its role in catchment storm runoff production at La Cuenca.

10.4   THE TOPOG_SBM MODEL

The Topog series of models are designed to predict the spatiotemporal hydrologic dynamics of small ($< 10\,km^2$) heterogeneous catchments. Topog_SBM is one of the several models in this series and was derived by hybridising elements of

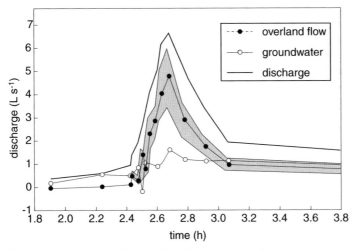

Figure 10.7. Hydrograph separation for a typical rainy-season event at La Cuenca, based on a three-component mixing model (for sake of clarity, the soil water component is not shown). The shaded area denotes the bounds of the 90 % confidence limits for the overland flow estimate.

the spatially explicit, fully dynamic model, Topog_dynamic (Vertessy et al., 1993, 1996; Dawes and Short, 1994; Dawes et al., 1997) and the aspatial, quasi-dynamic model, TOPMODEL (Beven and Kirkby, 1979; Beven, 1997). In brief, Topog_SBM consists of:

- the contour-based "streamtube" network for surface and subsurface flow routing, common to all Topog applications (see Figure 10.8),
- a simple bucket model for handling soil water fluxes in and between each element (as opposed to the Darcy–Richards approach described for previously reported versions of Topog), and
- a one-dimensional kinematic wave overland flow module for simulating surface runoff along the Topog "streamtubes".

Full details of the various components of Topog_SBM are given in Vertessy and Elsenbeer (1999). Figure 10.9 provides a schematic representation of the model, illustrating that it is capable of simulating infiltration excess, saturation excess and exfiltration (or return) overland flow, as well as lateral subsurface flow through the soil matrix. An underlying assumption of the model (borrowed from TOPMODEL) is that K_s is greatest at the soil surface and declines exponentially with depth through the soil profile. The model as used here does not explicitly represent pipe flow.

Topog_SBM is a fully distributed model, meaning that each spatial unit (or element) can be ascribed unique system properties if so desired. For each timestep (a five minute interval was adopted here) and each catchment element, the model computes water table depth, soil moisture storage, deep drainage loss, lateral subsurface flow, and overland flow height, velocity and discharge. As the

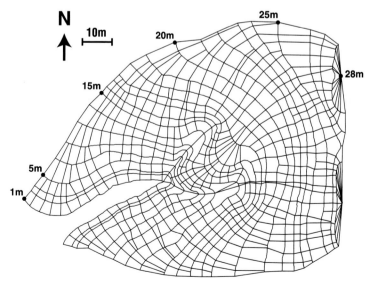

Figure 10.8. Element network computed by Topog for the La Cuenca catchment. Catchment area is 0.68 ha and mean element area is 10.1 m^2. (From Vertessy and Elsenbeer, 1999; reproduced with permission.)

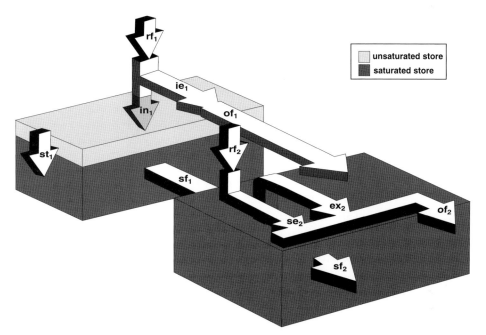

Figure 10.9. Schematic representation of the hydrologic processes modelled by Topog_SBM. (From Vertessy and Elsenbeer, 1999; reproduced with permission.) Symbol definitions are as follows: rf = rainfall, in = infiltration, st = transfer between unsaturated and saturated zone, ie = infiltration excess, se = saturation excess, ex = exfiltration, of = overland flow, sf = subsurface flow.

model is typically run for discrete storm events only, evapotranspiration processes are ignored. Our choice of a five minute timestep was dictated by the temporal discretisation of our rainfall input series. For such a small catchment, a smaller timestep (say one or two minutes) may have been preferable.

Aside from a rainfall series to drive the simulation and the topographic data used to derive the flow net, six model inputs must be specified for each element. These are soil depth (z), saturated soil water content (θ_s), residual soil water content (θ_r), saturated hydraulic conductivity at the soil surface (K_o), the rate of K_s decay through the soil profile (m) and the Manning roughness value (n). Each of these inputs may be considered to be uniform across the catchment, or ascribed on an element by element basis, to represent spatial variability across the catchment.

10.5   MODEL APPLICATION

The La Cuenca catchment was discretised into a network of 678 elements (Figure 10.8), resulting in a mean element area of 10.1 m^2 and a maximum element area of 35.9 m^2. The element slope averaged 0.43 m m^{-1} and ranged between 0.02 m m^{-1} on the floodplain and 1.92 m m^{-1} on the gully sideslopes. We represented the catchment floodplain (about 9 % of the catchment area) by allocating a low slope (0.02 m m^{-1}) to the bottom row of elements, which represent about 8 % of the catchment area.

We compared four different 'sets' of parameter values, which were distinguished by the manner in which soil hydraulic properties were represented. In set 1 (the "uniform" case) we applied the median value of the master K_o data distribution (2.3 m d^{-1}) to every element in the catchment. In set 2 (the "organised" case) we applied the median K_o value for land units B, C and D (1.2, 10.6 and 8.1 m d^{-1}, respectively) to all elements lying within each of these units. In set 3 (the "random" case), we randomly allocated deciles of the master cumulative frequency distribution of K_o values (see Figure 10.4d) across the whole catchment. In set 4 (the "random&organised" case), we randomly allocated deciles of the cumulative frequency distribution of K_o values for land units B, C and D (see Figures 10.4a, 10.4b and 10.4c), to elements lying within each of these units, respectively. It should be noted that we used a single random realisation of K_o values for sets 3 (random) and 4 (random&organised) rather than multiple realisations as has been used by, for example, Smith and Hebbert (1979), Freeze (1980) and Loague and Kyriakidis (1997). Because the measured variability of K_o is at a small scale relative to the size of the catchment, we believe that multiple realisations would produce similar results to the single realisation used here. If the scale of variability were larger relative to catchment size (such as the case of rainfall patterns in Chapter 6, pp. 133–4) we would have to use multiple realisations.

In all sets, soil depth, z, was fixed at 1.0 m for the whole catchment, as field observations did not reveal any significant variation in this quantity. Similarly θ_s and θ_r were fixed at 0.4 and 0.05, respectively, for all sets, again because little variability was evident in the field data gathered from the site. Preliminary model sensitivity analyses we have conducted suggest that θ_s and θ_r values (when systematically changed across their natural range) have a minor impact on model behaviour. Soil depth has a more significant effect, but the < 30 % variation over the La Cuenca catchment has a minimal effect on simulated runoff behaviour.

For sets discriminating between land units (sets 2-organised and 4-random& organised), the K_o decay parameter (m) was set to 0.07, 0.02 and 0.01 for land units B, C and D, respectively. For sets 1-uniform and 3-random, the mean of these three m values (0.03) was adopted.

A single event (event 12) was used to calibrate the model for all four sets. All simulations were initiated with a catchment wetness pattern derived from a "warm-up" simulation. The warm-up simulations involved applying a steady rainfall input equivalent to the observed pre-storm runoff rate; the run was terminated after the model produced a steady rate of runoff, equivalent to the pre-storm rate. On average, the warm-up run lasted about 100 days.

For each of the four parameter sets, it was possible to "fit" the model discharge hydrograph well to event 12, despite the fact that the observed hydrograph had a fairly complex shape. The Manning roughness parameters (n) obtained from these calibrations were 1.2, 1.1, 0.7 and 1.1 for sets 1, 2, 3 and 4, respectively. These n values are much higher than commonly reported in the literature, although Shen and Julien (1992) note that n can exceed 1.0 for extremely dense vegetation (their Table 12.2.1, p. 12.15). Acceptable hydrograph recessions could only be obtained by adopting such high n values. The model fit for set 4 (random&organised) is shown in Figure 10.10, illustrating that the height and timing of both runoff peaks, and the shape of the runoff recessions were faithfully reproduced by the model; similar quality fits were obtained for the other three sets.

We simulated 34 individual events for La Cuenca, chosen to span a wide range of rainfall totals, intensities and durations, and associated with varying antecedent soil moisture conditions (Figure 10.11). Rainfall totals varied between 10.2 and 82.5 mm (Figure 10.11a), with maximum five-minute intensities (I_5) ranging between 31.2 and 82.8 mm h^{-1} (Figure 10.11b). Pre-storm runoff rates varied between 0.005 and 0.37 mm h^{-1} (Figure 10.11c). Runoff totals varied between 1.3 and 44.8 mm (Figure 10.9d), and peak runoff rates ranged between 1.3 and 17.5 mm h^{-1} (Figure 10.11e). Times between the start and peak of runoff (time of rise) ranged between 20 and 125 min, and averaged 40 min (Figure 10.11f). The graphs showing the frequency distributions for rainfall and I_5 also display the distributions for a larger population of events (187 in total) which were associated with the overland flow frequency observations shown in Figure 10.3. This shows that the 34 events which we modelled were skewed towards higher rainfall magnitudes and intensities when compared to the larger population of storm events.

10.6 MODEL RESULTS

Below, we explore how the model, when calibrated on the single event, performed on the other 33 events using four different sets of input parameters.

10.6.1 Hydrograph Predictions

Figures 10.12 and 10.13 compare observed and predicted values of total runoff, peak runoff and time of rise for all 34 events, with predictions shown for each of the four parameter sets. Figure 10.12 shows observed values plotted against

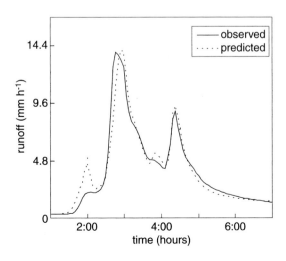

Figure 10.10. Observed and predicted runoff for event 12; the prediction is based on parameter set 4 (random&organised).

Table 10.3. Statistics for the 10 overlapping events for which observed and predicted overland flow frequencies were compared.

Event	Duration (min)	Pre-storm runoff (mm h^{-1})	Total rainfall (mm)	Maximum I_5 (mm h^{-1})
1b	90	0.112	28.0	36.0
4b	110	0.267	34.2	64.0
4c	550	0.364	83.5	96.0
5	140	0.267	31.8	33.6
7	280	0.275	30.6	36.0
8	270	0.257	39.9	64.8
16	220	0.163	14.3	26.4
17	150	0.078	20.4	48.0
21	80	0.131	19.5	45.6
22	75	0.219	14.2	30.0
Median	145	0.238	29.3	40.8
Median$_{34}$	128	0.125	31.4	51.4

The subscript 34 denotes that these are the median values for all 34 events which were simulated

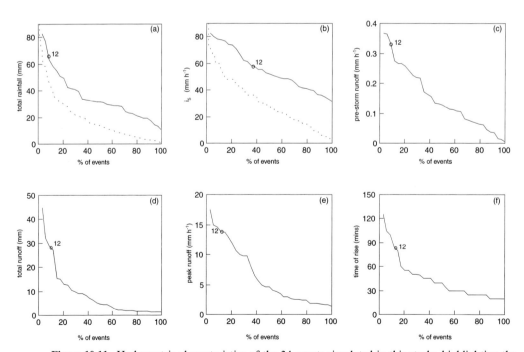

Figure 10.11. Hydrometric characteristics of the 34 events simulated in this study, highlighting the model calibration event (event 12). The dashed lines in (a) and (b) refer to the characteristics for the 187 events associated with the overland flow frequency analysis reported in Elsenbeer and Lack (1996).

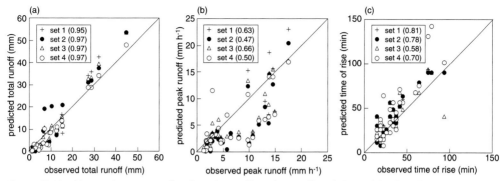

Figure 10.12. Observed versus predicted total runoff, peak runoff, and time of rise for all 34 events, for sets 1 (uniform), 2 (organised), 3 (random) and 4 (random&organised). The diagonal line in each plot is the 1:1 line.

predicted values, whereas Figure 10.13 compares the cumulative frequency distributions for the observed and predicted values. By examining Figure 10.12 we can gain a sense of model error for particular events and thus detect where the model fails. Figure 10.13 shows us how the frequency of predicted hydrograph properties compares to what was observed in the field; this is most relevant when considering the ability of the model to predict multiple events.

All four sets produced very good estimates of total runoff for most of the events, with r^2 values ranging between 0.95 and 0.97 (Figure 10.12a). The best total runoff estimates were obtained from set 4 (random&organised), particularly for the larger events. Figure 10.13a shows that all sets slightly underpredicted the distribution of runoff values for the smallest 65 % of events. It also shows that set 4 (random&organised) yielded the best cumulative distribution of total runoff volumes for the largest 35 % of events.

Peak runoff was simulated less well by all sets, with r^2 values ranging between 0.47 and 0.66 (Figure 10.12b). Generally, good peak runoff predictions were obtained for events with small runoff peaks ($< 6\,\mathrm{mm\,h^{-1}}$) and large runoff peaks ($> 12\,\mathrm{mm\,h^{-1}}$), though intermediate events were poorly simulated by all sets. For the larger events, peak runoff was predicted best by sets 3 (random) and

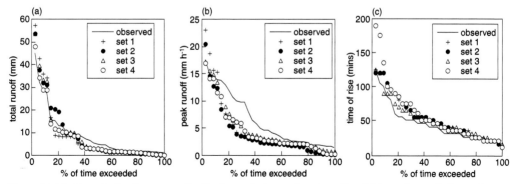

Figure 10.13. Cumulative frequency distributions of observed and predicted (a) total runoff, (b) peak runoff, and (c) time of rise, for all 34 events.

4 (random&organised), and worst by set 1 (uniform). Similar conclusions can be drawn from Figure 10.13b which shows that the predicted distribution of peak runoff values was always lower than observed, except for the largest 10 % of events. The largest discrepancies occurred between the 20th and 40th percentiles. Figure 10.13b also shows that set 4 (random&organised) predicted the most accurate cumulative distribution of peak runoff values, and that set 1 (uniform) yielded the worst results, although differences between the sets were small.

The time of rise predictions were generally good for all sets, with r^2 values ranging between 0.58 and 0.81 (Figure 10.12c). Set 3 (random) yielded the best cumulative distribution of time of rise values, being quite close to the observed distribution across the range of events (Figure 10.13c). Set 4 (random&organised) predicted much greater times of rise than were observed for the largest 10% of events.

In summary, reasonable catchment outflow hydrographs could be obtained for 34 events of varying magnitude and duration, using four different parameterisations of the model. This was achieved in spite of the fact that the model was calibrated on a single event (event 12), and that fairly simplistic initial moisture conditions were adopted in the simulations. Overall, sets 3 (random) and 4 (random&organised) yielded the best results and set 1 (uniform) yielded the worst.

10.6.2 Overland Flow Predictions

As noted earlier, one of our main aims was to simulate credible spatiotemporal patterns of surface runoff generation across the La Cuenca catchment. Figure 10.3 showed the frequency of overland flow occurrence at 72 detector sites for a total of 187 events. In Figures 10.11a and 10.11b, we showed that the distributions of total and maximum I_5 values of rainfall for our 34 events were much more skewed to "big events" than in the 187 events associated with the observed overland flow data set. There are two reasons for this. First, when selecting events to simulate, we tended to choose events that generated significant streamflow. Many of the 187 events sampled for overland flow frequency simply did not generate streamflow. Second, most of the 187 overland flow events were sampled early in the field campaign when mild drought conditions were prevailing, resulting in a greater than normal percentage of low-rainfall events. Ultimately, overland flow frequency amongst the 72 detector sites was only measured for 10 of the 34 events we simulated. These 10 "overlapping" events were used as the basis to compare observed and predicted spatial patterns of overland flow. The median event characteristics for these 10 events varied only slightly from those for the full 34 modelled, the median event duration being 145 min as opposed to 128 min, the median event rainfall being 29.3 mm as opposed to 31.4 mm, and the median maximum I_5 value being 40.8 mm h^{-1} as opposed to 51.4 mm h^{-1}.

Figure 10.14 compares observed and predicted frequency distributions of overland flow occurrence for the 10 "overlapping" events at La Cuenca, and illustrates the strong effect of soil property representation in the model. These

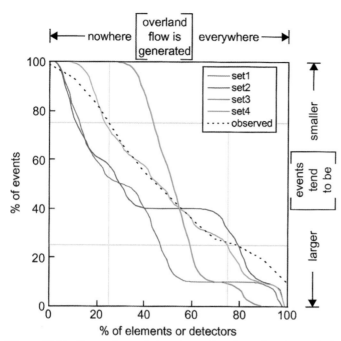

Figure 10.14. Comparison of the observed and predicted cumulative frequency distributions of overland flow occurrence across the catchment for 10 of the events simulated. Predictions are based on percentage of elements registering overland flow. Observations are based on the percentage of the 72 detectors registering overland flow. *Y*-axis values of 0 and 100 % equate to 'never' and 'always' in Figure 10.15. The curves shown were smoothed using a 5-point window. The horizontal and vertical grid lines define the interquartile ranges for the *x* and *y* axes.

data show what percentage of the model elements was predicted to generate overland flow. Also shown is the observed frequency distribution of overland flow at 72 detector sites for the same 10 events. Associated spatial patterns of overland flow frequency for sets 1 (uniform), 2 (organised), 3 (random) and 4 (random&organised) are shown in Figure 10.15.

The observed overland flow frequency distribution shown in Figure 10.14 indicates a near-linear pattern, with overland flow being generated at half of the detector sites for about half of the events. Only the tails of the distribution diverged from this linear pattern.

Using set 1 (uniform), the predicted pattern of overland flow development is strongly influenced by topographic factors, as soil properties are assumed to be uniform across the catchment (Figure 10.15). In this case, overland flow is concentrated in valleys and along the bottom contour, which we have represented as a floodplain by allocating low slope values to it. According to Figure 10.14, for set 1 (uniform) overland flow is generated over much less of the catchment area than is observed for almost all events. Using set 2 (organised), the influence of spatially variable soil is evident (Figure 10.15), with widespread occurrence of overland flow in landscape unit B, which has the lowest median K_o value (Figure 10.4). The associated frequency distribution for set 2 (organised) shown in Figure 10.14 indicates that overland flow occurs far more extensively than in set 1 (uniform)

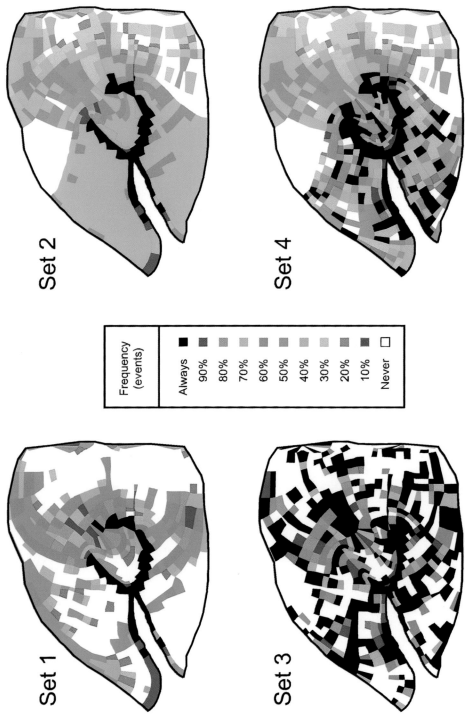

Figure 10.15. Predicted spatial patterns of overland flow frequency across the catchment for 10 events, as predicted by set 1 (uniform), set 2 (organised), set 3 (random), and set 4 (random&organised). (From Vertessy and Elsenbeer, 1999; reproduced with permission.)

Set 1

Set 2

Set 3

Set 4

Frequency
(events)

Always
90%
80%
70%
60%
50%
40%
30%
20%
10%
Never

for events ranked between the 10th and 40th percentile, but is of a similar pattern for all other events. Again, this frequency distribution differs significantly from that which has been observed in the field. In the case of set 3 (random), overland flow is generated in a random pattern across the catchment (Figure 10.15), though subtle topographic control is still evident. The pattern of runoff occurrence for set 3 (random) is dominated by extremes with a lot of elements showing no runoff and a lot with almost always runoff. This is because the mean soil conductivity is of a similar value to the precipitation intensities so, when randomness is introduced, those elements with lower K tend to "switch on" while those with higher K tend to "switch off". According to Figure 10.14, set 3 (random) predicts that overland flow is generated more widely than is observed for the smaller events (i.e. too many elements are "switched on"), and less widely than is observed for the wettest half of events (i.e. too many elements are "switched off"). A major failing of set 3 (random) is that it predicts overland flow to occur over at least 30 % of the catchment area for all events. Set 4 (random&organised) also displays a random pattern of overland flow generation, but not the extreme pattern of set 3 (random). The frequency of runoff occurrence is generally highest in land unit B and lowest in land unit C (Figure 10.15). This is a consequence of the K_o values in land unit B being almost an order of magnitude lower than those in land unit C (Figures 10.4a and 10.4b). Figure 10.14 shows that set 4 (random&organised) predicts an overland flow frequency distribution which is very similar to the observed distribution, particularly in the interquartile region. Beyond this region the model slightly overpredicts the occurrence of overland flow for the wettest events, and underpredicts its occurrence for the driest events.

Figure 10.16 provides for a visual comparison of the spatial patterns of observed runoff occurrence with the simulations of Figure 10.15. It is clear

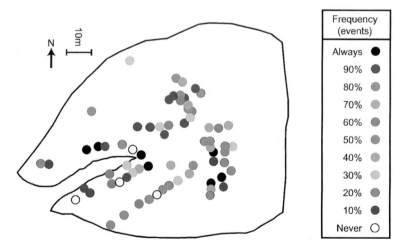

Figure 10.16. Observed pattern of overland flow frequency across the catchment for the 10 "overlapping" events.

that the observations do not display the topographic or soil-property induced spatial pattern seen in the simulations using set 1 (uniform) or set 2 (organised). The simulation for set 3 (random) is generally characterised by extremes of no runoff or always runoff being detected, but this pattern is not evident in the observations. The pattern of runoff occurrences in the observations is best matched by the pattern using set 4 (random&organised) where there is a full spectrum of occurrences simulated.

10.7   DISCUSSION AND CONCLUSIONS

Whilst La Cuenca is a tiny catchment by any standard, it is characterised by considerable spatial variability in soil hydraulic conductivity and complexity in storm runoff generation. A large body of hydrometric and hydrochemical data indicates that storm runoff production in this catchment is dominated by overland flow. The hydrometric evidence further indicates that the spatial pattern of overland flow is governed by the distribution of soil hydraulic properties and that subsurface pipe outlets are major point sources for overland flow generation. Hydrochemical evidence shows that the K/SiO_2 signature of overland flow is spatially variable, indicating that it arises from a variety of pathways. Again, subsurface pipes are believed to play an important role in overland flow generation and the chemical signature that overland flow assumes. As we showed earlier, the volumes of overland flow generated by a subsurface pipe pathway could be large, but this contention must be regarded as speculative because we only have direct volumetric measurements from a single pipe. In hindsight, it would have been wise to have continuously monitored flows emerging from the other five pipes noted within the catchment (Figure 10.3).

We have described a fully dynamic and distributed storm runoff generation model which was relatively simple to parameterise, but did not include the process of subsurface pipe flow. In fact, a pipe flow process could have been invoked in the model as such a capability exists in Topog_SBM. However, we chose to ignore this process for two reasons. First, whilst we knew of the locations of up to six pipe outlets, we had no knowledge of the pipe dimensions, nor their catchment area. Second, because we had flow data for only a single pipe, we had no means of evaluating model predictions of pipe flow dynamics.

Topog_SBM was applied to La Cuenca for a wide range of event conditions, using four different sets of soil hydraulic properties, each of which could be defended as legitimate representations of field data. In the simplest case, set 1 (uniform), we ascribed the median K_o value to all elements within the catchment. In the most complicated case, set 4 (random&organised), we ascribed deciles from three different distributions of K_o values, randomly, to elements residing within three different sub-areas (land units B, C and D). In this latter case we represented both the stochastic and deterministic variability in catchment soil properties, an approach rarely employed in distributed hydrologic modelling.

The four sets of model predictions of total runoff, peak runoff rate and the time of rise at the catchment outlet were compared against field observations for a total of 34 events. Relatively inferior hydrograph predictions were obtained when it was assumed that soil hydraulic properties did not vary in space and the median K_o value was ascribed to all catchment elements, set 1 (uniform). The best results were obtained using set 4 (random&organised), in which the model represented the measured spatial variability of K_o values within *and* between land units.

The discharge hydrograph differences between the four sets were relatively subtle, when compared to the predicted differences in spatiotemporal patterns of overland flow occurrence between sets. By randomly varying K_o across the catchment as a whole, but assuming no deterministic pattern in that variability, set 3 (random), the simulated pattern of overland flow occurrence changed radically but did not improve relative to sets 1 (uniform) and 2 (organised). By far the best results were obtained using the parameterisation for set 4 (random&organised), where measured spatial variability in K_o was represented within each individual land unit. Further modelling efforts at La Cuenca could compare multiple realisations of random K_o fields (e.g. Loague and Kyriakidis, 1997) to assess how much particular random patterns affect results, but given the small scale of variability in soil properties compared to the catchment scale, it is unlikely that the conclusions will differ appreciably.

From our results, we conclude that in order to get features of the spatial patterns of runoff occurrence correct, it is necessary to represent spatial variability in soil properties. Yet, it is rare to find such detailed soil property data as has been collected at La Cuenca, and in most modelling exercises of this kind the soil property inputs are probably guessed. In cases such as La Cuenca where the precipitation intensity is similar to the soil hydraulic conductivity, we suggest that it is still probably best to conduct the fitting with a randomised log-normal distribution of values such as those used in sets 3 (random) and 4 (random&organised), even if the distribution is entirely synthetic. On the basis of our findings we recommend representing stochastic variation in soil hydraulic property values, imposed on a deterministic pattern if multiple soil types are present within the area of interest.

Our strategy for setting initial conditions warrants discussion. As noted earlier, all of our event simulations were initialised with a soil moisture pattern derived from a "warm-up" run, in which a steady rate of rainfall, equivalent to the pre-storm runoff rate, was applied to the catchment. This approach was predicated on an assumption that the catchment was small enough to have drained to a *near* steady state condition prior to each storm. While the low conductivity subsoils would require several days to drain to a near steady rate, the more active upper 30 cm of soil could drain in a matter of hours. In Figure 10.11c we showed that the observed pre-storm runoff rates ranged between 0.005 and 0.37 mm h^{-1}, so our event simulations were based on a very broad range of initial moisture conditions. There were probably circumstances where the time interval between successive events was too short for adequate catchment drainage to have occurred. There were four occasions where more than 10 mm of rain fell within the 24 hours preceding the event to be modelled; conditions that might invalidate our assumption of a well drained catchment prior to the storm. Two of these events were simulated poorly and the other two were simulated quite well, including one where 32 mm of rain fell only nine hours prior to the start of the event simulated. Hence, poor hydrograph predictions cannot necessarily be blamed on errors in assumed initial conditions. Ideally, we would have selected events preceded by significant periods of no rain, thus reducing the possibility of errors in initial conditions. However, as it rains at La Cuenca on most days during periods of significant runoff generation (i.e. the wet season) this criterion proved impossible to satisfy.

We now briefly describe three possible ways in which to improve the performance of Topog_SBM in predicting the spatial patterns and temporal characteristics of runoff for the La Cuenca data set. These include an alternative way to prescribe initial moisture conditions, the incorporation of a fast subsurface flow path, and modifications to the soil water accounting scheme we have used.

Firstly, more realistic initial soil moisture conditions might be obtainable by letting the catchment drain from saturation until the pre-storm runoff rate has been attained. This would probably yield a moisture pattern more like the one that would occur under natural drainage between storms. However, as noted earlier, only some of the model error we have detected is attributable to initial moisture conditions, and the tendency in error is not at all systematic.

Secondly, by incorporating a rapid subsurface flow path into the model, we could represent the pipe network that has been observed to operate at La Cuenca during storm events (Elsenbeer and Vertessy, 2000). To some extent, the random pattern of K_o values adopted in sets 3 (random) and 4 (random&organised) have represented the effect of a pipe network by creating multiple point sources of overland flow generation in elements with low K_o values. However, a rapid subsurface flow path would result in the persistence of fast runoff after rainfall has ceased and allow us to relax our dependence on unrealistic roughness parameter (n) values to model hydrograph recessions correctly. A simple algorithm has been described by Bronstert and Plate (1997) which is within the Topog_SBM model but was not invoked in this study. However, as argued earlier, the geometry of the subsurface pipe network is unknown, and the parameters underpinning this sub-model would thus need to be treated as "black box" or calibration parameters.

Thirdly, some gains could be made by modifying the expression for K_s decay with depth, as has been advocated in recent TOPMODEL applications (Ambroise et al., 1996; Beven, 1997; Iorgulescu and Musy, 1997). The parabolic, linear and power function decay models that have been proposed as alternatives to the exponential decay model adopted in this study should be evaluated in Topog_SBM on La Cuenca for the same events studied here. It is possible that a more appropriate decay model would improve the hydrograph recession fits, thus permitting some relaxation of the high n values we were forced to use.

Concluding, we see a useful role for pattern comparisons in testing predictions of storm runoff dynamics in heterogeneous catchments from distributed models such as Topog_SBM. There are less complicated, and probably more accurate, modelling methods available if one's interests are confined to hydrograph generation. But, if the spatial pattern and magnitude of different runoff components must be ascertained, as is required in pollutant transport or landscape evolution modelling, then distributed models must be used. Though we regard it as virtually impossible to replicate the exact pattern and magnitude of overland flow across even the best parameterised and simple catchments, we do believe that it is possible to approximate their functional behaviour. To achieve this, we have argued that it is critical to represent spatial variability in soil hydraulic properties, both in a deterministic (pattern) sense and a random (variability) sense. Future research in storm runoff generation modelling should focus on improving ways of representing such variability as well as methods to prescribe initial moisture conditions.

ACKNOWLEDGEMENTS

This study was funded by the Cooperative Research Centre for Catchment Hydrology and Grant No. 21-39353.93 of the Swiss National Science Foundation. The first author also gratefully acknowledges a visiting fellowship from the OECD.

11

Shallow Groundwater Response at Minifelt

Robert Lamb, Keith Beven and Steinar Myrabø

11.1 INTRODUCTION

The spatial distribution of perched or shallow groundwater is widely recognised to be significant for physically realistic modelling of catchment runoff production, especially within humid regions and areas of shallow soils. The distribution of water stored as a dynamic, near-surface saturated zone has an important role in theories of runoff production embodying the concept of a variable source or response area, such as those of Hursh and Brater (1941) and Hewlett and Hibbert (1967). Changing spatial distributions of shallow saturated storage may also affect the dynamics of land–atmosphere fluxes (via supply of moisture to vegetation and the unsaturated zone) and water quality (by controlling the pathways and residence times of flows within the catchment).

In Scandinavia, water table fluctuations have been shown to control the runoff response of catchments where the saturated zone exists at a shallow depth in the soil, and is therefore able to respond quickly to precipitation. For example, Rodhe (1981) used isotope analysis in two catchments in Sweden to show that discharge from shallow groundwater storage could constitute a large proportion of the runoff during spring melt events. In two Norwegian catchments, Myrabø (1986, 1997) has used observations of patterns of surface saturation or subsurface groundwater levels to show that it is the dynamics of a shallow saturated zone that control runoff production from a variable response area.

Measured data from the Seternbekken Minifelt catchment study of Myrabø (1988) will be used in this chapter to test simulated spatial and temporal patterns of shallow groundwater, using the distributed model TOPMODEL (Beven and Kirkby, 1979; Beven et al., 1995), extending the work of Lamb et al. (1997, 1998a). TOPMODEL is based on an assumption that there is a unique relationship between local saturated zone storage (or storage deficit) and position. Here, position is expressed in terms of topography via the topographic index $\ln(a/\tan \beta)$ of Kirkby (1975) or topography and soils via the soils–topographic index

Rodger Grayson and Günter Blöschl, eds. *Spatial Patterns in Catchment Hydrology: Observations and Modelling* © 2000 Cambridge University Press. All rights reserved. Printed in the United Kingdom.

$\ln(a/T_0 \tan \beta)$ of Beven (1986). Formally, a is the upslope specific area contributing to flow through a point (dimension L), $\tan \beta$ is the plan slope angle, used to approximate the downslope hydraulic gradient in the saturated zone, and T_0 $[L^2T^{-1}]$ is the transmissivity of the soil profile when just saturated.

Distributed approaches to modelling saturated storage vary in complexity between the explicit physics of grid-based models such as variants of the Système Hydrologique Européen (SHE) (Bathurst et al., 1995; Refsgaard and Storm, 1995; Abbott et al., 1986), flow-strip representations such as Thales (Grayson et al., 1995) or the Institute of Hydrology Distributed Model (IHDM) (Calver and Wood, 1995), and the conceptual, 'quasi-physical' approach of TOPMODEL. As with the discussion in Chapter 3, no rigid system of model classification will be attempted here, not least because some models are capable of interpretation at several different levels.

Hydrological processes may be represented using different degrees of approximation and different model structures. The models mentioned above (amongst others) allow a "link to physical theory" (Beven et al., 1995) at the hillslope or catchment scale by simulating the changing spatial patterns of water storage, or storage deficit, over time. However, as argued throughout this book, compared to the total number of catchment hydrology studies using distributed models, there has been a general lack of attempts to test distributed simulations against observed data. As discussed in Chapter 1, in large part this has been because of a scarcity of suitable observations, in contrast to the much greater availability of rainfall and streamflow records.

Whereas the use of data from large numbers of boreholes is routine in regional groundwater modelling, fewer measurement sites have generally been available for spatially distributed modelling of shallower systems and hydrological response at the hillslope or small catchment scale. Probably the smallest catchment used in this context was a 2 m^2, artificial micro-catchment simulated using the model Thales (Moore and Grayson, 1991; Grayson et al., 1995). More typical field measurement densities were available for a 440 km^2 catchment where Refsgaard (1997, Chapter 13) compared observed water levels from eleven wells with levels simulated using the model MIKE-SHE. On the hillslope scale, observed piezometer data were compared to simulations made using the IHDM by Calver and Cammeraat (1993). Studies reporting tests of TOPMODEL concepts against observed shallow groundwater patterns will be described below.

The studies referred to have generally reported mixed results in reproducing observed water table patterns. Predictions are often reasonably good for some locations or on some occasions, but poor at other places or times. This can be attributed to the limitations imposed by model assumptions in representing spatially complex processes (Refsgaard, 1997) and the difficulty of estimating distributed model parameters, even when these have a clear physical interpretation in theory (Beven, 1989; Grayson et al., 1992b). Although TOPMODEL has physically meaningful parameters, in the work presented here, we have not attempted to fit these a priori using field measurements, but have instead used the exceptionally dense and extensive distribution of shallow groundwater measurements available for the Minifelt to estimate local parameter values by model inversion. In effect, TOPMODEL will be used as a distributed-parameter model, but one that is simple enough to be calibrated in the spatial domain using observed shallow groundwater levels, i.e. simple enough for inversion to be tractable.

Bedrock underlying the Minifelt is thought to be relatively impermeable, at least when considering the timescales of storm runoff responses, where it is the dynamics of the shallow saturated zone within the overlying soils that are important. This saturated zone is very shallow, generally only about one metre thick, with the water table less than one metre below the ground. The situation is therefore one of hillslope hydrology, rather than regional groundwater processes. Hence, we will consider groundwater levels measured with reference to the local ground surface, rather than as elevations relative to a fixed datum. The shallow nature of the system promotes a direct topographic influence on the saturated zone storage, which forms a convenient starting point for a simple distributed model. Unlike many regional groundwater problems, the topographic catchment boundary can be used as a very good approximation for the saturated zone flow divide. However, the local heterogeneity of soils in the Minifelt weakens the local influence of topography on the water table, and leads to a requirement for distributed soil parameters.

11.2   MEASUREMENTS AT THE SETERNBEKKEN MINIFELT

The Minifelt is a small (0.75 ha) natural catchment located in an area of pine woods about 10 km west of Oslo, Norway, at an altitude of approximately 250 metres above sea level. An intensive measurement campaign was established in 1986 to investigate runoff processes, as reported in detail by Myrabø (1988, 1997). Soil conditions in the Minifelt are dominated by Quaternary till deposits, with some bedrock outcrops, some areas of bog, and high organic content in places, especially in the top few centimetres. The maximum soil depth is about one metre. Saturated hydraulic conductivity was estimated by Myrabø (1997) to have a mean value of the order of $0.01 \, \mathrm{m \, h^{-1}}$, and to vary between 0.0072 and $0.29 \, \mathrm{m \, h^{-1}}$. Sampled soil grain sizes vary from 0.02 to 20.0 mm, and there are also many small boulders and macropores in the soil. Sampled total porosity varied between 40 % and 80 %.

Flows at the outlet of the Minifelt catchment were gauged at a V-notch weir where water levels were logged automatically. Precipitation, snowmelt and temperature were also gauged nearby. Average annual rainfall and potential evaporation are about 1000 mm and 600 mm, respectively. A recent view of the site is shown in Figure 11.1; vegetation is now somewhat denser than during the period of field measurements used in this work.

A dense network of instruments measuring water table depths was established in the catchment, as shown in Figure 11.2. Four observation wells of about 6 cm diameter were installed in different topographic settings, located in Figure 11.2 at the centres of the numbered circles. Water levels in these boreholes were mea-

Figure 11.1. View of part of the Seternbekken Minifelt catchment. (Photograph taken in 1999.)

sured using pressure transducers, and recorded by data loggers every hour. There are also 108 piezometers of about 2 to 3 cm diameter, in which manual observations of water levels were made on five occasions, spanning a wide range of conditions. On each occasion, a reading was made in every piezometer, all piezometers being read within a one-hour period, during which time the changes in level were not observed to exceed 5 cm. The locations of the piezometers are indicated in Figure 11.2 by triangle symbols. Surface elevations were surveyed at each piezometer location, with reference to a datum at the catchment outlet, and also at some points around the catchment boundary. Soil depths were also recorded at the piezometer locations.

Although snowmelt is an important part of the overall annual hydrological regime, it was not modelled in the present study, which concentrates on the subsurface responses to rainfall. Periods influenced by snowmelt have therefore been avoided.

The dynamics of the shallow groundwater control runoff production mainly through a dynamic, saturated source area. However, other runoff processes can also occur, including rapid lateral subsurface flux in macropores and coarse organic material close to the surface, pipeflow and saturated zone discharge into the stream.

11.3 MAPPING THE OBSERVATIONS

11.3.1 Terrain Data

Catchment-wide topographic data are needed to calculate the spatial distribution of $\ln(a/\tan\beta)$. In the case of irregularly distributed spot height data, inter-

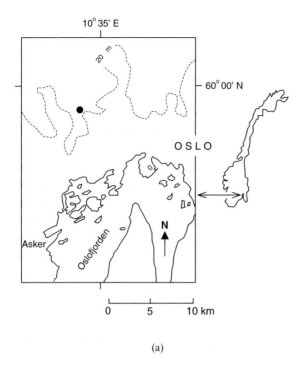

(a)

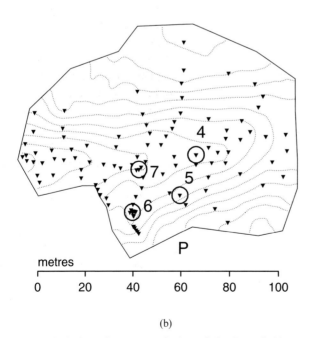

(b)

Figure 11.2. Location map and plan of the Seternbekken Minifelt catchment. Triangle symbols indicate piezometers, centres of numbered circles indicate location of continuously-logged bore-holes, P denotes the precipitation gauge. Dotted lines are topographic contours at 1 m intervals. The outlet is at the far left of the catchment. (After Lamb et al., 1998b.)

polation is therefore required to estimate elevations throughout the catchment (see discussion in Chapter 2, pp. 34–5). For this work, as in most TOPMODEL studies, topography was represented in the form of a grid-based digital terrain model (DTM). Interpolation of spot heights onto a regular grid raises the issue of the choice of grid element size, or grid resolution. This problem has been investigated in the TOPMODEL context (see, for example, Quinn et al., 1995; Wolock and Price, 1994; Saulnier et al., 1997b), but attention has generally centred on the effects of changing grid resolution on spatially-averaged model parameters and areally-integrated predictions. A number of studies have suggested that changes in grid resolution can be compensated for by adjustment of parameter values. Where explicit spatial predictions are to be made (and tested), grid resolution is also important because predictions at a given location will be a function of the local grid cell values of the topographic index $\ln(a/\tan\beta)$, which can change for different grid element sizes.

The link between TOPMODEL concepts and physical processes becomes difficult to sustain for grid cells that are large compared to hillslope lengths (Beven, 1997). For prediction of spatial patterns of the water table in the Minifelt, a fine grid resolution is therefore desirable, and can be supported by the available topographic data. However, there is a risk that a very high resolution DTM might contain significant false topographic features created during interpolation. Hence a $2\,m \times 2\,m$ regular grid was chosen, following the earlier analysis of Erichsen and Myrabø (1990), as a compromise to capture real topographic detail without creating artefactual features.

Similar arguments apply to the choice of interpolation algorithm. Although factors such as regularity of the original measurement sites, measurement density and coverage may influence this choice, it is also likely that availability of convenient software and preferences established through previous experience will play a part. In this case, a smoothed bilinear interpolation algorithm, also used by Erichsen and Myrabø (1990), was chosen after qualitative comparisons with other available algorithms and visual inspection of the catchment.

Contours derived from the DTM are shown in Figure 11.2. It will be seen that the piezometer and topographic survey locations were chosen to coincide with a number of hydrologically significant topographic features in the catchment, especially the main "valley" extending behind the outlet roughly along the horizontal axis. Also present are a number of slight spurs and hollows, areas in which piezometers were located.

Soil depth data from each piezometer location were also interpolated onto a $2\,m \times 2\,m$ grid using the bilinear algorithm, and are mapped in Figure 11.3. The pattern of soil depth, particularly along the main valley axis, shows a number of depressions in the bedrock elevation which produce small areas of deeper soils, separated by shallower sills. These features are verified from field observation, and are not merely artefacts of interpolation, as can often be the case with such "sink" features. However, a lack of many soil depth measurement points between the piezometers and the boundary does mean that the interpolated map is perhaps less reliable in these areas.

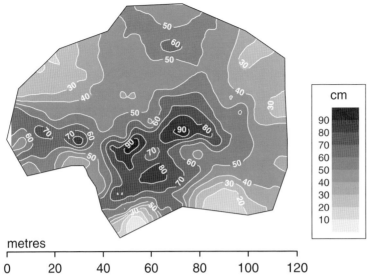

Figure 11.3. Interpolated map of soil depths in the Minifelt. Contours are plotted at 10 cm depth intervals.

Choices of grid resolution and interpolation method are subjective, based on experimentation with different options and assessment of the results given qualitative knowledge of the field situation (Chapter 2, pp. 45–9). In fact, it is recognised that the chosen grid resolution does not capture the detailed microtopography of the catchment, a factor that has to be considered when interpreting measured data and assessing model predictions. However, the interpolated DTM does capture hydrologically significant landscape features of the catchment that exist above the grid scale.

11.3.2   Mapping Groundwater Patterns

For each piezometer water level survey, $\bar{Q}$ was calculated as the average of the discharge at the outlet at the start and end of the measurement period. Surveys were carried out for $\bar{Q}$ equal to 0.1, 0.54, 0.61, 4.89 and 6.8 mm h^{-1}. All measurements were taken during recessions, but rain occurred while the observations corresponding to $\bar{Q} = 4.89$ and $\bar{Q} = 6.8$ mm h^{-1} were being recorded. The piezometer data were interpolated using the same bilinear algorithm and grid resolution applied to topography and soil depth. The observed spatial patterns are shown in Figure 11.4.

As discussed in Chapter 2, it should not be overlooked that interpolation is itself a form of modelling, and that interpolated maps can therefore only represent estimates of the pattern of water table depths. But despite the uncertainty introduced by interpolation, it is useful to represent these data as spatial patterns for comparison with the topographic and soil depth maps. In particular, comparison of the mapped water table data with the topographic contours shows areas on side slopes that saturate under wet conditions despite there being no

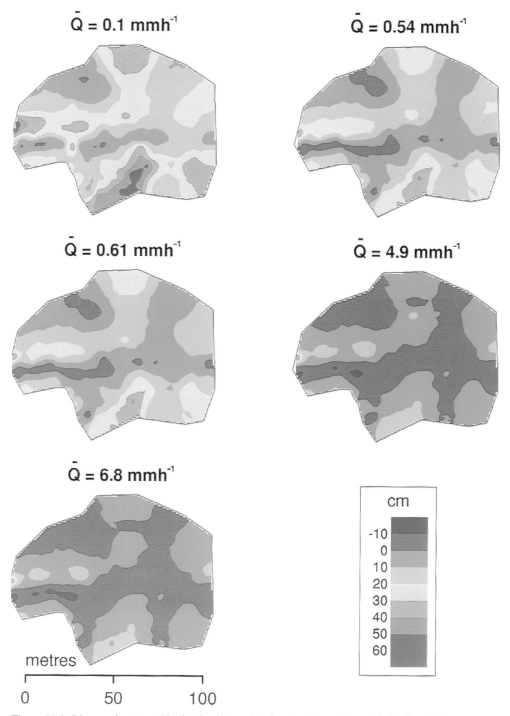

Figure 11.4. Measured water table depths, interpolated onto a 2 m × 2 m grid, for five discharges, time-averaged over each piezometer survey period. Depths are in cm (positive downwards).

apparent slope convergence. In such locations, the water level observations reflect spatial variations in soil properties or small topographic features not captured by the DTM.

There is some correspondence between the pattern of soil depth and water table depths close to the outlet, best seen for $\bar{Q} = 0.1\,\mathrm{mm\,h^{-1}}$. It would appear from the mapped data that the reductions in soil depth approximately along the valley axis affect the water table, which rises towards the surface just upslope of the small areas where soil depth decreases. One hypothesis put forward to explain this is that the reduction of soil depth, and likely consequent reduction in soil profile transmissivity, combined with a very low topographic slope angle, creates local conditions that favour exfiltration (or "run-on") from the saturated zone, even under fairly dry conditions.

11.4 TOPMODEL

TOPMODEL was introduced by Beven and Kirkby (1979) as a quasi-physical rainfall-runoff model, able to simulate the distribution of a dynamic storm runoff source area on the basis of a topographic control on saturated zone storage. Recent reviews of TOPMODEL concepts and applications have been provided by Beven et al. (1995), Beven (1997) and Kirkby (1997). A complete derivation of TOPMODEL theory will be omitted here, but may be found in the references cited above. Here, we will concentrate on assumptions invoked in making distributed water table depth predictions using TOPMODEL concepts.

TOPMODEL provides a simple, yet physically meaningful model of basic hillslope and catchment scale runoff processes, at least in relatively humid conditions and where soils are shallow relative to slope lengths (allowing the assumption that the local saturated zone gradient is approximated by the surface slope). It can be shown (e.g. Kirkby, 1997) that TOPMODEL derives directly from physical principles under an assumption that the rate of flux produced in the saturated zone quickly becomes spatially uniform for any change in a uniform input (or recharge) rate. This assumption has also been referred to as the "quasi-steady state assumption" because it implies a spatially uniform transition between steady state saturated zone profiles for a given discrete change in the uniform recharge rate between successive time steps. The dynamics of the saturated zone are thus represented as a succession of steady states (Beven, 1997).

The difference, at any point in the catchment, between the local storage deficit due to gravity drainage D [L] and the areal average deficit $\bar{D}$ is described in TOPMODEL by the equation

$$\frac{\bar{D} - D}{m} = \left[\ln\frac{a}{\tan\beta} - \lambda\right] - \left[\ln(T_0) - \overline{\ln(T_0)}\right] \tag{11.1}$$

where λ is the areal average of $\ln(a/\tan\beta)$, $\overline{\ln(T_0)}$ is the areal average of $\ln(T_0)$ and m[L] is a parameter controlling a vertical change in soil profile transmissivity with depth (see equation (11.2) below). The logarithmic terms in equation (11.1) enter because of an assumption that the transmissivity T [L^{2}T^{-1}] of the soil profile decreases exponentially as a function of storage deficit and T_0, the transmissivity when the soil is just saturated, such that the local subsurface lateral flux q $[L^2T^{-1}]$ is given by the equation

$$q = T \cdot \tan \beta = \left[T_0 \cdot \exp\left(\frac{-D}{m}\right) \right] \cdot \tan \beta \qquad (11.2)$$

When combined with a continuity equation, (11.2) has been found by Kirkby (1988, 1997) to satisfy, to a good approximation, the assumption of spatial uniformity of flux production in the saturated zone.

Beven et al. (1995) show how a numerical integration of (11.2) at the base of slopes along the channel network results in the following exponential lumped storage equation for the total saturated zone specific discharge:

$$Q = Q_0 \cdot \exp\left(\frac{-\bar{D}}{m}\right) \qquad (11.3)$$

where the intercept parameter $Q_0 = \exp(\overline{\ln(T_0)} - \lambda)$ and Q has dimensions $[LT^{-1}]$.

In equation (11.1), it may be noted that the position-dependent parameters $\ln(a/\tan \beta)$ and $\ln(T_0)$ are separated. If, as is often the case, the soil transmissivity is assumed to be uniform, then $\ln(T_0) = \overline{\ln(T_0)}$ everywhere and the right-hand term vanishes. However, it is easily seen that equation (11.1) can be written in terms of the soils–topographic index $\ln(a/T_0 \tan \beta)$ if knowledge of the variation of T_0 in space is available. To date, TOPMODEL applications have not used distributed soil transmissivities estimated directly by field measurements. Such measurements are difficult to interpret, as the natural variation of soil properties may be considerable, and it is difficult to match the scale of measurements with the model grid scale. Also, inference of the saturated soil profile transmissivity T_0 requires either depth sampling of (lateral rather than vertical) hydraulic conductivities, or reliance on the assumption of a known and fixed relationship between conductivity and depth (to allow integration of the conductivity from a measured surface value to the base of the soil).

Alternatively, local values of T_0 and hence $\ln(a/T_0 \tan \beta)$ can be estimated by model inversion, given knowledge of D. Although deficits are difficult to measure per se, observations of the depth $z[L]$ to the water table can be used, provided that a relationship is assumed between z and D. To keep the number of parameters as small as possible, most studies have assumed a simple linear scaling between the depth to the water table and storage deficit, such that

$$z = \frac{D}{\Delta \theta} \qquad (11.4)$$

where the dimensionless effective porosity, or storage coefficient, $\Delta \theta$ represents the "readily drainable" fraction of the pore space between 'field capacity' and saturation, and is assumed constant with depth.

Both soil transmissivity and effective porosity are likely to vary spatially in a catchment and can, in principle, be represented in a spatially distributed manner in TOPMODEL. Furthermore, Saulnier et al. (1997a) have shown that m can also be distributed in space. However, allowing three saturated zone parameters to be distributed in space would increase the number of degrees of freedom in fitting to distributed water table data. The m parameter, if spatially constant, can be related directly to an integrated variable, streamflow, using (11.3) and is therefore conveniently treated as a lumped parameter in catchment runoff studies, even though a physically more complete description of the catchment might allow m to vary. Variations in soil properties may be more readily associated with variations in transmissivity (which may in any case implicitly account for differences in soil depth) and effective porosity. Although a number of studies have allowed T_0 to vary in space, but assumed a constant value for effective porosity (Lamb et al., 1997, 1998a; Seibert et al., 1997), the effective porosity can be calibrated where, as in this case, data are available to describe the dynamics of the water table. An implication of (11.1) is that points in a catchment having the same values of $\ln(a/\tan \beta)$ and $\ln(T_0)$ are predicted to respond identically, in terms of storage deficit, to changing recharge. Successive simulated water table profiles will therefore be drawn at different depths, but in parallel with each other.

The moisture status of the unsaturated zone can be simulated in a distributed manner using the saturated zone storage deficit for any value of topographic index (and soil parameters) as the lower boundary condition on some model for the unsaturated zone. To represent the unsaturated zone in a simple manner, consistent with the overall level of simplification in TOPMODEL, unsaturated zone storage was calculated here based on a simple combination of root zone storage and a vertical time delay, as described by Beven et al. (1995).

The saturated zone in TOPMODEL can be derived (Kirkby, 1997) as a simplification of an ensemble of parallel, variable width flow strips, represented by the equation of continuity and a Darcian flow law (with fixed hydraulic gradients, assumed to equal local topographic slope). This formulation is not a fully 2D model in that there are no exchanges between adjacent flow strips, but can be thought of as a simply-distributed kinematic model. The assumptions made in TOPMODEL, especially the assumption of spatially uniform recharge, permit straightforward analytical solution, although at the expense of somewhat simplified dynamics. For a comparison of TOPMODEL with an explicit, grid-based model for topographically driven subsurface flow, see Wigmosta and Lettenmaier (1999).

Surface water storage and overland flow occur in the Minifelt and present a problem in formulating a minimally-parameterised distributed model. A key difficulty is that the surface water can arise through a combination of processes, namely extension of the saturated zone above the surface (as exfiltration), ponding in sub-grid scale topographic depressions and lateral extension of the "channel" as stream levels rise. Calibration of TOPMODEL against observed streamflows has produced very good simulations without any explicit model for overland flow (Lamb et al., 1997), by treating surface water essentially in the same way as the saturated zone. Physically, this approach represents a great simplification of processes, but has the advantage of parsimony in that no roughness coefficients, wave velocities or time delay parameters need to be calibrated. However, the simplifying assumptions do have to be carried through to the analysis of distributed water table depths. This is the example of trade-offs discussed in Chapter 3.

Despite the ability to simulate distributed responses, the focus of most TOPMODEL studies has been on areally-integrated simulation of runoff. However, a number of studies have tested distributed aspects of the TOPMODEL concepts (Ambroise et al., 1996; Moore and Thompson, 1996; Jordan, 1994; Burt and Butcher, 1985) or used a distributed parameter approach without testing (Coles et al., 1997). Two recent studies, in the Minifelt (Lamb et al., 1997) and another small Scandinavian catchment (Seibert et al., 1997), have reported tests of TOPMODEL in simulating extensive shallow water table depth observations. In both studies, predictions obtained by the simple TOPMODEL concepts were often in error locally, but could be improved by estimation of local parameters; Seibert at al. (1997) used the water level observations to fit local values of a "groundwater index", equivalent to $\ln(a/T_0 \tan \beta)$ whilst Lamb et al. (1997) explicitly estimated local values of log-transmissivity, but with spatially constant $\Delta\theta$. The following sections describe extensions to this work to investigate the estimation of local values for both $\Delta\theta$ and $\ln(T_0)$, and to test the predictive performance of the resulting "spatially calibrated" model.

11.5 ESTIMATION OF $\Delta\theta$ AND $\ln(T_0)$ USING TIME SERIES OBSERVATIONS

Calibration of TOPMODEL was approached in several stages. Firstly, 'global' (i.e. spatially-constant) parameters were estimated by fitting against observed flow series from a period of six weeks in 1987, as described by Lamb et al. (1997). Then, local values of the soil parameters $\Delta\theta$ and $\ln(T_0)$ were calibrated against measured water table depth data, initially using logged borehole water levels from the 1987 period, then using two of the five piezometer surveys. The calibration against logged borehole data was used to look at temporal dynamics of the boreholes, while the separate calibration against piezometer surveys was used to look at spatial patterns. Effective porosity, $\Delta\theta$, affects the dynamics of water level changes, and was calibrated using differences in water levels with respect to simulated storage deficits. Transmissivity, $\ln(T_0)$, was then treated, in effect, as a "correction factor" to adjust simulated water levels up or down to match observations.

A random search procedure was used by Lamb et al. (1997) to estimate values for the TOPMODEL saturated zone parameters m and Q_0 by maximisation of the Nash and Sutcliffe (1970) efficiency statistic (NSE) calculated on the difference between observed and simulated streamflow. Because $\overline{\ln(T_0)}$ is the only unknown factor in Q_0, calibration of Q_0 effectively provides a first estimate of the mean log-transmissivity, arrived at independently of any local values. The calibrated parameter values of m and $\ln(T_0)$ are 3.5 mm and 0.27 (T_0 in m/h). The fit of simulated and observed flows over the calibration period, which contained several rainfall events, including one large storm, was visually very good, with NSE $= 0.9$ (see Lamb et al., 1997, Figure 6), even without an explicit model for overland flow.

The parameters calibrated using flow data were applied in (11.1) to simulate time series of the local storage deficit D at each of the four logged borehole locations. To transform the simulated deficits to water table depths for comparison with the observations, (11.4) was applied, requiring estimation of the effective porosity parameter $\Delta\theta$ for each borehole.

However, both $\ln(T_0)$ and $\Delta\theta$ affect the simulated water table depth. To estimate $\Delta\theta$ and $\ln(T_0)$ independently, it is necessary to resolve this dependency. By rearranging (11.1), it is possible to write expressions for the catchment mean storage deficit in terms of both uniform soil parameters and distributed soil parameters, such that

$$\hat{D} + m \cdot \Lambda = \bar{D} = (\Delta\theta \cdot z) + m \cdot \Lambda - m \cdot \left(\ln(T_0) - \overline{\ln(T_0)}\right) \qquad (11.5)$$

where

$$\Lambda = \ln\frac{a}{\tan\beta} - \lambda \qquad (11.6)$$

The left-hand side of (11.5) states that the mean storage deficit in the catchment can be expressed as a function of topography and a local deficit $\hat{D}$, where $\hat{D}$ is simulated using the parameters calibrated by fitting against observed flow data (the "hat" notation is used here to emphasise that this term is a *simulated* storage deficit). The same mean deficit is also expressed on the right-hand side of (11.5) as a function of topography, the difference between mean and local log-transmissivities, and a local deficit, $D(z) = (\Delta\theta \cdot z)$, estimated as a function of the observed water table depth z.

If it is assumed that $\hat{D} = D(z)$, then (11.5) will be true only if $\ln(T_0) = \overline{\ln(T_0)}$. Any difference between the simulated storage deficit $\hat{D}$ and the estimated deficit $(\Delta\theta \cdot z)$ can thus be attributed to differences between the local transmissivity and the global transmissivity, scaled by m, provided there are no significant timing errors in the simulated storage deficit series.

Equation (11.5) can be rearranged to eliminate $\bar{D}$ and Λ, leading to an expression for the difference between the mean and local soil transmissivity parameters,

$$\overline{\ln(T_0)} - \ln(T_0) = \frac{\hat{D} - (z \cdot \Delta\theta)}{m} \qquad (11.7)$$

For the four boreholes in the Minifelt, small differences were found between the timing of the responses of simulated deficits and observed water levels to rainfall events, but the onset of recession periods was very nearly simultaneous (errors of only one or two hours) for both time series. It was therefore concluded that (11.5) and (11.7) could reasonably be applied during recession periods.

Assuming that the difference between local and mean log-transmissivities does not change over time, and that the parameters m and $\Delta\theta$ are also constant in time, (11.7) can be used to express $\Delta\theta$ in terms of the simulated storage deficits and observed water levels at times t and $t + \Delta t$, where

$$\Delta\theta = \frac{\hat{D}_{t+\Delta t} - \hat{D}_t}{z_{t+\Delta t} - z_t} \qquad (11.8)$$

and the subscripts denote time.

In applying (11.8), the time interval Δt was set to be five hours and t was chosen to coincide with a prolonged recession period such that values of z over the interval $t + \Delta t$ were centred about the mean water table depth in each borehole. Values for $\Delta\theta$ were estimated in this way for boreholes 4 ($\Delta\theta = 0.06$), 5 ($\Delta\theta = 0.04$), and 6 ($\Delta\theta = 0.04$), but not for borehole 7 because the observed data at this location were of poorer quality, owing to instrument malfunction. Simulated water levels are plotted along with the observed levels for boreholes 4, 5 and 6 in Figure 11.5.

Variations in simulated water table depths shown in Figure 11.5 appear to be similar in amplitude to the variations in the observations, with the exception that the simulated water table does not extend as far above the surface as the observed water level at borehole 4, although the timing of simulated surface saturation is approximately correct. This difference arises because, when applying (11.4) to

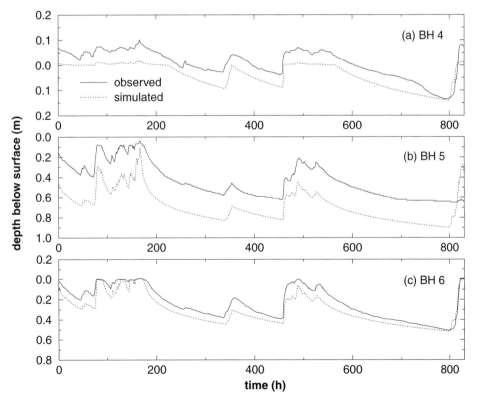

Figure 11.5. Simulated and observed water levels in boreholes (BH): 4 (a); 5 (b); and 6 (c) using local estimates of $\Delta\theta$ based on analysis of logged borehole data.

transform from simulated storage deficits to water table depths, the parameter $\Delta\theta$ was set to be equal to one for negative simulated deficits (i.e. when the simulated water table would be above the surface) to reflect the theoretical transition of porosity from a value less than one in the soil to exactly one in air.

However, it can be seen from Figure 11.5a that this theoretical condition, when applied within the current model structure, is not consistent with the data observed at borehole 4, where $\Delta\theta < 1$ would give rise to an improved fit between simulation and observations. There are two physical reasons for this finding. One is that an abrupt transition to $\Delta\theta = 1$ above the surface oversimplifies the complex and continuous transition in the uppermost layers of the soil, or in dense mossy vegetation and organic litter just above the soil surface. Perhaps a greater influence on the observed water levels is the accumulation of water in topographic features that are not properly represented in the catchment DTM. Allowing $\Delta\theta$ to be less than one above the surface leads to improved simulated water levels, but it must be recognised that the parameter then becomes less physically meaningful, and would be functionally compensating for errors and simplifications in the model.

The model results shown in Figure 11.5 are clearly biased, this being particularly notable for borehole 5, where the simulated water table is approximately 0.2 m below the observed level for much of the series. This bias can be attributed to a difference between local and mean log-transmissivities. Once $\Delta\theta$ is known, the local log-transmissivity $\ln(T_0)$ can be estimated by rearranging (11.7). For each borehole, (11.7) was therefore applied with the same simulated deficits and observed water table depths as used to estimate $\Delta\theta$. The resulting estimated local transmissivities for boreholes 4, 5 and 6 were $T_0 = 0.81\,\mathrm{m^2\,h^{-1}}$, $T_0 = 0.14\,\mathrm{m^2\,h^{-1}}$ and $T_0 = 0.68\,\mathrm{m^2\,h^{-1}}$ respectively. The simulated local water table depths were then revised, using the local values of T_0 and $\Delta\theta$. These are shown in Figure 11.6, plotted along with the observed water levels and the original predicted levels, based on the global transmissivity formulation. It can be seen that the local transmissivities effectively correct for much of the bias in the original simulated levels.

11.6 ESTIMATION OF SPATIALLY DISTRIBUTED $\Delta\theta$ AND $\ln(T_0)$ USING SPATIAL OBSERVATIONS

11.6.1 Spatial Predictions for Uniform Soil Properties: The "Global Parameter" Model

Simulations of storage deficits in the network of 108 piezometers were carried out for each of the five sets of observed water table depths. This was done in the space domain only by application of (11.1), under an assumption that the discharge on each occasion represented drainage from the saturated zone alone, allowing the observed discharges to be used to calculate the catchment mean storage deficit in each case after rearranging (11.3). Although some of the piezo-

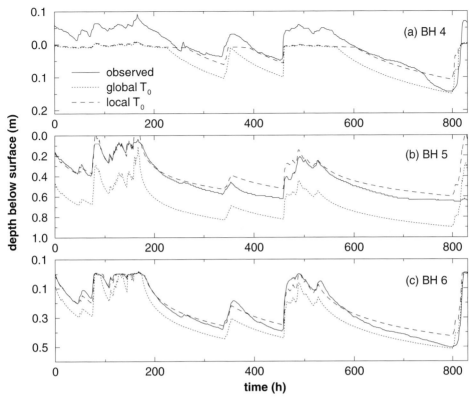

Figure 11.6. Simulated and observed water levels in boreholes: 4 (a); 5 (b); and 6 (c) using local estimates of $\Delta\theta$ and $\ln(T_0)$ based on analysis of logged borehole data.

meters indicate surface saturation (i.e. $z < 0$), use of the measured flow rates to approximate the saturated zone discharge has been justified (Lamb et al., 1997) on the basis that there was no rainfall during the measurement periods for the three data sets for which $\bar{Q} \leq 0.61\,\mathrm{mm\,h^{-1}}$, and that even under wetter conditions, the ponded areas represent in large part extensions of the saturated zone, arising due to local reductions in transmissivity, leading to exfiltration (or "return flow") accumulating on the surface in local depressions. Further discussion of this approximation will be found at the end of this section.

Water levels at each piezometer location were first simulated using the global transmissivity $\overline{\ln(T_0)}$ and an average value for $\Delta\theta$ of 0.05, which is the mean (to one significant figure) of the estimates for boreholes 4, 5 and 6. The simulated water levels can be considered as "validation" results for the global parameter case, where the lumped saturated zone parameters were calibrated solely on the basis of time series data. Simulated water levels are plotted as crosses in Figure 11.7, for each of the five values of $\bar{Q}$, as a function of the local $\ln(a/\tan\beta)$ index value. Also plotted as circles are the observed water levels. It can be seen that the simple "global parameter" model is not able to explain the local variations in water levels, the implication being that topography alone is insufficient to account for spatial variations in the depth to the water table.

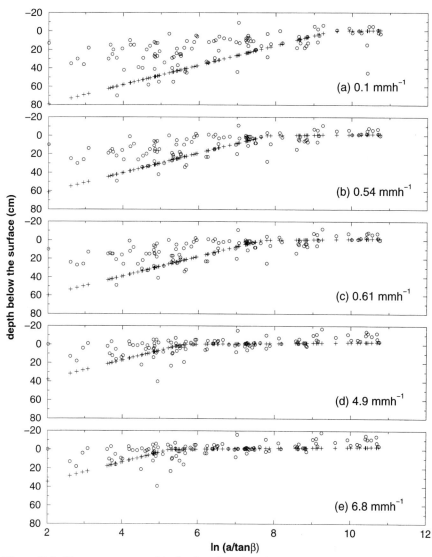

Figure 11.7. Piezometer water table depths, plotted against $\ln(a/\tan\beta)$ for the "global parameter" model ($\Delta\theta$ and $\ln(T_0)$ both constant in space). Discharges for the five data sets increase as indicated from graph (a) to graph (e). Circles are observations, crosses are simulated depths.

11.6.2   Spatial Predictions for Distributed Effective Porosity: the "Distributed $\Delta\theta$" Model

Spatially distributed values of $\Delta\theta$ were then calibrated as follows, using the piezometer surveys for $\bar{Q} = 0.1$ and $\bar{Q} = 0.61 \, \text{mm h}^{-1}$. Equation (11.8) was applied to estimate $\Delta\theta$ at each of the 108 piezometers, where instead of specifying a time interval between t and $t + \Delta t$, the values of $\hat{D}$ and z corresponded to $\bar{Q} = 0.1$ and $\bar{Q} = 0.61 \, \text{mm h}^{-1}$. These conditions were chosen to avoid using water table measurements made at the two higher averaged flow rates, thus reducing as far as possible any overestimation of the "true" saturated zone discharge and effectively integrating the estimate of $\Delta\theta$ over a vertical soil depth of the order of 10 cm, a much greater interval than that used in the case of boreholes 4, 5 and 6. It will be realised that the centre of this interval was also closer to the surface in places where the water table tends to be close to the surface.

The spatial pattern of $\Delta\theta$ is shown in Figure 11.8. The range between 0.02 and 0.40 encompasses 90 % of the values, with a mean of 0.16. This may be compared to a range of $0.1 < \Delta\theta < 0.2$, estimated from measurements in the field of drainable water content and total porosity. The largest calibrated value was 0.96. Calibrated values of $\Delta\theta$ vary most in areas where the water table tends to be closest to the surface. In areas where the water table is generally deeper, no large calibrated values were obtained.

The spatial pattern of estimated $\Delta\theta$ values mapped in Figure 11.8 shows some similarities with the patterns of observed water levels and surface elevations. Particularly notable is the appearance of high values of $\Delta\theta$ along the main axis of the valley, both close to the outlet and at the catchment boundary. These are also areas where the water table rises above the surface, even at relatively low flows (see Figure 11.4). Large estimated values of $\Delta\theta$ are to be expected in these areas because of the use of data recorded close to or at the surface. However, small estimated values of $\Delta\theta$ also occur in wet areas close to the catchment outlet.

It would be tempting to use the spatial patterns of soil depth, topography and water table depths to postulate some more general relationship between these variables and $\Delta\theta$. However, any similarities between the patterns of these "independent" variables and the spatial distribution of $\Delta\theta$ seem only to be very

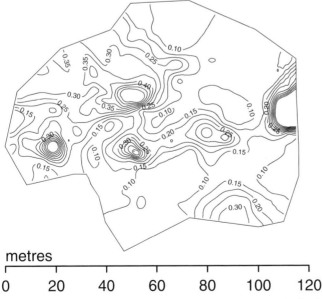

Figure 11.8. Spatial pattern of interpolated local values of effective porosity, $\Delta\theta$ obtained from calibration using measured patterns of depth to water table (for $\bar{Q} = 0.1$ and $0.61\,\mathrm{mm\,h^{-1}}$).

localised. When examined across the whole catchment, $\Delta\theta$ was found to be very poorly correlated ($R^2 = 0.2$) with the topographic index $\ln(a/\tan\beta)$, and soil depth ($R^2 < 0.1$).

By applying the local values of $\Delta\theta$ to each piezometer location, revised simulated water table depths were calculated, retaining the assumption of spatially constant transmissivity. These simulated levels are shown in Figure 11.9, where the data are plotted as a function of $\ln(a/\tan\beta)$. The data plotted in Figure 11.9b, d and e are effectively "validation" results for the distributed-$\Delta\theta$ model since these data were not used in the calibration of the distributed $\Delta\theta$ values. Data shown in Figure 11.9a and c were used for calibration; however, it is worth

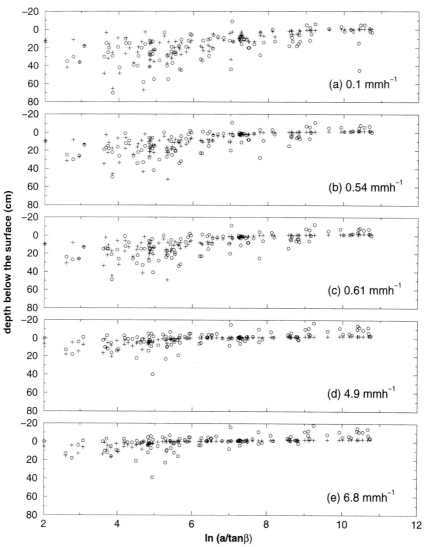

Figure 11.9. Simulated and observed piezometer water table depths for the "distributed $\Delta\theta$" model ($\ln(T_0)$ constant in space). Circles are observations, crosses are simulated depths.

noting again in this context that the local values of $\Delta\theta$ were calibrated on the basis of *differences* between these two piezometer surveys.

Comparison with Figure 11.7 shows that estimation of local values for $\Delta\theta$ has introduced a degree of spatial variation into the predicted water levels that is similar to the variation in the observations. Qualitatively then, the "distributed $\Delta\theta$" model appears more realistic than the original model, although there are still many errors in the predicted water levels.

The root mean squared error (RMSE) for each set of simulated water levels was computed for both the "distributed $\Delta\theta$" and "global parameter" models (see Table 11.1 below). Lower RMSE values for the distributed $\Delta\theta$ model confirm the visual impression of an improvement in the simulated water levels.

11.6.3   Estimation of Distributed Transmissivities

Given knowledge of local values of $\Delta\theta$, (11.7) was applied to calibrate the local log-transmissivity at each piezometer. The simulated storage deficit $\hat{D}$ was calculated for $\bar{Q} = 0.61\,\mathrm{mm\,h^{-1}}$, representing the median of the range of flow rates under which the piezometer data were collected. The resulting values of $\ln(T_0)$ are plotted as a contour map in Figure 11.10, and range from -10.3 to 24.4, with a mean of 2.5. However, 92% of the estimated values lie within the range $-5.0 < \ln(T_0) < +5.0$, for which the mean value of 0.77 is much closer to the global log-transmissivity calculated from the calibrated value of Q_0 ($\ln(T_0) = 0.27$).

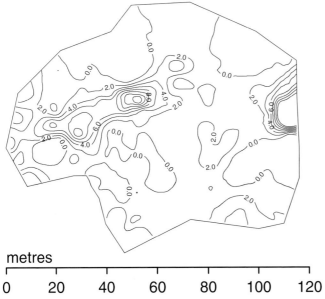

metres

0   20   40   60   80   100   120

Figure 11.10. Interpolated spatial pattern of local log-transmissivity $\ln(T_0)$ based on calibration using the measured pattern of depth to water table for $\bar{Q} = 0.61\,\mathrm{mm\,h^{-1}}$.

Physically, it might be expected that $\ln(T_0)$ would be related to soil depth, because, for any two locations having identical soil characteristics but different soil profile depths, the lateral transmissivity when saturated would be greater for a deeper profile. However, no clear relationship was found between local $\ln(T_0)$ and soil depth (correlation coefficient of -0.2). It can be seen from Figure 11.10 that the interpolated spatial pattern of $\ln(T_0)$ is complex and not easy to interpret, although there may be a weak correspondence between areas of low transmissivity and areas shown in Figure 11.4 to be wettest. An alternative explanation could be that the local transmissivity parameter might act to compensate for errors in estimating the "true" upslope contributing area from a DTM. Such compensation could occur because of the interaction between $\ln(a)$ and $\ln(T_0)$ implied by the right-hand side of (11.1).

11.6.4   Spatial Predictions for Local Soil Properties: the 'Distributed Parameter' Model

Water table depths were simulated using the "distributed parameter" model (local estimates of $\Delta\theta$ and $\ln(T_0)$) and are shown in Figure 11.11, plotted as a function of $\ln(a/\tan\beta)$ along with the observed water levels. Because the data sets for $\bar{Q} = 0.1\,\mathrm{mm\,h^{-1}}$ and $\bar{Q} = 0.61\,\mathrm{mm\,h^{-1}}$ were those used to calculate the distributed parameters, it is not surprising that the simulated water table depths for these two mean discharges fit very well to the observations (Figure 11.11a and c). However, this outcome was, again, not inevitable, given that $\Delta\theta$ is computed using differences between the two data sets, whereas $\ln(T_0)$ is calculated solely on the basis of the $\bar{Q} = 0.61\,\mathrm{mm\,h^{-1}}$ measurements (Figure 11.11c).

Also notable is the very close fit (RMSE $= 1.2\,\mathrm{cm}$) between simulated and observed water table depths for the $\bar{Q} = 0.54\,\mathrm{mm\,h^{-1}}$ data (Figure 11.11b), which were not used in calibration. However, for wetter conditions ($\bar{Q} = 4.89$ or $6.8\,\mathrm{mm\,h^{-1}}$) it can be seen from Figure 11.11d and e that the distributed parameter model performs less well, leading to errors at a few locations that are as large as those produced by the original, "global parameter" TOPMODEL.

Despite these errors, the RMSE results in Table 11.1 suggest that the distributed parameter formulation still performs better in predicting water table depths for the wetter catchment conditions than either of the simpler models tested. The model formulations are also compared in Table 11.2, which gives the frequency (as a percentage of all piezometers) for which the local parameter models performed better than the global case, irrespective of the magnitude of errors. These results confirm the improvement in predictions made using distributed parameters.

Spatial patterns of the simulated depth to water table were constructed by interpolating TOPMODEL predictions, using distributed parameters, at each piezometer site onto a $2\,\mathrm{m} \times 2\,\mathrm{m}$ grid, using the same algorithm adopted for the observed data. The simulated patterns are shown in Figure 11.12 and can be compared to the observed water table depths in Figure 11.4. There is a very

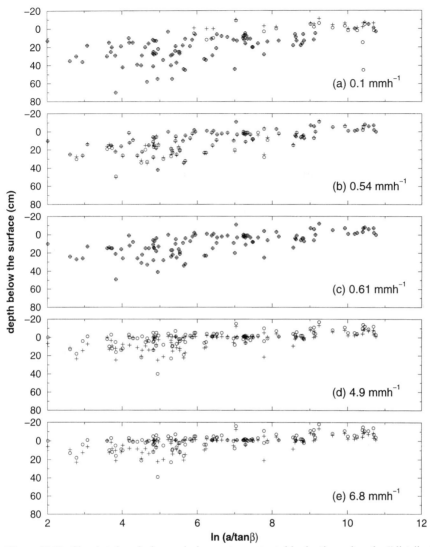

Figure 11.11. Simulated and observed piezometer water table depths, using the "distributed parameter" model ($\Delta\theta$ and $\ln(T_0)$ both spatially distributed). Circles are observations, crosses are simulated depths.

high level of agreement between the simulated and observed patterns. Recall that the observed patterns for $\bar{Q} = 0.1$ and $0.61\,\mathrm{mm\,h^{-1}}$ were used for calibration of the spatially variable $\Delta\theta$ and $\ln(T_0)$ so these patterns would be expected to be almost identical (which they are). The observed patterns of depth to water table are not appreciably different for the values of $\bar{Q} = 0.54$ and $0.61\,\mathrm{mm\,h^{-1}}$, reflecting the minor difference in magnitude of $\bar{Q}$ (i.e. the patterns are essentially for the same runoff magnitude, albeit measured at a different time). The simulated patterns are therefore also not appreciably different from each other or the observed patterns. The simulated patterns for $\bar{Q} = 4.9$ and $6.8\,\mathrm{mm\,h^{-1}}$ do differ

Table 11.1. Root mean squared errors (cm) for spatial water table simulations. Parentheses indicate data sets used for calibration of local values of ln (T_0)

	RMSE		
Q (mmh^{-1})	Global parameter[1]	Local $\Delta\theta$[2]	Distributed parameter[3]
0.10	22.5	10.4	5.5
0.54	15.0	9.4	1.2
0.61	14.7	9.1	(1.2×10^{-5})
4.89	9.7	7.5	5.7
6.80	9.1	7.5	5.7

[1] Spatially constant $\Delta\theta$ and ln(T_0)
[2] Local values of $\Delta\theta$, spatially constant ln(T_0)
[3] Local values of $\Delta\theta$ and ln(T_0)

from the observations with the width of the saturated area in the main drainage line being over-predicted (i.e. too wet) and the depth to water table around the boundary tending to be over-predicted (i.e. too dry). The general pattern is still well simulated because this is dominated by the spatial pattern of the calibrated $\Delta\theta$ and ln(T_0). These parameters impose a basic pattern which is the same for each wetness level; it just moves "up and down" depending on the overall wetness. The key to such good agreement between Figures 11.12 and 11.4 is the availability of enough detailed spatial observations to calibrate the spatial patterns of $\Delta\theta$ and ln(T_0).

As noted above, spatial parameter calibration and simulation were carried out under an assumption that the recorded (averaged) flows during each piezometer survey were discharged from the saturated zone alone. This assumption might overestimate the true saturated zone discharge for the wetter conditions. However, it is not at all certain what proportion of the flows recorded during piezometer surveys should be assumed to have reached the outlet via a surface "storm runoff" route. Further complications arise because of macropore and pipe flows, leading to rapid sub-surface responses, combined with relatively

Table 11.2. Frequency table for comparison of model formulations. Parentheses indicate data sets used for calibration of local values of ln (T_0)

Mean discharge (mmh^{-1})	0.10	0.54	0.61	4.89	6.80
% of points where distributed $\Delta\theta$ model is better than global model	64	52	51	27	22
% of points where distributed parameter model is better than global model	94	93	(100)	67	64
% of points where distributed parameter model is better than distributed $\Delta\theta$ model	90	94	(100)	58	57

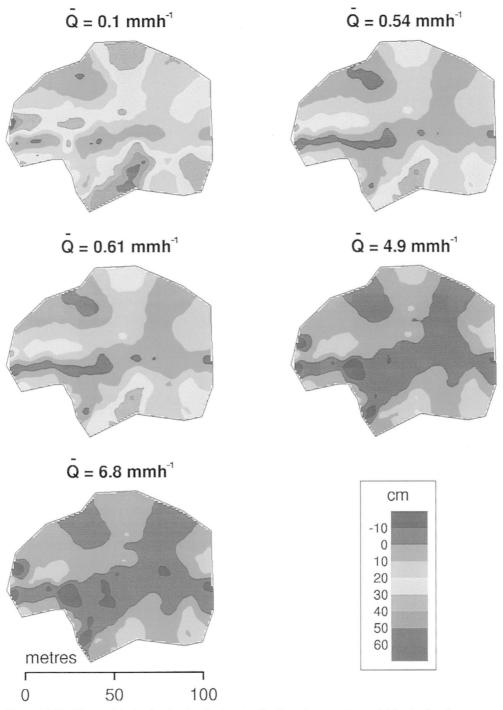

Figure 11.12. Water table depths simulated using the distributed parameter model for the five time-averaged flow rates corresponding to the observed water table patterns shown in Figure 11.4. Depths are in cm (positive downwards).

low topographic gradients in the valley bottom, and the role of sub-grid scale topography. The preceding results could therefore be altered by different assumptions about the partitioning of the gauged runoff, although the principles of the calibration approach are not affected.

To provide a simple test of implications of runoff partitioning, the saturated zone discharge was adjusted for the two water table surveys ($\bar{Q} = 4.89 \, \text{mm h}^{-1}$, $\bar{Q} = 6.8 \, \text{mm h}^{-1}$) during which rainfall was observed. Flows were reduced by an amount equal to the product of the rainfall during the survey and the proportion of the catchment that was saturated (as estimated from the interpolated water table maps in Figure 11.4). This straightforward partitioning caused the assumed flow rates to be reduced by up to approximately one-third of the recorded value. The reduced flow rates lead to reductions in saturated zone discharge, implying slightly increased catchment average storage deficits. Local deficits therefore increased uniformly, and calibrated transmissivities decreased slightly to compensate. However, the effects on parameter estimates were only very slight. Water levels were then simulated using the distributed parameter formulation; root mean squared errors increased by no more than 11%, compared to the results obtained for the original saturated zone flow rates.

11.7   UNCERTAINTY ESTIMATION AND SPATIALLY DISTRIBUTED DATA

The formulations of TOPMODEL developed in this chapter represent a very simple approach to distributed hydrological modelling although establishing the spatially variable values of $\Delta\theta$ and $\ln(T_0)$ required a substantial amount of spatial data. One of the main motivations for adopting a simple approach is to obtain a model that has as few unknown parameters as possible. This has the advantage of reducing the number of degrees of freedom present if the model has to be fitted to observations of "output" or "internal state" variables, which are often more accessible than physical measurements of parameters.

However, even using a simple and parsimonious model structure, there may still be considerable uncertainty about the values of some or all of the model parameters. A number of studies (e.g. Freer et al., 1996; Franks et al., 1997; Lamb et al., 1998b) have used Monte Carlo methods with variants of TOPMODEL to reveal that there may be multiple sets of parameter values that lead to similarly acceptable simulations, judged in terms of various objective functions. This behaviour, which is well known as "non-uniqueness" or "equifinality", appears to be a generic problem, and has been noted in a number of contexts, such as groundwater modelling (Neuman et al., 1980), catchment modelling (Binley and Beven, 1991; Duan et al., 1992), hydraulic floodplain inundation modelling (Romanowicz et al., 1994), and in predictions of water quality (Beck, 1987; Zak et al., 1997).

Parameter uncertainty resulting from the presence of equifinality can be expected to lead to uncertainty in predictions. Uncertainty can also be expected as a result of the difference in hydrological models between the scale of model solutions and the scale at which physical measurements can be made, as a result of errors in measurements, and of errors in the translation of real processes to the mathematical constructs used in all conceptual or physical models (also see discussion on pp. 19–22, 27 (Chapter 2) related to the matching of process, measurement and model scales, and pp. 70–4 (Chapter 3) on model complexity).

Uncertainty about TOPMODEL predictions in the Seternbekken Minifelt has been investigated by Lamb et al. (1998b). This work used the Generalised Likelihood Uncertainty Estimation procedure (GLUE) of Beven and Binley (1992), a Bayesian Monte Carlo method that can be applied to investigate how different sets of observed data constrain parameter and simulation uncertainty, i.e. to see how useful the detailed spatial measurements of piezometric heads are as compared to other types of information such as runoff, or time series of heads for a small number of piezometers. GLUE is a fairly generic procedure and is easily implemented for nonlinear models of arbitrary complexity, given sufficient computing resources. In the context of distributed subsurface hydrological modelling, GLUE can be related to Monte Carlo methods as described by Peck et al. (1988). For comparison, a comprehensive treatment of Bayesian parameter estimation in groundwater modelling has been given by Neuman and Yakowitz (1979), whilst Cooley (1997) has compared various methods for estimating confidence intervals for a nonlinear regression model of a hypothetical groundwater system. The GLUE procedure will not be described in detail here, but, briefly, involves random generation of a large number, M, of independent sets of model parameter values (typically $M \geq 10,000$). For each parameter set, a model simulation is performed, and the value of a goodness-of-fit function or "likelihood measure" calculated. This likelihood measure may be a function of one or more observed variables. Parameter sets for which the likelihood measure falls below a specified threshold are rejected as unacceptable simulators of the observed data.

For the Seternbekken catchment, Lamb et al. (1998b) used the following function to define the likelihood measure L for the ith parameter set Θ_i, given observations Y, such that

$$L(\Theta_i | Y) = \exp\left(-W \cdot \frac{\sigma_e^2}{\sigma_o^2}\right) \tag{11.9}$$

where σ_e^2 is the variance of the simulation errors, σ_o^2 is the variance of the observations and W_i is a weight. Equation (11.9) has the property that when N different sets of observations $Y_1, Y_2, \ldots, Y_N$ are combined using the principle of Bayes' theorem, the resulting "updated" value of L, is given by

$$L(\Theta_i | Y_{1,\ldots,N}) = \frac{1}{C} \cdot \exp\left[-\left(W_1 \cdot \frac{\sigma_{e,1}^2}{\sigma_{o,1}^2} + W_2 \cdot \frac{\sigma_{e,2}^2}{\sigma_{o,2}^2} + \cdots + W_N \cdot \frac{\sigma_{e,N}^2}{\sigma_{o,N}^2}\right)\right]$$

$$\tag{11.10}$$

where the scaling factor

$$C = \sum_{i=1}^{M} L\left(\Theta_i | Y_{1,\dots,N}\right) \tag{11.11}$$

The exponential term in (11.9) implies additive combination of individual error variance terms within the exponent of (11.10). This is an important feature if Θ_i successfully reproduces some sets of observations but is very poor at simulating others; it may be considered undesirable in this case to reject Θ_i, but this would almost certainly happen if the individual variance ratios were multiplied together directly. The weight given to the ith set of observations is controlled by choosing a value for W_i.

Equation (11.10) was applied in the case of the Minifelt to examine the changing statistical distributions of L conditioned on different combinations of flow data, borehole water level time series and spatial patterns of piezometer water table depths. Uncertainty bounds can be constructed for any simulated variable using the array of values of L conditioned on any combination of observed data sets as follows: First, the simulations at every ordinate (time step or location) are sorted in terms of magnitude, and the corresponding distribution for L is found. Then, uncertainty bounds are drawn at values of the simulated variable corresponding to selected quantiles of the distribution of L, in this case the 10th and 90th percentiles.

The GLUE procedure was applied using the simple, "global parameter" formulation of TOPMODEL to estimate the uncertainty about simulations of the spatially distributed water table depths at every piezometer location and to assess the utility of different sorts of data (catchment discharge, piezometric levels from a few logged boreholes, levels from the 108 piezometers) in constraining uncertainty. The main reasons for not using either of the distributed (local) parameter formulations were the computational demands involved in generating sufficient parameter sets to sample independently from a wide range of possible values for local parameters at each of over 100 locations, and then storing the resulting simulations.

Figure 11.13 shows, for three of the piezometer data sets, the uncertainty bounds computed using spatially constant parameters, conditioned firstly on a combination of time series observations (flows and logged borehole water levels), secondly on the errors in simulating the 108 spatially distributed piezometer water table depths and, thirdly, on a combination of the time series and spatially distributed data. The flow and borehole data were given equal weighting with the piezometer data in (11.10).

It can be seen from Figure 11.13 that, for the drier parts of the catchments (small $\ln(a/\tan\beta)$), uncertainty bounds conditioned on the patterns of piezometer observations (dashed lines) are slightly narrower than the bounds conditioned on time series data (dotted lines) while, for the wetter parts of the catchments (large $\ln(a/\tan\beta)$), they are wider. This indicates that in the drier parts of the catchments that are closer to the ridges, patterns of depth to the water table are indeed

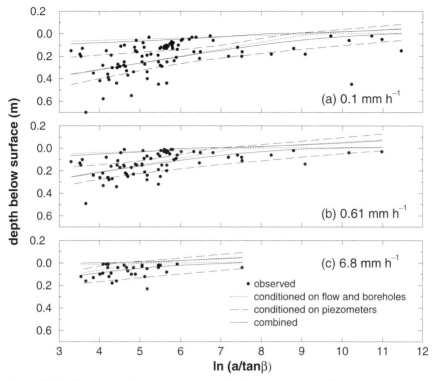

Figure 11.13. Uncertainty bounds on simulated piezometer water table depths compared to observations for time-averaged flows: (a) $\bar{Q} = 0.1\,\text{mm}\,\text{h}^{-1}$; (b) $\bar{Q} = 0.61\,\text{mm}\,\text{h}^{-1}$; and (c) $\bar{Q} = 6.8\,\text{mm}\,\text{h}^{-1}$. (After Lamb et al., 1998b.)

valuable. However, in the gullies their value is significantly lower than that of the dynamic information of streamflow and water table. This may be due to surface water ponding in which case the piezometers do not contain much information on the characteristics of catchment response. The combined case in Figure 11.13 (solid line) shows narrower bounds than either of the other two cases as the simulations are constrained by both patterns and time series data. However, this does not necessarily mean that constraining the parameters by both patterns and time series data improves the accuracy of the distributed model. In fact, the bounds of the combined case exclude a large proportion of the observations, especially where the water table is high. In order to interpret this it is important to realise that the total uncertainty of a model simulation is the sum of uncertainty in the parameters, uncertainty in the inputs and uncertainty in the model structure. The uncertainty bounds in Figure 11.13 only reflect uncertainty in the parameters, assuming that the effects on the function L of data and model structure errors are similar for all parameter sets. While we assume that uncertainty in the inputs is relatively small, it is likely that if model structural uncertainty were included in the GLUE procedure, the bounds would significantly change to cover the larger scatter in the observed borehole data. Indeed, the TOPMODEL concepts impose a relatively high degree of structure on the system, as they are based on quasi one-dimensional (in the map view) flow redistribution. In order to obtain wider uncertainty bounds that may be more consistent with the data in Fig. 11.13, one would therefore also have to vary the model structure, in addition to the parameter values.

Some indication of the effects on local uncertainty of knowledge of water levels at a number of points can be gained by examining simulation uncertainty bounds for water levels in the four logged boreholes. These are shown in Figure 11.14 for three of them, where two sets of bounds are drawn, the first (denoted "conditioned on local data") being conditioned separately on the observations in each individual borehole, the second being conditioned on a combination of the errors in simulated water levels for the three boreholes together.

Figure 11.14 shows that the uncertainty about simulated water levels is comparable to that seen in Figure 11.13 for the piezometer water table depths. The uncertainty bounds computed for the logged boreholes, conditioned separately on the data from each location, do not vary greatly over time. This is in contrast to the more dynamic observed levels. The apparent lack of response may be a consequence of the simplification of subsurface dynamics inherent in the TOPMODEL formulation. However, individual parameter sets can be

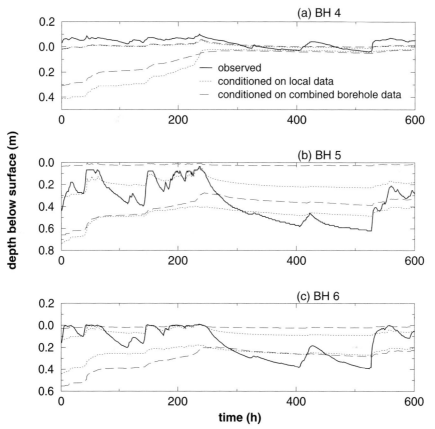

Figure 11.14. Uncertainty bounds on simulated water level time series in boreholes: (a) 4; (b) 5 and (c) 6. (After Lamb et al., 1998b.)

found to simulate the water level data in each borehole quite well, as illustrated in Figure 11.5. Other factors must therefore be considered to explain the rather "flat" uncertainty bounds. It was found by Lamb et al. (1998b) that the immediate reason for the lack of realistic dynamics in the borehole uncertainty bounds was uncertainty about the value of $\Delta\theta$, brought about by the degree of interaction in the model structure between parameters. This is consistent with the interpretation of Figure 11.13. Inspection of Figure 11.14 shows that taking account of observed water levels in all logged boreholes (dashed lines) generally tends to increase the uncertainty about simulated water levels, with the exception of the first part of the record for borehole number 4. This suggests that the addition of information from three points in the catchment does not constrain the uncertainty about predictions made using the simple, global parameter model at a given point but widens the range of possible parameter values and therefore the range of possible simulation results. This finding is likely to reflect the limitations of the model structure, suggesting that good simulations cannot be obtained at different points using the same parameter values everywhere.

11.8 SUMMARY

Observed spatial patterns of water table depth based on measurements at 108 locations have been used here to carry out spatial calibration of the distributed rainfall-runoff model TOPMODEL. The model calibration was formulated in three ways, as follows:

1. A "global parameter" case, where soil parameters have the same, areally averaged value everywhere and the distribution of saturated zone storage deficits is determined by topography alone.
2. A "distributed effective porosity ($\Delta\theta$)" case, where a catchment average effective transmissivity is used such that topography alone controls the distribution of saturated zone storage deficits, but the pattern of water table depths also depends on locally-estimated values of $\Delta\theta$.
3. A "distributed parameter" case, where the distribution of saturated zone storage deficits is a function of topography and locally-estimated values of the log-transmissivity, $\ln(T_0)$. In this case, the final distribution of water table depths also depends on locally-estimated values of $\Delta\theta$.

The local parameters were 'back-calculated' using a subset of the available spatial data. There is a risk that this method, like most model calibration procedures, may introduce an element of circular reasoning; parameters estimated using a given set of observations should be expected to lead to good simulations of the same data. Testing of the model on data not used for calibration is a typical response to this problem, and one that has been followed here (i.e. split sample testing – see Chapter 3, p. 76 and Chapter 13, p. 340 for further discussion).

Spatial patterns of the calibrated local parameters were difficult to interpret. Comparison with maps of topography, soil depth and observed water table depths suggested some weak correspondence between the variables in limited areas. However, correlation analyses showed there to be no general relationships between these observed variables and the calibrated local parameters.

Despite the difficulty in explaining physically the patterns of the calibrated parameters, it has been shown that estimation of local values greatly improves the simulation of spatial patterns of water table depths over a range of conditions. The results for the root mean squared errors (Table 11.1) of simulations using the three TOPMODEL formulations showed clearly that the distributed parameter model gives rise to relatively accurate simulations of water table patterns within what remains a simple model structure. This was made possible by the large amount of information on spatial patterns of water table depth which enabled the spatial calibration to be carried out.

Although the global parameter formulation of TOPMODEL is quite successful in simulating streamflows from the Minifelt catchment, it has been shown that a topographic control on the spatial distribution of storage is not a complete description for this catchment; spatial predictions made using the same soil parameter values everywhere were in fact wrong in most places, leading to a high degree of predictive uncertainty. The patterns of water table levels have allowed us to develop a spatially-calibrated model while retaining the original, simple structure. Even so, it has to be recognised that model assumptions still differ in some important respects from physical processes in the Minifelt. A further challenge will be to reconcile such differences but this will not be possible with the present model structure. TOPMODEL as used here greatly simplifies the spatial dynamics of catchment responses. In particular, overland flow is not treated explicitly, macropore flow is not represented and the downslope dynamics of the saturated zone are represented as a succession of steady states.

Despite these simplifications, subsurface water level responses in specific locations can be simulated reasonably well, and the pattern of the water table can be reproduced realistically when local soil parameters are used. An important question is whether or not the improved local predictions represent a meaningful improvement in modelling capability. Ideally, confidence in the distributed parameter model formulation would be enhanced by a physically convincing explanation for the values (and patterns) of the inferred, local parameters. Unfortunately, this was not possible. Likely reasons include that the conceptual definition of model parameters is not the same as the physical definition of measured properties, and that model parameters may be compensating for the effects of processes not explicitly represented in the model structure, possibly at sub-grid scales.

Results from the GLUE procedure suggest that whilst local or "point" data can help to constrain predictive uncertainty for distributed water levels, the interaction between parameters can still produce rather wide uncertainty bounds. This is also a consequence of limitations in the model structure which made it difficult to determine the value of different types of data in constraining uncertainty.

Assessment of confidence in distributed models should include testing against spatial data, both within an uncertainty framework and, where parameters can be calculated directly, in a manner similar to the "split-sample" approach. We have been fortunate to have access to detailed spatial data, allowing both these approaches to be explored. It is to be hoped that the collection of such distributed data sets will continue, both for the development of conceptual understanding of hydrological systems and for the refinement of distributed models.

12

Groundwater–Vadose Zone Interactions at Trochu

Guido Salvucci and John Levine

12.1 INTRODUCTION

The chapters in this book document sources of spatial variability of hydrologic fluxes and moisture storage in catchments. Some variability reflects that of the atmospheric forcing, and some results from the interaction of that forcing with spatially varying soil, vegetation, and topographic properties. Part of the variability due to interaction, for example that arising from the dependence of soil moisture on soil hydraulic properties during infiltration, may change from storm to storm depending on storm intensity and duration (e.g. Salvucci, 1998). Patterns that arise from the spatial distribution of groundwater–vadose zone interactions, however, have a persistent nature due to the long timescales of groundwater redistribution (see, e.g., Tóth, 1966). The existence, nature and cause of these patterns in the Trochu catchment of Alberta, Canada, are the focus of this chapter.

Throughout the chapter it is assumed that the dominant mode of this interaction (with respect to impact on spatial patterns) is the dependence of surface fluxes on the position of the water table relative to the land surface. This assumption is explored by coupling an equilibrium model (Salvucci and Entekhabi, 1995) that estimates long-term average water table dependent surface fluxes to a groundwater flow model, and then comparing the model results with patterns of recharge and discharge measured by Tóth (1966). The equilibrium model is particularly well suited for comparison with Tóth's measurements because the latter are largely based on field observations of natural time-integrators of sub-surface flow conditions (e.g. presence of salt precipitates). These measurements are a mix of quantitative and observational indicators that provide a spatial picture of long-term recharge and discharge locations.

The water table position relative to the ground surface is assumed to represent the dominant mode of interaction because it impacts on the partitioning of rainfall in two important ways: 1) by bounding the moisture profile, and 2) by

Rodger Grayson and Günter Blöschl, eds. *Spatial Patterns in Catchment Hydrology: Observations and Modelling* © 2000 Cambridge University Press. All rights reserved. Printed in the United Kingdom.

creating a potential source of capillary rise to the root zone. In areas of shallow water table, the bounding of the moisture profile promotes runoff (Dunne and Black, 1970b) and the potential for continuous capillary rise (Gardner, 1958) maintains evapotranspiration at potential rates long after other parts of the landscape dry out. The position of the water table, in turn, depends on the spatial distribution of recharge, capillary rise, and surface water contacts (e.g. springs and lakes). This interdependence of vadose zone and groundwater flows can be viewed either as a consequence of coupled soil moisture and groundwater dynamics, or simply as a constraint imposed by mass conservation; that is, averaged over many wetting and drying cycles, the divergence of the groundwater flow field, which depends on the groundwater pressure distribution and is thus reflected in the water table topography, must balance the net of inputs from and losses to the vadose zone.

12.1.1   Relation to Previous Studies

The central theme of this chapter, the continuity and interdependence of groundwater and vadose zone flows, was recognised and explored in a series of papers by Tóth (1962, 1963, and 1966). Therein a comprehensive analysis of the spatial structure of recharge is detailed, and a theory of regional groundwater flow that accounts for losses through the vadose zone as discharge from the aquifer system is developed. In the third paper, Tóth describes how chemical, biological and piezometric observations can be used to map the spatial distribution of aquifer recharge and discharge areas. In the first two papers he provides methods by which the groundwater flow equation can be solved, for a given fixed water table, in order to predict these patterns.

Despite the recognition of this interdependence three decades ago, many models today either treat groundwater and vadose zone flow systems in isolation, or at most treat conditions at the boundaries between them as fixed quantities. When interactions at the boundaries are ignored, however, potentially important feedbacks are not allowed to occur. For example:

1. Vadose zone analyses that assume a condition of gravity drainage at the bottom of a soil column (e.g. Milly and Eagleson, 1987) may predict recharge to groundwater in excess of what the underlying aquifer can transmit;
2. Climate models that incorporate one-dimensional land surface parameterisations (e.g. Rosenzweig and Abramopoulos, 1997) and ignore lateral groundwater redistribution may fail to simulate large low-lying areas where moisture is evaporated long after higher areas dry out, thus underestimating evaporation and overestimating the sensitivity of evaporation to model parameters;
3. Groundwater studies that take vadose zone inputs as independent of the groundwater flow regime (e.g. Danskin, 1988) can predict artificially high water tables, and those that fix the water table a priori and diagnose

vadose zone fluxes from the groundwater divergence field (e.g. Stoertz and Bradbury, 1989; Ophori and Tóth, 1989) may predict recharge in excess of annual precipitation; and

4. Catchment models that account for near-surface lateral saturated flow (e.g. Beven and Kirkby, 1979; Hatton et al., 1995), but fix the hydraulic gradient to reflect surface topography, constrain the spatial variability of lateral redistribution to be accounted for solely by changes in transmissivity. These models cannot account for the regional groundwater circulations that maintain riparian zones (which in turn affect evaporation and streamflow).

In summary, critical rate-limiting processes governing the local hydrologic cycle may be overlooked when applying methods in which the coupling at the water table is not specifically considered. This can have a large impact on the spatial estimates of ET, recharge and surface saturation.

There is great difficulty in evaluating models and assessing the importance of coupling because the relevant data is generally not available. The unusually detailed observations of Tóth (1966) provide one of few data sets useful for this purpose. In the following section we first describe the data and then present a model that will enable the issue of coupling to be explored.

12.2 METHODS

12.2.1 The Study Area and Available Data

The catchment chosen for this study (Trochu) is located in southern Alberta in the Canadian plains (Figure 12.1). This site was chosen because of the extensively documented fieldwork undertaken in the area by Tóth (1966), and because the poorly drained prairie topography emphasises the importance of three-dimensional groundwater circulation in determining the water balance (Levine and Salvucci, 1999a). The Trochu catchment is 16 km^2 in area with low relief (Figure 12.2). The maximum change in elevation from the water divide to the outlet is approximately 100 m and the maximum slope is approximately 6 %. The general flow direction in the catchment area is west to east. The average slope from the farthest point in the catchment to the outlet is under 2 %.

The soils in the study catchment are identified as thin black soils (combinations of silt, sand and gravel) developed on glacial drift material (Bowser et al., 1951). The bedrock consists primarily of nearly horizontal layers of sandstone and siltstone with some discontinuous layers of claystone and shale (Carlson, 1969; Tóth, 1966). The vegetation in the area is primarily cultivated rapeseed, alfalfa, and forage grasses with some small patches of aspen and willow trees along the ridge lines. Precipitation and temperature (New et al., 1999) both peak in the summer months (Figure 12.3), and the frost-free period typically begins in May and ends between September and October (Tóth, 1966). Average annual precipitation is approximately 440 mm.

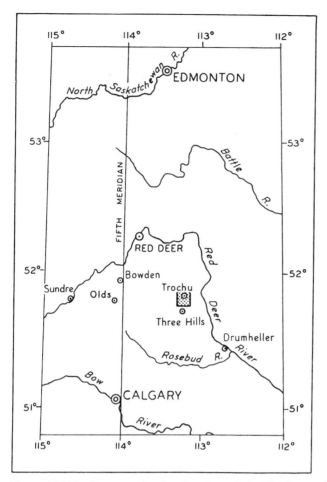

Figure 12.1(a). Site map showing location of the Trochu catchment in the Canadian Plains. (Reprinted from Tóth, 1966; reproduced with permission.)

Tóth (1966) studied this catchment extensively in an attempt to determine whether or not a correlation exists between physiographic features (e.g. vegetation types, presence of salt precipitates, moist depressions, well levels) and the direction of groundwater movement. In this work he provides a table of 152 field observations, 48 of which fall within the Trochu catchment. These observation points are labeled on Figure 12.2 (and the subsequent maps in this chapter) as R for recharge, D for discharge, C for creek bed and I for intermediate.

The categories were determined by applying Tóth's (1966) criteria for evaluating surface observations and wells as follows. Sites classified as recharge through chemical analysis are those where ground or surface water testing showed low concentrations of dissolved minerals. Discharge sites were assumed where high concentrations of dissolved minerals were present. Observations of vegetation and surface salt deposition were used together to classify otherwise dry observation points as either discharge or recharge points (Figure 12.2). Where phreatic vegetation is present (e.g., slough grass, sedges and rushes)

Figure 12.1(b). Photograph of the Trochu catchment.

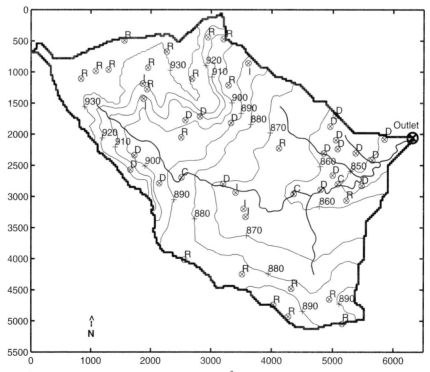

Figure 12.2. Surface elevation map of the 16 km^2 Trochu catchment with Tóth's observation points. Observation points are labelled as follows: R = Recharge, D = Discharge, I = Intermediate, C = Creek bed. Elevation contours in metres. Grid labeling in metres.

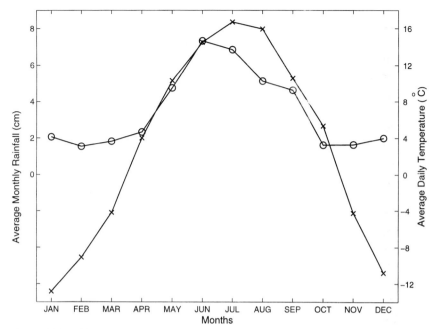

Figure 12.3. Thirty year mean (1961–1990) of monthly temperature (crosses) and precipitation (circles) for the Trochu region. (Data from New et al., 1999.)

without the presence of salt precipitates, the indication is that surface water recharges in this area. Phreatophytes with salt precipitates or salt precipitates alone indicate significant evaporation of groundwater, and so these points are classified as discharge points. Where two sets of criteria (i.e. water chemistry, vegetation, etc.) yielded different classifications, the point is labeled with both classifications.

Springs, seeps and flowing shotholes were all classified as discharge locations. Creek beds can be either gaining or losing reaches and so are not classified as either recharge or discharge. Piezometric classification was based on head to surface elevation comparisons. Where the head was near ($< 3\,\mathrm{m}$) or above the surface elevation, a well was classified as being a discharge observation. Where head in a well was significantly lower than the surface elevation ($> 10\,\mathrm{m}$), the piezometric determination was that the well was in a recharge zone. Between the two extremes wells were classified as being in intermediate zones. This criteria was not directly indicated in Tóth (1966), but rather inferred by comparing his reported measurements of depth to water against his final recharge–discharge map. Well head data was used as the sole determinant (i.e. without collocated chemical or botanical indicators) for approximately ten percent of the locations.

The resulting map (Figure 12.2) indicates a general pattern of recharge in the highlands (e.g. north-west and south-east borders of the basin and the north-west trending ridge in the northern third of the basin), and discharge in the low-lying areas, especially near the basin outlet. It is important to remember that in this gentle prairie topography, mapped discharge locations include both surface contacts where liquid water seeps out of the ground and areas of persistent evapotranspiration of capillary rise. In the following section a model is described that accounts for the continuity and interdependence of such groundwater and vadose zone flows, and thus should be well-suited to reproduce the observed patterns.

12.2.2 Model Review

The position of the water table is dependent on the convergence (divergence) of groundwater flow, the amount of water being lost (gained) at the saturated/unsaturated interface, and the location of direct aquifer–surface water contacts (seeps, springs, lakes, etc.). The loss (gain) at the water table interface depends on the partitioning of fluxes in the vadose zone.

The vadose zone receives water from rainfall and capillary rise, and loses water through evapotranspiration and recharge. The rates of capillary rise, recharge, infiltration and evapotranspiration are all influenced by the soil moisture profile, depth to the water table, soil characteristics, and surface meteorological forcing.

Changes in vadose zone characteristics (soil moisture, matric potential, etc.) occur over short time and length scales, whereas the characteristic temporal and spatial scales for groundwater flow are generally larger. To ease the resulting computational burden of full saturated–unsaturated numerical simulations (e.g. Freeze, 1971; Paniconi and Wood, 1993), Salvucci and Entekhabi (1995) built on Eagleson's (1978a–f) work to develop an equivalent steady state solution to the Richards equation bounded by a water table and driven by climate statistics (mean storm duration, intensity and frequency).

The climate statistics enter the model to parameterise the probability distributions of boundary conditions at the soil surface. Derived distribution techniques are used to average the resulting moisture fluxes over the storm–interstorm timescale, therefore providing an estimate of the time-averaged soil water flow through the vadose zone. This time–averaged flow, which may be downward recharge or upward capillary rise, forms a groundwater divergence boundary condition that is used to drive the spatially distributed groundwater model MODFLOW (McDonald and Harbaugh, 1996). As is discussed further below, the resulting distribution of water table depths influences the predicted vadose zone flow, and thus iteration is required to find the spatial distribution of water table depths for which the saturated flow divergence and vadose zone recharge are at equilibrium.

It is assumed in the model that interstorm evaporation and transpiration are driven by potential evaporation, but are influenced by both the mean root zone moisture content and by interstorm sources (capillary rise) and losses (recharge) below the root zone. The soil storage and infiltration capacities determine surface runoff through storage excess and infiltration excess mechanisms. The storage capacity of the soil is determined by integrating the soil moisture deficit from the surface to the water table. The infiltration capacity is governed by gravity and matric potential gradients through a two-term Philip (1957) equation. The expected values of these soil and moisture dependent fluxes are derived subject to random precipitation events by integrating the event fluxes over probability distributions of storm intensity, duration and intermittency.

The fluxes through the vadose zone are determined by finding the equivalent steady state soil moisture profile that yields closure to the surface water budget. This profile is used as an initial condition for determining the infiltration and evapotranspiration capacity of the surface. It provides a coupling to groundwater by matching the time-averaged flux (recharge or discharge) and pressure head at the mean position of the water table. Salvucci and Entekhabi (1994a,b) show, by comparison with a long-term finite element simulation, that the equivalent steady state moisture profile solution closely approximates the long-term mean vadose zone flux over a wide range of soil texture and climate conditions.

An example solution, used below to drive MODFLOW over the Trochu catchment, is illustrated in Figure 12.4 (the climate and soil parameters used in this example are discussed later). Note that the dependence of the water budget

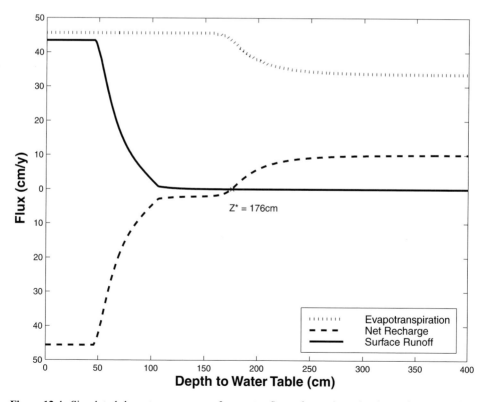

Figure 12.4. Simulated, long-term mean surface water fluxes for various depths to the water table. Silt-loam soil. $Z^* = 176$ cm. The first 45 centimeters above the water table are tension saturated. (From Levine and Salvucci, 1999a; reproduced with permission.)

on the depth to the water table (Z_w) occurs over a finite range from zero to the depth at which net recharge equals the maximum recharge rate. For this soil (silt–loam), the range of water table dependence is approximately 250 cm. There is a depth to the water table (Z^*) for each of the soil textures tested where the net recharge equals zero. At this depth, the long-term mean of transient groundwater losses due to capillary rise are balanced by the long-term mean of intermittent gains through recharge.

As the depth to groundwater decreases from Z^* to the tension saturated zone (for the silt–loam soil between 176 cm and 45 cm), runoff increases, evapotranspiration increases up to the potential rate, and the net flux across the vadose zone–water table interface is upward. Runoff (solid line in Figure 12.4) increases due to a reduction in both the infiltration capacity and the storage capacity that occurs for higher initial soil moisture content and reduced depth to the saturated zone. Evapotranspiration (dotted line in Figure 12.4) increases to the climate limited rate because capillary rise from shallow water tables is large enough to replenish all water lost to evapotranspiration, even over long interstorm periods. As the depth to saturation increases from Z^*, evapotranspiration decreases, and recharge increases to balance the net of infiltration over the reduced evapotranspiration. The decrease in evapotranspiration occurs primarily because moisture supplied to the root zone by capillary rise decreases with increasing depth to water table.

Note that the water table dependence of vadose zone fluxes disappears for depth to water table greater than that at which both evaporation and net recharge reach their asymptotic values. This implies that water table–vadose zone coupling is insignificant in determining net recharge outside this range of depths.

The long-term mean of the net flux across the water table (the dashed line plotted in Figure 12.4) is used as an input to the groundwater flow model, MODFLOW, which is run in steady-state mode. This flux can be either positive for deep water tables (recharge) or negative for high water tables (discharge/capillary rise). This dependence provides a feedback between the surface water balance model and the groundwater flow model (Figure 12.5) whereby high water tables lose water to, and deep water tables gain water from, the vadose zone. The methodology for coupling the water table dependent recharge/discharge flux to the groundwater flow model is detailed in Levine and Salvucci (1999a).

12.2.3 Model Modifications

The vadose zone model presented in Salvucci and Entekhabi (1995, 1997) was modified to account for winter precipitation and snowmelt. Snowmelt is divided between storage excess runoff and infiltration according to the storage capacity of the vadose zone. Cold season evaporation is assumed negligible. The infiltration from snow melt is added to the flux to groundwater, assuming that its effect is mainly to increase soil moisture, and thus recharge, as a single pulse during a period of low evaporation.

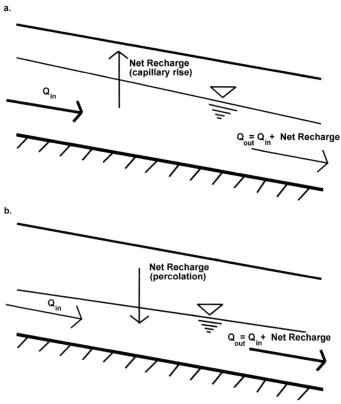

Figure 12.5. Equilibrium conditions at discharging (a) and recharging (b) sections of a hillslope. The net recharge is positive (negative) for areas where the water table is deeper (shallower) than the depth at which mean annual groundwater flux is zero for the simulated soil and climate conditions. (From Levine and Salvucci, 1999a; reproduced with permission.)

The mean bare soil evaporation equation derived in Salvucci and Entekhabi (1995, 1997) has also been modified to account for vegetation. The interstorm transpiration is modelled as a two-stage process (unstressed and stressed), with the transition time dependent on soil type, soil moisture and rooting depth (Levine and Salvucci, 1999b). As for the other event-based fluxes in the model, the mean transpiration is calculated by integrating over the probability distributions of the time between storms. Because the modelled transpiration is more efficient than bare soil evaporation, warm season recharge is negligible and the source of deep-water recharge in Figure 12.4 is mainly melted winter precipitation. As will be seen in the coupled model runs below, lateral groundwater redistribution of this winter recharge makes up the moisture deficit (evaporation – rainfall – surface runoff) in shallow water table areas throughout the rest of the year.

Groundwater discharge at the surface is simulated by distributing drains over the surface using the drain package in MODFLOW (McDonald and Harbaugh, 1996). The stream networks predicted by the models were drawn using a flow accumulation algorithm with weights determined from the sum of groundwater discharge and surface runoff predicted at each cell.

12.2.4 Scenarios and Model Parameters

Following are short descriptions of estimated model parameters, discretisation choices, and soil–bedrock combinations chosen for model runs. Results, discussions and conclusions are documented in the subsequent sections.

The groundwater model was run with one soil layer and six bedrock layers at a horizontal grid spacing of 30 metres. The parameters in each layer were spatially uniform. The groundwater divide was assumed to coincide with the surface divide. The vertical spacing was variable, with the top layer thickness set to the depth of unconsolidated material (derived from Carlson, 1969) and the lower layers adjusted such that the modelled impermeable bottom was reached without any layer being more than 1.5 times as thick as the layer above it. The modelled impermeable aquifer bottom was placed at 500 m above sea level (an average depth of 385 m below the surface). Following Tóth (1966), the bedrock is approximated as a single homogeneous-isotropic unit. Drains were placed just below the surface in each column in order to simulate springs and seeps if the water table intersects the land surface at equilibrium conditions. All water leaving the saturated zone via the drains is assumed to exit the catchment as stream flow. A test run with 10 metre horizontal grid spacing yielded nearly identical results, most likely because the gentle topography of the prairie catchment is adequately described by 30 metre data.

The climate statistics required as input to the model (Table 12.1) were derived from the long-term record (29 years) for Lacombe, Canada (52°28′ N, 113°45′ W) provided by Environment Canada. This is the closest meteorological recording station with long-term precipitation and potential evaporation records. The potential evaporation data are pan evaporation multiplied by a single site-wide adjustment factor. All climatic variables are assumed to be spatially uniform.

Results are presented for three simulations (Table 12.2) which include two soil types (silt-loam and clay-loam) over low conductivity bedrock and one soil type (silt-loam) over medium conductivity bedrock. The bedrock conductivities tested (Table 12.3) were chosen to cover the range of values estimated by Tóth (personal communication) in studies carried out over the Ghostpine and Three Hills Creek areas for the Edmonton and Paskapoo geologic formations. The soil types were chosen as representative of a silt-loam (similar to the local soil) and a clay-loam to demonstrate the effect that soil type has on surface/aquifer coupling. The

Table 12.1. La Combe climate parameters used in the simulations

Mean time between storms*	3.60 days
Mean storm duration*	0.92 days
Mean storm intensity*	1.08 cm/d
Mean evaporation*	0.37 cm/d
Winter precipitation (snow water equivalent)	9.96 cm

* Storm and evaporation statistics are for the 154-day average snow and frost-free period.
Source: Data from Environment Canada. Period of record is 1963 to 1992.

Brooks and Corey (1966) soil hydraulic parameters used to represent the soils are listed in Table 12.4. For simplicity the area is modelled with complete vegetation cover with an effective rooting depth of 45 cm. The model was also run for a two-dimensional cross-section in order to illustrate the variety of scales of circulation (local, intermediate and regional) making up the flow system. The two-dimensional results are presented in Levine and Salvucci (1999a).

12.3 RESULTS AND DISCUSSION

12.3.1 Impact of Soil and Aquifer Hydraulic Parameters on Spatial Patterns Induced by Saturated Unsaturated Zone Coupling

The influence of the coupling between vadose zone flux and depth to the water table is evident in the maps of simulated net recharge (Figures 12.6, 12.7 and 12.8). Case I (Figure 12.6) shows strong recharge (red) and discharge (blue) and many springs (white). Case II (Figure 12.7) shows strong recharge (red), discharge (blue), and springs (white) occurring over a smaller percentage of the area, and more extensive intermediate areas (yellow to pale orange) over which coupling strongly influences the height of the water table. Case III (Figure 12.8) is dominated by weak recharge (orange) and discharge (pale green) zones with extensive intermediate (yellow to pale orange) zones. The higher intensity of the recharge and discharge simulated in case I (and to a lesser extent case II) occurs because the higher conductivity allows greater flow through the aquifer, which in turn allows Z_w to remain significantly below Z^* over large areas.

Table 12.2. Numbering of the case studies

Case number	Soil type	Bedrock type
Case I	Silt-loam	Medium conductivity
Case II	Silt-loam	Low conductivity
Case III	Clay-loam	Low conductivity

Table 12.3. Bedrock Conductivity

Bedrock type	Saturated conductivity (cm/d)
Low conductivity	0.2
Medium conductivity	2.0
High conductivity	20

Table 12.4. Brooks and Corey parameters

Soil Type	Saturated conductivity (cm/d)	Pore size distribution index	Effective porosity	Bubbling pressure head (cm)
Clay-loam	3	0.44	0.45	−90
Silt-loam	30	1.2	0.35	−45
Sand-loam	300	3.3	0.25	−25

Source: Bras (1990)

Reducing the soil conductivity and increasing the soil water retention ability (i.e. comparing case II, Figure 12.7 to case III, Figure 12.8) decreases the strength of the recharge (red to orange) and discharge (dark blue to pale blue and green) and yields larger midline areas (light orange and yellow). This compensation

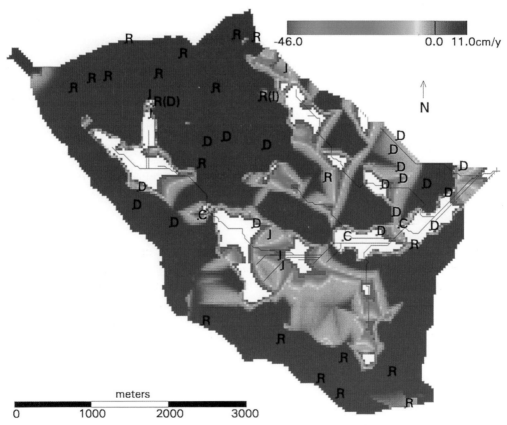

Figure 12.6 Case I: Simulated equilibrium recharge, capillary rise, spring locations, and surface drainage. Colour bar scales from maximum recharge (red, +11 cm/year) to maximum capillary rise (blue, −46 cm/year). Seeps are denoted by white pixels. Observation points are labelled as follows: R = Recharge, D = Discharge, I = Intermediate, C = Creekbed. (From Levine and Salvucci, 1999a; reproduced with permission.)

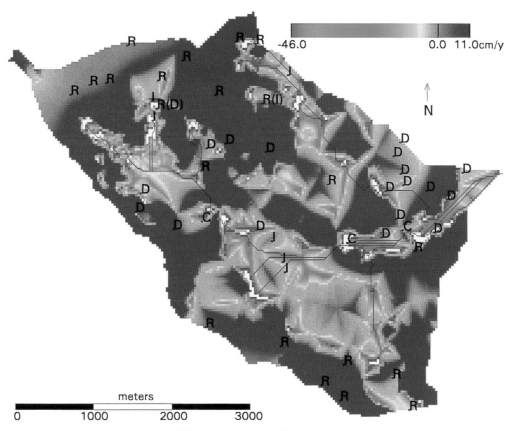

Figure 12.7. Case II: Simulated equilibrium recharge, capillary rise, spring locations, and surface drainage. Colour bar scales from maximum recharge (red, +11 cm/year) to maximum capillary rise (blue, −46 cm/year). Seeps are denoted by white pixels. Observation points are labelled as follows: R = Recharge, D = Discharge, I = Intermediate, C = Creek bed. (From Levine and Salvucci, 1999a; reproduced with permission.)

occurs because the coupling effects for the clay-loam soil are active over a greater depth than for the silt-loam soil, requiring a greater depth to the water table to maintain recharge. The greater depth of interaction enables the feedback mechanism to regulate the water table height more strongly by restricting the saturated depth and hydraulic head in the relatively conductive soil layer, thus restricting the ability of the groundwater to drain away recharge. The reduced drainage potential, in turn, drives the water table back toward the equilibrium distance (Z^*) from the land surface.

Holding the soil type constant and increasing the bedrock conductivity (i.e. comparing case I, Figure 12.6 to case II, Figure 12.7) reduces the water table height, increases the strength and areal extent of recharge (red) and discharge (blue and green) zones, and shrinks intermediate zones (yellow to pale orange). The decrease in water table height under topographic highs weakens the impact that vadose zone–water table coupling has on the extent and strength of recharge. The decreased coupling under surface highs results in a larger amount of water

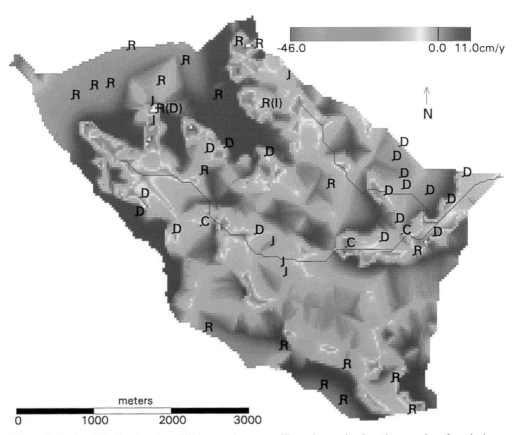

Figure 12.8. Case III: Simulated equilibrium recharge, capillary rise, spring locations, and surface drainage. Colour bar scales from maximum recharge (red, +11 cm/year) to maximum capillary rise (blue, −46 cm/year). Seeps are denoted by white pixels. Observation points are labelled as follows: R = Recharge, D = Discharge, I = Intermediate, C = Creek bed. (From Levine and Salvucci, 1999a; reproduced with permission.)

entering the flow system. This increased groundwater flow cannot be removed solely through evaporation in discharge areas, and thus surface seeps and springs develop where the water table is at the land surface. A good example of this is illustrated in case I (Figure 12.6) where springs (white) developed in most model predicted discharge areas, strong recharge (red) extends over large areas of the catchment, and only small zero net flux areas (pale orange) appear.

The significance of local variations in topography on the simulated spatial distribution of recharge and discharge areas can be seen in the more strongly coupled cases (II and III) (Figures 12.7 and 12.8). Note, for example, the discharge areas predicted at the base of steep slopes and in local convergence areas in the western portion of the study area. These areas are marked by a convergence of the local topography (Figure 12.2) and groundwater flow. The effect of the convergence of flow is to bring the water table closer to the surface, lowering the hydraulic gradient (slowing the rate of groundwater flux), and increasing the potential for discharge (through evaporation, springs or runoff production).

Case II (Figure 12.7) qualitatively captures many of the recharge and discharge points in the catchment. Well captured are the discharge points near the outlet of the catchment (eastern tip), the discharge areas along zones of groundwater convergence (west central portion of the catchment), and the recharge points near the higher elevation catchment boundaries (northern, southwestern and southern boundaries). Poorly captured are the discharge points that occur along the side of the ridge forming the upper southwestern boundary, the discharge areas that occur at the eastern edge of the ridge that extends southeastward from the upland area at the northern boundary of the catchment, and the two recharge points that occur in the small valley that runs north from just west of the southern tip of the catchment (lower right corner).

Case III (Figure 12.8) captures the discharge areas that occur along the base of the ridge that runs south-east from the north-western boundary of the catchment to just above the centre of the catchment. These areas are missed by cases I (Figure 12.6) and II (Figure 12.7). This is an area where surface slope changes abruptly from a downhill pitch to more level ground (Figure 12.2). The water table in this region is fairly close to the surface for all the simulations, but for the silt-loam cases it is still below Z^*, and thus is predicted to receive water from the vadose zone (Figure 12.4). Only in case III is Z^* great enough that the simulated mean flux at the saturated/unsaturated interface is toward the surface.

But case III fails to identify the recharge points located in the relatively flat upland area in the northern portion of the catchment (pale orange, Figure 12.8). This results from a competition between the relatively large depth to water table (Z_w) over which net recharge is negative for clay-loam soil ($Z^* = 370$ cm) and the relatively shallow Z_w (i.e. water table close to the ground surface) needed to provide the gradient and transmissivity for driving groundwater flow through the low conductivity aquifer. Together these effects act to restrict the strength and extent of recharge zones, as shown for simple hillslopes in Salvucci and Entekhabi (1995).

In case I (Figure 12.6), the higher lateral conductivity reduces the head gradient necessary to laterally drain groundwater recharge, resulting in a larger depth to the water table. This lower water table decreases the area over which vadose zone–water table coupling occurs. It results in larger and stronger recharge areas (red/orange), almost non-existent midline areas, very small (but strong) discharge areas (blue), and the development of significant areas of direct aquifer discharge to the surface (white). This case fails to capture the discharge areas at the transitions from steep hillsides to flat valley bottoms, but captures well the strong discharge along the centre of the lower valleys in the catchment (as does case II in general).

All of the model simulations miss the discharge point closest to the southwestern boundary, approximately half-way up the catchment. This point is located midway down a hillside and may be the result of a discontinuity in the soil or aquifer material that creates a very localised flow system.

The matches between the model and field estimates for all three cases are summarised in Table 12.5. This table lists the percentage of observations for which the equilibrium model estimated the same direction of flow (recharge or

Table 12.5. Percentage of recharge and discharge points for which equilibrium model estimates match estimates based on field observations

Observation type	Case I	Case II	Case III
R	100 %	86 %	57 %
D	67 %	78 %	89 %

discharge) as was estimated from the field observations. As might be expected there is a tradeoff in which catchment parameters that lead to good estimates of recharge location underpredict areas of discharge and vice-versa. In part this results from the nature of the spatial structure; that is, recharge in one area of the basin impacts discharge in others. Note also that the only spatially distributed parameters in the model application were surface and bedrock topography. Calibration to individual field observations could be made by adjusting local soil hydraulic properties, but without independent measurements of these properties the significance of such calibration would be questionable.

12.3.2   Stream Networks Generated by Equilibrium Model

The model was also tested by comparing simulated to observed stream networks. The stream locations were predicted using a threshold-based accumulation algorithm of surface runoff and spring flow. The predicted stream network (shown in Figures 12.6, 12.7 and 12.8) thus reflects the surface and groundwater dynamics to a much greater degree than simple contributing area methods (e.g. algorithms that assume uniform runoff production across a basin).

The stream locations predicted for case II (Figure 12.7) are nearby to Tóth's (1966) three creek bed observation points and broadly correspond to the locations of streams plotted on the Canadian Centre for Mapping topographic map for the area (which, unfortunately, is at a scale where what constitutes a stream is somewhat subjective). In comparison with the mapped streams, the model overpredicts stream development in the area of groundwater convergence that runs north-south in the western portion of the catchment, and on the relatively flat areas in the southern third of the central area. Differences in stream locations generated from the simulated output may also be due to errors in the DEM, terrain analysis, or mapped locations.

Note also that the stream lengths and patterns change with the changes in patterns of recharge and discharge (Figures 12.6, 12.7, and 12.8). Case III (Figure 12.8) has the lowest overall groundwater flow, very little surface runoff, and high evaporation, which together result in the development of only a small stream network. The sources of the two streams that develop in this case are located in areas of strong groundwater convergence and high water tables and not in areas of weaker convergence. Case I (Figure 12.6) yielded a smaller stream network than case II (Figure 12.7) because the deeper water table of case I (resulting from moderate bedrock conductivity) concentrates discharge in the lowest portions of the catchment. These results indicate that for this climate, high drainage densities would be expected in areas with highly conductive soils (to restrict evaporation) and low permeability bedrock.

Because of limitations in the mapped stream locations, the comparisons between mapped and predicted streams are of limited value. However, the strong sensitivity of the network shape and extent on soil and bedrock properties suggests that such comparisons could be useful for model calibration in situations where the streams are mapped, based on direct field observations.

12.3.3 Benefits and Drawbacks of Vadose Zone–Groundwater Coupling in Modelling the Spatial Patterns of Hydrologic Fluxes

In order to test whether water table coupling is a significant determinant in the spatial structure of recharge/discharge zones, or if the structure is constrained mainly by topography, a simulation was run holding the water table at the surface. As shown by Stoertz and Bradbury (1989) and others, holding the water table as a fixed boundary condition allows recharge to be estimated as the flow divergence at the surface of the aquifer. Note, however, that Stoertz and Bradbury (1989) proposed using the actual water table in such applications, while in our case the water table was held at the land surface (as in Ophori and Tóth (1989)). This was done because the water table data available from Tóth (1966) were too sparse.

Because holding the water table at the surface determines a spatial distribution of net recharge without consideration of vadose zone flow processes, comparing the skill of this method with the skill of the equilibrium model runs provides a simple test of the importance of two-way coupling. Furthermore, if it was found that the uncoupled method was able to represent the observed pattern of recharge and discharge, then topographic analysis alone could provide a useful estimate of spatial patterns of groundwater flow in a similar fashion to the way TOPMODEL (Beven and Kirkby, 1979) estimates spatial patterns in runoff (see Chapter 11).

The resulting patterns of recharge and discharge (Figure 12.9) do vary in general accordance with the observations of Tóth (1966). The amount of recharge necessary to maintain the water table can be much higher than the climate potentials. For example, recharge rates greater than 300 cm/yr are required to balance groundwater divergence in some locations (e.g. the deep red areas). Lowering the conductivity of the bedrock until the recharge areas have physically realistic intensities (i.e. less than annual precipitation) causes the disappearance of spatial patterns in recharge and discharge and eliminates almost all lateral flow (Figure 12.10).

Note that the fixed water table method also overemphasises the impact of small changes in topography by forcing the water table to reflect too closely the surface topography. This creates a more irregular recharge–discharge field, and

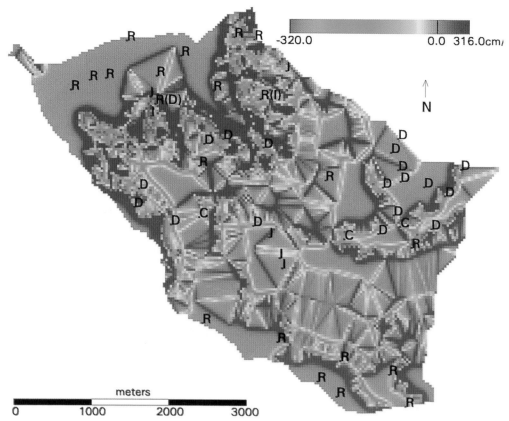

Figure 12.9. Simulated equilibrium recharge, capillary rise, spring locations, and surface drainage for an uncoupled model run with water table held at the surface. Soil and bedrock parameters as in Case II. Colour bar scales from maximum recharge (red, +316 cm/year) to maximum capillary rise (blue, −320 cm/year). Observation points are labelled as follows: R = Recharge, D = Discharge, I = Intermediate, C = Creekbed.

even highlights the triangular irregular networks formed in converting the surface elevation contour map to a digital elevation model.

As discussed in the introduction, it is not surprising that uncoupled groundwater models are susceptible to predicting recharge rates outside of reasonable bounds. On the other hand, coupled models such as that presented here suffer numerous limitations as well. Here the purpose was to show how observed patterns of recharge and discharge can inform modelling, but if the model were to be used for detailed predictions of spatially distributed catchment behaviour, one would need many more parameters in order to describe the meteorological forcing, vegetation, and unsaturated hydraulic properties. Other limitations (specific to the equilibrium model presented here) include the inability to simulate transient groundwater response and the inability to account for vertical soil heterogeneity in the vadose zone. This latter limitation could be addressed by replacing the Eagleson-type equilibrium water balance function (plotted in Figure 12.4) with the long-term mean flux partitioning predicted by a Richards-based soil–vegetation–atmosphere transfer model driven by long time

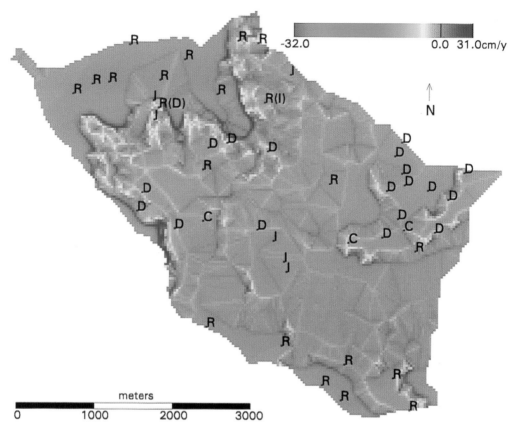

Figure 12.10. Simulated equilibrium recharge, capillary rise, spring locations, and surface drainage for an uncoupled model run with water table held at the surface. Soil is clay-loam, bedrock conductivity lowered to 0.02 cm/day in order to restrict recharge to reasonable values. Colour bar scales from maximum recharge (red, +31 cm/year) to maximum capillary rise (blue, −32 cm/year). Observation points are labelled as follows: R = Recharge, D = Discharge, I = Intermediate, C = Creek bed.

records of meteorological forcing. Such a hybrid approach would not be limited by soil homogeneity assumptions or other simplifying approximations that must be made to derive the analytical flux capacity relations in the Eagleson (1978a–f) and Salvucci and Entekhabi (1995) models. Whether or not relaxing these assumptions would lead to better model results, or simply more parameters to estimate, is an open question.

12.4 CONCLUSIONS

12.4.1 Modelling Spatial Patterns of Hydrologic Fluxes

The importance of accounting for two-way groundwater–vadose zone interaction when modelling the spatial distribution of surface fluxes has been demonstrated through a comparison against Tóth's (1966) field observations and coupled and uncoupled model results. Allowing the position of the water table to influence both the surface water balance and the groundwater divergence field, and then constraining these two fluxes to balance over the long-term mean, defines an equilibrium condition for the catchment system. Past and current research in surface and groundwater hydrology (e.g. Tóth, 1963; Freeze and Witherspoon, 1966; Eagleson, 1978a–f; Stoertz and Bradbury, 1989; Salvucci and Entekhabi, 1994a,b, 1995; Kim et al., 1999) has demonstrated and/or supposed that this equilibrium state forms an estimate of long-term mean conditions. Under this coupled and time-averaged condition, water balance partitioning in any one part of a basin can influence the partitioning at distant points. This behaviour imparts strong diagnostic value to spatially distributed field observations.

12.4.2   Comparison with Observations

The simulated recharge and discharge patterns match field observations of both recharge and discharge best for case II, but not as well for the more and less permeable conditions of cases I and III. Case I had more permeable bedrock and allowed the groundwater to reside deeper in the ground. This minimised saturated–unsaturated zone coupling and resulted in better prediction of recharge areas but underprediction of discharge areas. Case III had less permeable bedrock and more retentive (clayey) soils. The reduced permeability forced the water table closer to the surface (to increase hydraulic gradients and transmissivity), and thus brought more of the catchment area under saturated–unsaturated coupling. As a result, this case underpredicted recharge areas but captured most observed discharge areas.

Together the results indicate that coupling is an important factor in determining the spatial structure of recharge/discharge zones. Model estimates of the location of recharge and discharge with the water table shape determined by topography alone (i.e. without two-way interaction) also matched the field observations reasonably well, but appeared, in a qualitative sense, unrealistically heterogeneous and could be achieved only with physically unrealistic recharge intensities. In contrast, the fluxes predicted by the coupled model are within the bounds imposed by the climate forcing (i.e. model predicted recharge is less than rainfall, and model predicted discharge, except in cases of surface water contact, is less than potential evaporation). This condition is not met by the uncoupled model results where the water table was held fixed at the surface.

12.4.3   Impacts of Saturated–Unsaturated Zone Coupling on Catchment Modelling

Conclusions specific to the modelling experiments, which may be transferable to other catchments, can be summarised as follows:

- The overall effect of the vadose zone feedback mechanism (at a point) is to increase recharge if the depth to the water table is greater than equilibrium depth Z^* and to increase capillary rise if it is less than Z^*.
- The spatial patterns of recharge/discharge are dependent on both the aquifer permeability and the depth (of order Z^*) over which strong surface coupling exists. The patterns are more dependent on the local surface topography when strong coupling exists (Figures 12.7 and 12.8), and on the regional topography when weaker coupling exists (Figure 12.6).
- Low bedrock conductivity limits lateral redistribution of water and requires the water table to rise under topographic highs in order to drain recharging water. Higher water tables, however, restrict net recharge through both enhanced capillary rise to the root zone and increased runoff generation. Thus limited transmissivity increases the horizontal extent over which coupling at the water table plays a role in keeping the flow system in balance.

This work has shown that an equilibrium, coupled groundwater–vadose zone model is able to represent the long-term spatial patterns of recharge and discharge in the prairie landscape of the Trochu catchment. This conclusion was made possible by the extraordinarily detailed field data of Tóth (1966) that was unusual not only in its spatial detail, but also in the nature of the measurements. These were generally integrators of long-term response, such as vegetation and soil or water chemistry, and so were ideal for the testing of the equilibrium-style model used in this study. Indeed, without data of this type, testing the spatially distributed predictions from equilibrium models would be impossible.

ACKNOWLEDGEMENTS

This work was supported by NSF grant EAR-9705997. We thank J. Tóth, B. Rostron, and C. Mendoza (all from University of Alberta) for leading us on a site visit and providing their insights on groundwater hydrology in prairie environments.

PART THREE

IMPLICATIONS

13

Towards a Formal Approach to Calibration and Validation of Models Using Spatial Data

Jens Christian Refsgaard

13.1   INTRODUCTION

Spatially distributed models of catchment response tend to be highly complex in structure and contain numerous parameter values. Their calibration and validation is therefore extremely difficult but at the same time essential to obtain confidence in the reliability of the model (Chapter 3). Traditionally, calibration and validation has been mainly based on a comparison of observed versus simulated runoff at the outlet of the catchment of interest, but it has been pointed out numerous times that this is a very weak test of the adequacy of a model. Ideally, the internal behaviour of the models, in terms of simulated patterns of state variables and model output, should be tested, but examples of such internal tests are only relatively recent. In 1994 Rosso (1994, pp. 18–19) pointed out that "Conventional strategies for distributed model validation typically rely on the comparison of simulated model variables to observed data for specific points representing either external boundaries or intermediate locations on the model grid . . . . Traditional validation based on comparing simulated with observed outflows at the basin outlet still remains the only attainable option in many practical cases. However, this method is poorly consistent with spatially distributed modeling . . . ". More recently, encouraging work has been done on demonstrating how observed spatial patterns of hydrologically relevant variables can be used for internal tests. Indeed, the case study chapters of this book (Chapters 6–12) have clearly illustrated the enormous value of detailed spatial data in developing and testing distributed representations of catchment hydrological processes. These chapters have used a plethora of different data types and models and are representative of the progress in this area within the scientific community. However, typically these studies have been performed in small, well instrumented research catchments. The models often have been developed or modified by the group of people who did the data collection and also were the users of the results; and the purpose of the model development was to obtain insight into spatial catchment processes and process interactions.

These conditions are quite different in practical applications. In a practical case, that is, when a model is used and/or developed to assist in making management or design decisions in a particular catchment, the catchments are usually much larger and one can often only rely on data from the standard network that are not nearly as detailed as those in research catchments. Often the standard data are of unknown and/or undocumented quality. This tends to make model calibration and validation in particular cases significantly more difficult and less accurate. In many practical cases there is then an issue of whether, with the given data, a spatially distributed model of catchment response can at all be considered to reliably portray catchment behaviour.

In practical applications, model users often use model codes they have not developed themselves and data that are provided by different agencies. It is sometimes not clear how reliable the code is and sometimes it is unclear to the user how the code exactly works. The lack of field experience also makes it more difficult to appreciate which processes operate in the catchment and what is the best model approach for representing them. The large scale of catchments often considered in practical applications of distributed models tends to cause scale problems similar to those discussed in Chapter 3 of this book. For example, it is not uncommon to use Richards' equation for elements that are as big as 500×500 m – this is an area that is larger than the size of the whole catchment in many of the case studies in this book which have shown an enormous complexity that goes far beyond the processes represented by Richards' equation. The fact that model users, model builders, data providers, and clients are different groups and have differences in terminology, creates further problems. Currently there appears to exist no unique and generally accepted terminology on model validation in the hydrological community and the many different and contradictory definitions used tend to be confusing.

Finally, in practical applications, the purpose of the modelling is to make predictions rather than to gain insight into spatial catchment behaviour. What is considered a useful model for understanding catchment behaviour is not necessarily useful for practical applications. Many case studies in the chapters of this book used comparison of observed and simulated patterns not only to calibrate models and ensure that they are working for the right reasons, but also to identify from the pattern comparison, processes that the model cannot handle very well. These may be the subject of future research work. On the other hand, the situation is quite different in the practical case. What is needed in this case is a reliable model for the projected conditions of model application and the insights obtained are only important to the extent they can be used to improve model performance and/or interpretation of the results in terms of management or design decisions. It is important to realise that the type of model application i.e. investigative versus predictive, has profound implications on both model structure (predictive models often having a simpler structure) and model calibration/validation (predictive models often having a better defined range of applicability). The validation and calibration of distributed models in practical applications is therefore quite different from that in research type applications.

Unfortunately, in practical projects there is often not very thorough model testing due to data and resource constraints. It is not uncommon for predictions of spatial patterns to be made with models that have not been properly tested in terms of their spatial behaviour. For example, Kutchment et al. (1996) applied a distributed physically-based model to simulate the 3315 km^2 Ouse basin in the UK. They calibrated their model against runoff data only, but stated also that the model can give "hydrologically meaningful estimates of internal values". Due to lack of data and lack of tests on internal variables, this statement appears as the authors' own perception rather than a documented fact. This problem has been recognised by some authors who are a little more circumspect about the performance of distributed models. Jain et al. (1992) applied a distributed physically-based model to the 820 km^2 Kolar catchment in India, where the runoff data comprised the only available calibration and validation data. They concluded that "The resulting final model calibration is believed to give a reasonably good physical representation of the hydrological regime. However, a preliminary model set-up and calibration... resulted in an equally good hydrograph match, but on the basis of apparently less realistic soil and vegetation parameter values. Thus, it may be concluded that a good match between observed and simulated outlet hydrographs does not provide sufficient guarantee of a hydrologically realistic description." However, the practical problems for which distributed models need to be applied remain, so the challenge is to better use the information available to us and to seek additional information, to help strengthen the confidence we can have in simulated responses from distributed models.

It is clear that proper validation and calibration of distributed models of catchment behaviour is of the utmost importance. This is obviously an uncertain endeavour. The primary role of model calibration and validation is to obtain a realistic assessment of this uncertainty – of what confidence we can place on the predictions of our model. In this chapter, these issues are addressed by proposing a framework for model validation and calibration. Also, issues of terminology will be clarified to develop a common language, and data issues relevant to distributed catchment models in practical applications will be discussed. The validation framework and data considerations will be illustrated in a case study for the 440 km^2 Karup catchment. The chapter concludes with a discussion on possible interactions between model builders, model users, and clients that could improve the understanding and treatment of uncertainty in practical applications of spatially distributed catchment models.

13.2   SOURCES OF SPATIAL DATA FOR PRACTICAL APPLICATIONS

Distributed hydrological models require spatial data. In this chapter, two different terms are used for the data, depending on its type: *parameter values* are those that do not vary with time while *variables* are time dependent. In a traditional model application the parameter values and the driving variables (typically climate data) are input data, while the other variables are simulated by the model.

An overview of typical data types and sources is given in Table 13.1 together with a characterisation of the typical availability of data from traditional sources and the potential for operational use of remote sensing data. Basically, traditional data sources provide point or vector data. Even for many of the exceptions to this, such as digital elevation maps or soil maps, the spatial data are inferred from originally measured point data. In general, it is possible to obtain such spatial data on catchment characteristics for use as model parameter input and by assuming relationships between these data and model parameters (e.g. between soil type and soil hydraulic properties, between vegetation types and water use etc.), model parameter values can be estimated (albeit with an accuracy determined by the validity of the assumed relationships – see discussion in Chapter 2, pp. 23–4, 41). It is also generally possible to get hydroclimatological time series for driving the model and for checking the overall catchment runoff. However, there is almost always a lack of data to check the detailed spatial patterns of internally simulated variables such as soil moisture, actual evapotranspiration and water depths. The only source for such data that can be characterised as realistic on scales above plots and small experimental catchments, is remote sensing data.

Remote sensing data have for a couple of decades been described as having a promising potential to supply spatial data to distributed hydrological models, e.g. Schulz (1988), Engman (1995) and De Troch et al. (1996). However, so far the success stories, at least in operational applications, are in practice limited to mapping of land use and snow cover, whereas scientific/technological breakthroughs are still lacking for assessing soil moisture, vegetation status, actual evapotranspiration and water depths. Chapters 5 and 6 in this book give examples of research applications where progress is clearly being made on representing variables such as soil moisture and saturated source areas, but as yet these methods are not available for practical application.

With progress made in recent years at the research level, we may foresee operational applications of remote sensing for practical modelling within the next decade in areas such as:

- Assessments of water depths and inundation areas at larger scales (> 1 km length) to be used for flood forecasting and flood mapping. This can today be done during cloud-free periods by use of thermal data, and appears promising in the future by use of SAR data. Furthermore, new high-resolution (few metres) visible satellite data are also promising.
- Assessment of vegetation and soil moisture status at field scale and above to be used for crop forecasting, irrigation management and meteorological forecasting.
- Assessment of vegetation status at field scale and below for supporting precision agriculture.
- Improved accuracy of RADAR derived precipitation.

A key point to remember about remote sensing data is that it is a surrogate measure – i.e. it depends on a relationship between properties of emitted or

Table 13.1. Spatial data used in distributed catchment modelling

Data type	Function in model	Typical traditional data source	Typical availability of traditional data	Potential for operational use of remote sensing data
Topography	Parameter – input data	Maps, DEMs	Very good	Interferometry
River network	Parameter – input data	Maps, derived from DEMs	Very good	
Geology	Parameter – input data	Geological surveys, maps	Good	
Soil	Parameter – input data	Maps, national databases	Good	
Land use/vegetation	Parameter – input data	Maps	Good	Well proven (Landsat, Spot, etc.)
Climate data (precipitation, temperature, wind speed, etc.)	Variable – input data	Meteorological databases	Usually exist, but some times difficult to get access to	Potential for rainfall data (Meteosat) Weather radar data in operational use, but quantitative accuracy so far not good
Snow cover	Variable – simulated	Ad hoc point measurements	Very seldom	Well proven
Soil moisture	Variable – simulated		Very seldom and only point values	Potential (microwave and SAR data), but so far no encouraging operational use
Vegetation status (leaf area, root depth)	Variable – input data or simulated		Very seldom and only point values	Potential, encouraging, but so far no operational use
Actual evapotranspiration/ surface temperature	Variable – simulated		Very seldom and only point values	Potential
Water depth/inundation area	Variable – simulated	River gauging stations	Only at river gauging stations	Potential

reflected radiation and a particular feature of interest such as soil moisture content. It is not a direct measure so, as with parameters like soil hydraulic properties estimated from soil type, the accuracy of derived measures is a function of the quality of the surrogate relationships (see Chapter 3, pp. 41–5).

13.3   ISSUES OF TERMINOLOGY

13.3.1   Background

Before presenting a practical methodology for model calibration and validation, it is worth reflecting on the more fundamental question of whether models at all can be validated, and issues of terminology. Konikow and Bredehoeft (1992) argued that the terms validation and verification are misleading and their use should be abandoned: "...the terms validation and verification have little or no place in ground-water science; these terms lead to a false impression of model capability". The main argument in this respect relates to the anti-positivistic view that a theory (in this case a model) can never be proved to be generally valid, but can on the contrary be falsified by just one example. De Marsily et al. (1992) argued in a response to Konikow and Bredehoeft (1992) for a more pragmatic view: "...using the model in a predictive mode and comparing it with new data is not a futile exercise; it makes a lot of sense to us. It does not prove that the model will be correct for all circumstances, it only increases our confidence in its value. We do not want certainty; we will be satisfied with engineering confidence." Part of the difference of opinion relates to interpretations of the terminology used.

Konikow (1978) and Anderson and Woessner (1992) use the term verification with respect to the governing equations, the code or the model. According to Konikow (1978) a model is verified "if its accuracy and predictive capability have been proven to lie within acceptable limits of errors by tests independent of the calibration data". The term model verification is used by Tsang (1991) in the meaning of checking the model's capability to reproduce historical data. Anderson and Woessner define model validation as tests showing whether the model can predict the future. As opposed to the authors above, Flavelle (1992) distinguishes between verification (of computer code) and validation (of site-specific model). Oreskes et al. (1994), using a philosophical framework, state that verification and validation of numerical models of natural systems theoretically is impossible, because natural systems are never closed and because model results are always non-unique. Instead, in their view models can only be "confirmed". Within the hydraulic engineering community attempts have been made to establish a common methodology (IAHR, 1994). The IAHR methodology comprises guidelines for standard validation documents, where validation of a software package is considered in four steps (Dee, 1995; Los and Gerritsen, 1995): conceptual validation, algorithmic validation, software validation and functional validation. This approach concentrates on what other authors call code verification, while schemes for validation of site-specific models are not included.

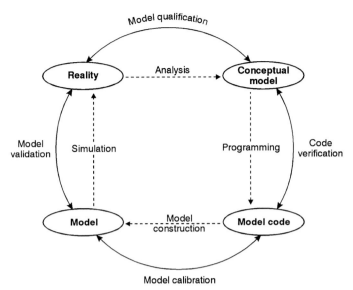

Figure 13.1. Elements of a modelling terminology and their interrelationships. Modified after Schlesinger et al. (1979).

The terminology and methodology proposed below has evolved from a background of more than twenty years' experience with research, development and practical applications of hydrological models. The proposed terminology and methodology is aimed at being pragmatic and does not claim to be in full accordance with scientific philosophy. Thus, it operates with the terms verification and validation, which are being used on a routine basis in the hydrological community, although with many different meanings. On the other hand, the term model validation is not used carelessly here but within a rigorous framework where model validation refers to site specific applications and to pre-specified performance (accuracy) criteria. Thus, in agreement with past practical experience and in accordance with philosophical considerations, a model code is never considered generally valid.

13.3.2 Definition of Terminology

The following terminology is inspired by the generalised terminology for model credibility proposed by Schlesinger et al. (1979), but modified and extended to suit distributed hydrological modelling. The simulation environment is divided into four basic elements as shown in Figure 13.1. The inner arrows describe the processes which relate the elements to each other, and the outer arrows refer to the procedures which evaluate the credibility of these processes. The most important elements in the terminology and their interrelationships are defined as follows:

Reality. The natural system, understood here as the hydrological cycle or parts of it.

Conceptual model. A conceptual description of reality, i.e. the user's perception of the key hydrological processes in the catchment and the corresponding simplifications and numerical accuracy limits which are assumed acceptable in the model in order to achieve the purpose of the modelling.

Model code. Generic software program.

Model. A site-specific model established for a particular catchment, including input data and parameter values.

Model construction. Establishment of a site-specific model using a model code. This requires, among other things, the definition of boundary and initial conditions and parameter assessment from field data.

Simulation. Use of a validated model to gain insight into reality and obtain predictions that can be used by water managers.

Model qualification. An estimate of the adequacy of the conceptual model to carry out the desired application within the acceptable level of accuracy.

Code verification. Substantiation that a model code is in some sense a true representation of a conceptual model within certain specified limits or ranges of application and corresponding ranges of accuracy.

Model calibration. The procedure of adjustment of parameter values of a model to reproduce the response of a catchment under study within the range of accuracy specified in the performance criteria.

Model validation. Substantiation that a model within its domain of applicability possesses a satisfactory range of accuracy consistent with the intended application of the model.

Performance criteria. Level of acceptable agreement between model and reality. The performance criteria apply both for model calibration and model validation.

In the above definitions the term conceptual model should not be confused with the word conceptual used in the traditional classification of hydrological models (lumped conceptual rainfall-runoff models).

13.4 PROPOSED METHODOLOGY FOR MODEL CALIBRATION AND VALIDATION

13.4.1 Modelling Protocol

The protocol described below is a translation of the general terminology and methodology defined above into the field of distributed hydrological modelling. It is furthermore inspired by the modelling protocol suggested by Anderson and Woessner (1992), but modified concerning certain steps.

The protocol is illustrated in Figure 13.2 and described step by step in the following.

1. The first step in a modelling protocol is to *define the purpose* of the model application. An important element in this step is to give a first assessment of the desired accuracy of the model output.
2. Based on the purpose of the specific problem and an analysis of the available data, the user must establish a *conceptual model*.
3. After having defined the conceptual model, a suitable computer program has to be selected. In principle, the computer program can be prepared specifically for the particular purpose. In practice, a *code* is often *selected* among existing generic modelling systems. In this case it is important to ensure that the selected code has been successfully verified for the particular type of application in question.
4. In case no existing code is considered suitable for the given conceptual model a *code development* has to take place. In order to substantiate that the code solves the equations in the conceptual model within acceptable limits of accuracy a *code verification* is required. In practice, code verification involves comparison of the numerical solution generated by the code with one or more analytical solutions or with other numerical solutions.
5. After having selected the code and compiled the necessary data, a *model construction* has to be made. This involves designing the model with regard to the spatial and temporal discretisation, setting boundary and initial conditions and making a preliminary selection of parameter values from the field data. In the case of distributed modelling, the model construction generally involves reducing the number of parameters to calibrate (i.e. reducing the "degrees of freedom", Chapter 3, pp. 75–6) e.g. by using representative parameter values for different soil types.
6. The next step is to define *performance criteria* that should be achieved during the subsequent calibration and validation steps. When establishing performance criteria, due consideration should be given to the accuracy desired for the specific problem (as assessed under step 1) and to the realistic limit of accuracy determined by the field situation and the available data (as assessed in connection with step 5). If unrealistically high performance criteria are specified, it will either be necessary to modify the criteria or to obtain more and possibly quite different field data.

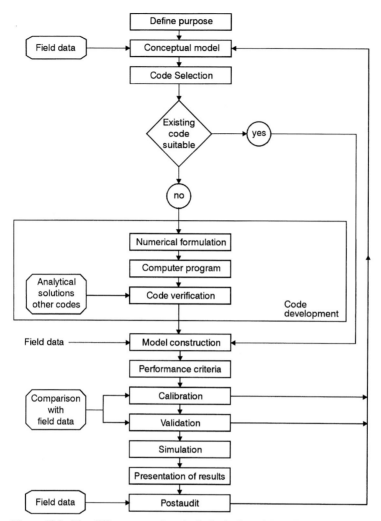

Figure 13.2. The different steps in a hydrological model application – a modelling protocol. (From Refsgaard, 1997; reproduced with permission.)

7. *Model calibration* involves adjustment of parameter values of a specific model to reproduce the observed response of the catchment within the range of accuracy specified in the performance criteria. It is important in this connection to assess the uncertainty in the estimation of model parameters, for example from sensitivity analyses.

8. *Model validation* involves conduction of tests which document that the given site-specific model is capable of making sufficiently accurate predictions. This requires using the calibrated model, without changing the parameter values, to simulate the response for a period other than the calibration period. The model is said to be validated if its accuracy and predictive capability in the validation period have been proven to lie within acceptable limits or to provide acceptable errors (as specified in

the performance criteria). Validation schemes for different purposes are outlined below.

9. *Model simulation* for prediction purposes is often the explicit aim of the model application. In view of the uncertainties in parameter values and, possibly, in future catchment conditions, it is advisable to carry out a predictive sensitivity analysis to test the effects of these uncertainties on the predicted results (see Chapter 11 for one such procedure).

10. *Results* are usually *presented* in reports or electronically, e.g. in terms of animations. Furthermore, in certain cases, the final model is transferred to the end user for subsequent day-to-day operational use.

11. An extra possibility of validation of a site-specific model is a so-called *postaudit*. A postaudit is carried out several years after the modelling study is completed and the model predictions can be evaluated against new field data.

13.4.2   Scheme for Construction of Systematic Model Validation Tests

Distributed hydrological models contain a large number of parameters, and it is nearly always possible to produce a combination of parameter values that provides a good agreement between measured and simulated output data for a short calibration period. However, as discussed in Chapter 3, this does not guarantee an adequate model structure nor the presence of optimal parameter values. The calibration may have been achieved purely by numerical curve fitting without considering whether the parameter values so obtained are physically reasonable. Further, it might be possible to achieve multiple calibrations or apparently equally satisfactory calibrations based on different combinations of parameter values (see also Chapter 11). Ideally, the ultimate purpose of calibration is not to fit the calibration data but to fit reality. If the other error sources, including the effects of non-perfect model structure and data uncertainties, are not somehow considered, there is a danger of overfitting.

In order to assess whether a calibrated model can be considered valid for subsequent use it must be tested (validated) against data different from those used for the calibration. According to the methodology established above, model validation implies substantiating that a site-specific model can produce simulation results within the range of accuracy specified in the performance criteria for the particular study. Hence, before carrying out the model calibration and the subsequent validation tests, quantitative performance criteria must be established. In determining the acceptable level of accuracy a trade-off will, either explicitly or implicitly, have to be made between costs, in terms of data collection and modelling work, and associated benefits of achieving more accurate model results. Consequently, the acceptable level of accuracy will vary from case to case, and should usually not be defined by the modellers, but by the water resources decision makers. In practice, however, the decision maker often only influences this important issue very indirectly by allocating a project budget and requesting the modeller to do as good as possible within this given frame.

The scheme proposed below is based on Klemeš (1986b) who states that a model should be tested to show how well it can perform the kind of task for which it is specifically intended. The four types of test correspond to different situations with regard to whether data are available for calibration and whether the catchment conditions are stationary or the impact of some kind of intervention has to be simulated.

The *split-sample test* is the classical test, being applicable to cases where there is sufficient data for calibration and where the catchment conditions are stationary. The available data record is divided into two parts. A calibration is carried out on one part and then a validation on the other part. Both the calibration and validation exercises should give acceptable results. This approach was taken in Chapters 6, 7, 10 and 11.

The *proxy-basin test* should be applied when there is not sufficient data for a calibration of the catchment in question. If, for example, streamflow has to be predicted in an ungauged catchment Z, two gauged catchments X and Y within the region should be selected. The model should be calibrated on catchment X and validated on catchment Y and vice versa. Only if the two validation results are acceptable and similar can the model command a basic level of credibility with regard to its ability to simulate the streamflow in catchment Z adequately.

The *differential split-sample test* should be applied whenever a model is to be used to simulate flows, soil moisture patterns and other variables in a given gauged catchment under conditions different from those corresponding to the available data. The test may have several variants depending on the specific nature of the modelling study. If, for example, a simulation of the effects of a change in climate is intended, the test should have the following form. Two periods with different values of the climate variables of interest should be identified in the historical record, such as one with a high average precipitation, and the other with a low average precipitation. If the model is intended to simulate streamflow for a wet climate scenario, then it should be calibrated on a dry segment of the historical record and validated on a wet segment. Similar test variants can be defined for the prediction of changes in land use, effects of groundwater abstraction and other such changes. In general, the model should demonstrate an ability to perform through the required transition regime.

The *proxy-basin differential split-sample test* is the most difficult test for a hydrological model, because it deals with cases where there is no data available for calibration and where the model is directed to predicting non-stationary conditions. An example of a case that requires such a test is simulation of hydrological conditions for a future period with a change in climate and for a catchment where no calibration data presently exist. The test is a combination of the two previous tests.

Examples of the four tests are given by Refsgaard and Knudsen (1996), and Styczen (1995) provides an example of a test procedure based on the same principles for the validation of pesticide leaching models for registration purposes. A general point related to all the tests, is that if the accuracy of the model for the validation period is much worse than for the calibration period, it is an indication of "overfitting"; that is, there is likely to be a problem with the model structure causing the parameters to be specific to the conditions used for calibration. The ratio of accuracy during the calibration period to accuracy during the validation period is sometimes used as a measure of the degree of overfitting.

It is noted that, according to this scheme, a distributed model cannot claim a predictive capability in simulation of spatial patterns unless it has been specifically tested for this purpose. Thus, if a model has only been validated against discharge data, which is very commonly the case, there is no documentation on its predictive capability with regard to, for example, simulation of spatial patterns of soil moisture and groundwater levels at grid scales. Claims on predictive capabilities with regard to, for example, soil moisture variation in time and space, require successful outputs from a validation test designed specifically with this aim. When designing validation tests the following additional principles must be taken into account:

- The *scale* of the measurements used must be appropriate for the scale of the model elements (see Chapter 2, pp. 19–20). The scales need not be identical, but the field data and the model results must be up/downscaled, so that they are directly comparable.
- The *performance criteria* must be specified, keeping the spatial patterns in mind.
- The *validation test* must be designed in accordance with a special emphasis on the spatial patterns and the distributed nature of the model.

For illustrative purposes, a hypothetical example is given in the following. Suppose that one purpose of a model application is to predict the patterns of soil moisture in the topsoil, and that the validation data consist of a remote sensing based SAR data set with a 30 m spatial resolution at four times during a given period. Suppose that the SAR data set has been successfully calibrated/validated against ground truth data and that the error can be described statistically (mean, standard deviation, spatial correlations). Suppose finally that the hydrological model uses a horizontal grid of 60 m and can match the vertical depth of the topsoil measured by the SAR. The above three principles could be implemented as follows:

- *Scaling*. The comparisons between model and data should be carried out at a minimum scale of 60 m. The data could also be aggregated to larger scales if that were sufficient for the subsequent model application. In any case the error description of the SAR data must be corrected to correspond to the selected scale, implying that the standard deviation of the error is reduced due to the aggregation process (see Chapter 2, p. 23).
- The *performance criteria* could, for example, be chosen to reflect various aspects such as:

- Capability to describe correct overall levels. This may include criteria on comparison of mean values and standard deviations of SAR data and model results.
- Capability to describe spatial patterns. This may include comparison of correlation lengths or division of the entire area into subareas and comparison of statistics within each subarea.
- Capability to describe temporal dynamics. This may include criteria on comparison of SAR and model time series for selected areas, such as regression coefficient and model efficiency (Nash and Sutcliffe, 1970).

In general, the numerical values of the performance criteria should depend on the uncertainty of the data, as described by the error statistics, and the purpose of the modelling study.

- The *type of validation test* should be decided from the same principles as outlined above. For instance, if SAR data were available for just part of the model area, a proxy-basin split sample test can be applied where model results, calibrated without SAR data, were compared in the overlapping area. The performance criteria would then be assumed to indicate how well the model will perform for the area not covered by SAR data.

13.4.3   Use of Spatial Data

Distributed hydrological models are structured to enable the spatial variations in catchment characteristics to be represented by having different parameter and variable values for each element. Often model applications require several thousands of elements, meaning that the number of parameters and variables could be two or three orders of magnitude higher than for a lumped model of the same area. Obviously, this generates different requirements for lumped and distributed models with regard to parameterisation, calibration and validation procedures.

A critique expressed against distributed models by several authors concerns the many parameter values which can be modified during the calibration process. Beven (1989, 1996) considers models which are usually claimed to be distributed physically-based as in fact being lumped conceptual models, just with many more parameters. Hence, according to Beven (1996) a key characteristic of the distributed model is that "the problem of overparameterisation is consequently greater".

To address this problem in practical applications of distributed models, it is necessary to reduce the "degrees of freedom" by inferring spatial patterns of parameter values so that a given parameter only reflects the significant and systematic variation described in the available field data. This approach is exemplified by the practice of using representative parameter values for individual soil types, vegetation types or geological layers along with patterns of these types (see also the discussion in Chapter 3, pp. 75–6 and examples of using this approach in Chapters 6, 9 and 10). This approach reduces the number of free parameter coefficients that need to be adjusted in the subsequent calibration procedure. The following points are important to consider when applying this approach (Refsgaard and Storm, 1996):

- The parameter classes (soil types, vegetation types, climatological zones, geological layers, etc.) should be selected so that it becomes easy, in an objective way, to associate parameter values. Thus the parameter values in the different classes should, to the highest possible degree, be assessable from available field data.
- It should explicitly be evaluated which parameters can be assessed from field data alone and which need some kind of calibration. For the parameters subject to calibration, physically acceptable intervals for the parameter values should be estimated.
- The number of real calibration parameters should be kept low, both from a practical and a methodological point of view. This can be done, for instance, by fixing a spatial pattern of a parameter but allowing its absolute value to be modified through calibration.

Reducing the number of free parameters in this way helps to avoid methodological problems in the subsequent phases of model calibration and validation. An important benefit of a small number of free parameters adjustable through calibration is that the whole parameter assessment procedure becomes more transparent and reproducible. The quality of the results, however, depends on the adequacy and accuracy of the imposed patterns. It is also important to consider to what extent the imposed pattern dominates the pattern of the simulated output (see Chapter 3, p. 76, and Figures 6.13, 9.12 and 10.15). Another example is the use of Thiessen polygons for representing spatial precipitation patterns where it is possible that soil moisture may be dominated by precipitation quantities making the simulated spatial patterns just reflect the Thiessen polygons. It may often be required to aggregate both model results and field data to a scale where the imposed pattern does not dominate.

Thus, the challenge for the hydrologist has been expanded beyond the task of tuning parameter values through calibration to the art of defining hydrologically sound methods for reducing the number of parameters to be calibrated.

13.5 CASE STUDY

The above methodology is illustrated step by step in the study described in Refsgaard (1997). The first seven steps are summarised with rather brief descriptions in Section 13.5.2, while the model validation step is addressed more thoroughly in Section 13.5.3.

13.5.1 The Karup Catchment

The 440 km^2 Karup catchment is located in a typical outwash plain in the western part of Denmark. From a geological point of view the area is relatively homogeneous, consisting of highly permeable sand and gravel with occasional lenses of moraine clay. The depth of the unsaturated zone varies from 25 m at the eastern groundwater divide to less than 1 m in the wetland areas along the main river. The aquifer is mainly unconfined and of glacial deposits. The thickness of the aquifer varies from 10 m in the western and central parts to more than 90 m to the east. The catchment has a gentle sloping topography and is drained by the Karup River and about 20 tributaries.

The Karup catchment has been subject to several hydrological studies (e.g. Stendal, 1978; Miljøstyrelsen, 1983; Styczen and Storm, 1993) and a comprehensive database exists both for surface water and ground water variables. The catchment area and the measurement sites referred to in the following are shown in Figure 13.3.

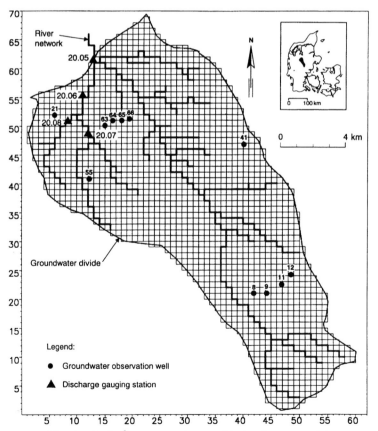

Figure 13.3. The 440 km^2 Karup catchment with the river network in a 500 m model grid together with the location of the discharge gauging stations and groundwater observation wells referred to in the text. (From Refsgaard, 1997; reproduced with permission.)

13.5.2   Establishment of a Calibrated Model – the First Steps of the Modelling Protocol

Step 1. Definition of Purpose

The overall objectives of the case study are to illustrate the parameterisation, calibration and validation of a distributed model and to study the validation requirements with respect to simulation of internal variables and to changing spatial discretisation. In this context the purpose of the model is to simulate the overall hydrological regime in the Karup catchment, especially the spatial pattern of discharges and groundwater table dynamics.

Step 2. Establishment of a Conceptual Model

The assumptions made regarding the hydrological system are described in detail in Refsgaard (1997). The main components of the conceptual model can be characterised as follows:

- The groundwater system is described by an unconfined aquifer comprising one main aquifer material with the same hydraulic parameters throughout the catchment and five minor lenses with distinctly different parameters. The aquifer system is modelled as two-dimensional.
- The unsaturated zone is described by one-dimensional vertical flows. The soil system is via maps and profile descriptions described by two soil types characterised by different hydraulic parameters.
- Four vegetation/cropping classes are assumed: agriculture, forest, heath and wetland.
- The main river system and the tributaries which could be accommodated within the 500 m spatial model discretisation are included in the model. The wetland areas are assumed to be drained by ditches and tile drainpipes. The stream–aquifer interaction is assumed to be governed by the head differences in the river and the main aquifer and controlled by a thin, low permeability layer below the riverbed.
- Daily values, averaged over the catchment, of precipitation, potential evapotranspiration and temperature are used.

Step 3. Selection of Model Code

The code selected for the study was MIKE SHE (Refsgaard and Storm, 1995). In the present case the following modules were used: two-dimensional overland flow (kinematic wave), one-dimensional river flow (diffusive wave), one-dimensional unsaturated flow (Richards' equation), interception (Rutter concept), evapotranspiration (Kristensen and Jensen concept), snowmelt (degree-day concept) and two-dimensional saturated flows (Boussinesq).

Step 4. Code Verification

As MIKE SHE is a well proven code with several verification tests as well as many large scale engineering applications, including prior tests on the present area and on similar cases, no additional code verification was required in this case.

Step 5. Model Construction

The details regarding spatial discretisation of the catchment, setting of boundary and initial conditions and making a preliminary selection of parameter values from the field data are described in Refsgaard (1997). The number of parameters to be calibrated was reduced to 11 by, for example, subdividing the domains based on soil classes with uniform soil parameters in each subdivided area. Three of the parameters related to the aquifer properties and stream-aquifer interaction, while the eight remaining ones were soil hydraulic parameters. Thus the degrees of freedom in describing the spatial pattern of ground water levels in practice reduces to three parameters that can be fitted through calibration. One of the costs of such simplification is that one (spatially constant) value for aquifer hydraulic conductivity may not be sufficient to adequately describe the spatial patterns in groundwater flows and groundwater levels. A previous calibration of groundwater transmissivities for the same aquifer (Miljøstyrelsen, 1983) suggests that the transmissivities vary substantially more than can be explained by the variation in aquifer thickness in the present model.

Step 6. Performance Criteria

The performance criteria were related to the following variables:

1. Discharge simulation at station 20.05 Hagebro (the outlet of the catchment) with a graphical assessment of observed and simulated hydrographs supported by the following two numerical measures:

    - average discharges of observed and simulated records, OBS_{ave} and SIM_{ave}, and
    - model efficiency, R^2, calculated on a daily basis (Nash and Sutcliffe, 1970).

2. Groundwater level simulations at observation wells 21, 41 and 55 located in the downstream part of the catchment and also used by Styczen and Storm (1993) plus observation wells 8, 9, 11, 12 representing a cross-section at the upstream part of the catchment.

These criteria were used for the calibration and the first part of the validation tests. For the second part of the validation tests, focussing on the capability to describe internal variables and spatial patterns, additional criteria were defined (see below).

Step 7. Model Calibration

Most of the parameter values were assessed directly from field data or transferred from experience in other similar catchments. The remaining eleven parameter values were assessed during calibration through a trial-and-error process. The model calibration was carried out on the basis of data for the period 1971–

74. Altogether, the calibration results are of the same accuracy as the results in Styczen and Storm (1993), and are, as such, considered acceptable.

13.5.3 Model Validation

The validation tests have been carried out in two steps:

- *Step 8a*. Validation on the same station/wells as used for the calibration.
- *Step 8b*. Validation on additional data representing internal variables not utilised during the calibration process.

The validation tests were designed in accordance with the three guiding principles, outlined in Section 13.4.2. The *validation test type* is a traditional split-sample test for Step 8a and a proxy-basin split-sample test for Step 8b.

Scaling. The discharge data are aggregated values integrating the runoff over the respective catchments. This applies both to the field data and the model simulations, so no scaling inconsistency occurs here. The groundwater level data from observation wells are point data as opposed to the model simulations which represent average values over 500 m grids. The groundwater levels are known to vary typically by 1–2 m over a 500 m distance (Stendal, 1978). Hence, observed and simulated groundwater levels cannot be expected to match more closely than 1–2 m with regard to levels but somewhat better with regard to temporal dynamics.

The *performance criteria* take the various aspects into account as follows:

- The overall levels are expected to match within 10 % with regard to average discharge and within 2 m with regard to groundwater levels. The 2 m criteria should be seen in view of a typical variation in observed groundwater levels of $1\frac{1}{2}$–2 m within 500 m (Stendal, 1978).
- The temporal dynamics is expected to match reasonably well. No specific, numerical criteria have been identified for this purpose, but the visual inspection will focus on amplitude and phasing of the annual fluctuations.
- The capability to describe internal spatial patterns has been tested by using additional data for the following stations, for which data were not used at all during the calibration process:
  - discharge values at the three stations 20.06 Haderup (98 km^2), 20.07 Stavlund (50 km^2) and 20.08 Feldborg (17 km^2) (Figure 13.3).
  - groundwater tables at observation wells 63, 64, 65 and 66, located in the area between the main river and the tributary with the three discharge stations 20.06, 20.07 and 20.08 (Figure 13.3).

Key results from this validation test are shown in Figures 13.4 and 13.5. These results from one discharge station and seven groundwater observation wells were comparable to the results from the calibration period and have been assessed as acceptable. Hence the model has now been successfully validated for simulation

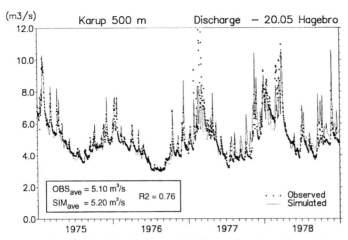

Figure 13.4. Simulated and observed discharge for the entire catchment for the validation period together with figures for average observed and simulated flows, OBS_{ave} and SIM_{ave}, and model efficiency on a daily basis, R^2. (From Refsgaard, 1997; reproduced with permission.)

of catchment discharge and groundwater levels in these seven observation wells with the expected accuracy similar to those shown in Figures 13.4 and 13.5.

The interesting question now is how reliable is the model for simulation of internal variables and spatial patterns. This was addressed during step 8b. Results from the first 28 months of the validation period, where data are available for all the above stations, are seen in Figures 13.6 and 13.7 for discharge and groundwater tables, respectively. As can be seen from the hydrographs, the water balance, and the model efficiency, the simulation results are significantly less accurate than for the calibrated stations. The simulated discharges at the three tributary stations are significantly poorer than for the calibrated station 20.05 in two respects. Firstly, there is a clear underestimation of the baseflow level and the total runoff for the three tributary stations, where the 10 % accuracy on the water balance performance criteria is not fulfilled for any of the three stations. Secondly, the simulation shows a significantly more flashy response than the observed hydrographs. The simulated groundwater tables (Figure 13.7) show correct dynamics, but have problems with the levels. The groundwater level error at well no. 64 is just above the 2 m specified as the accuracy level in the performance criteria on groundwater levels. Taking into account that the gradient between wells 64 and 65, which are located in two neighbouring grids, is wrong by about 3 m, the model simulation of groundwater levels can not be claimed to have passed the performance criteria in general. The primary reason for the differences in baseflow levels appear to be that the internal groundwater divide between the main river and the main tributary is not simulated correctly, with the result that the three tributary stations according to the model are draining smaller areas than they do in reality.

From the conducted validation tests, it may be concluded that the model cannot be claimed to be valid for discharge simulation of subcatchments, nor

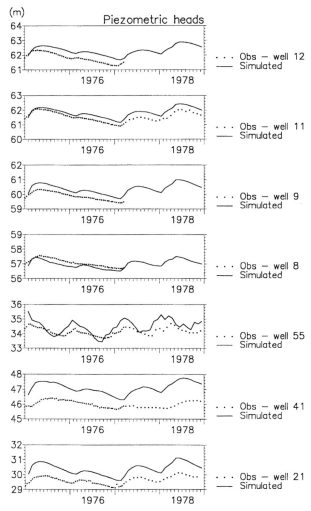

Figure 13.5. Simulated and observed piezometric heads at seven well sites for the validation period. The locations of the wells are shown in Figure 13.3. (From Refsgaard, 1997; reproduced with permission.)

for groundwater levels in general over the entire catchment area. Following the methodology represented in Figure 13.2, we should now re-look at the model conceptualisation and model construction, and perform additional calibration and tests to improve confidence. To do this we would need additional data. This could be more discharge data from other subcatchments, or additional data on groundwater levels to derive more detailed spatial patterns of ground water response. It would be hoped that using new and more detailed data will improve the model so that it meets the desired performance criteria. In the context of a practical application, more resources would need to be obtained to carry out these improvements. This is where the interaction with managers regarding acceptable accuracy and available budgets becomes critical.

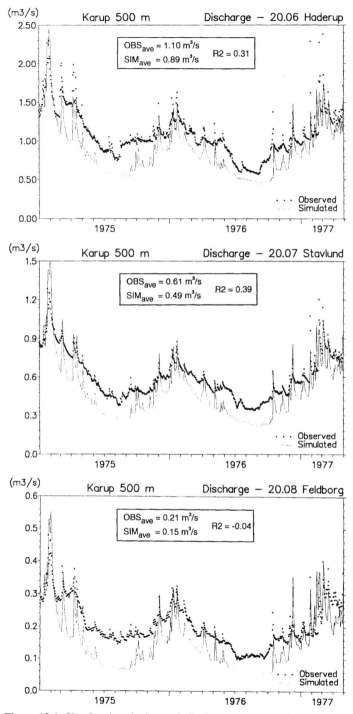

Figure 13.6. Simulated and observed discharges, average flows, SIM$_{ave}$ and OBS$_{ave}$, and model efficiencies, R^2, from the validation period for three internal discharge sites 20.06 (98 km^2), 20.07 (50 km^2) and 20.08 (17 km^2), which have not been subject to calibration. The locations of the discharge stations are shown in Figure 13.3. (From Refsgaard, 1997; reproduced with permission.)

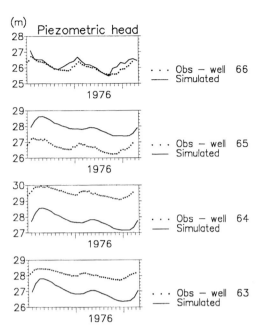

Figure 13.7. Simulated and observed piezometric heads from the validation period for four well sites for which no calibration has been made. The locations of the wells are shown in Figure 13.3. (From Refsgaard, 1997; reproduced with permission.)

13.6 DISCUSSION AND CONCLUSIONS

The need for model validation in distributed hydrological modelling was discussed in Chapter 3. Loague and Kyriakidis (1997) also concluded that the hydrologists need to establish rigorous model evaluation protocols. Gupta et al. (1998) argued that the whole nature of the calibration problem is multi-objective with a need to include not only streamflow but also other variables.

While some attention has been paid to systematic validation of lumped hydrological (rainfall-runoff) models (e.g. WMO, 1975, 1986, 1992), very limited emphasis has so far been put on the far more complicated task of validation of distributed hydrological models, where spatial variation of internal variables also has to be considered. Based on a review of some of the few studies focussing on validation of distributed models with respect to internal variables and multiple scales, the following conclusions can be drawn:

- Distributed models are usually calibrated and validated only against runoff data, while spatial data are seldom available.
- In the few cases, where model simulations have been compared with field data on internal variables, these test results are generally of less accuracy than the results of the validation tests against runoff data.
- Authors, who have not been able to test their models' capabilities to predict internal spatial variables, often state that their distributed models provide physically realistic descriptions of spatial patterns of internal variables.

In summary, it is possible to simulate spatial patterns at a quite detailed level and produce nice colourgraphics results; but due to lack of field data it is in general not possible to check to which extent these results are correct. This fact is pre-

sently one of the most severe constraints for the further development of distributed hydrological modelling. It is believed that although predictions of spatial
patterns both by distributed models and by remote sensing are subject to considerable uncertainties, the possibilities of combining the two may prove to be of
significant mutual benefit. There is an urgent need for more research on this
interface. A recent example of such an exercise by Franks et al. (1998), who
combined SAR based estimates of saturated areas with TOPMODEL simulations, shows encouraging results.

A particular area, where limited work has been carried out so far, is on the
establishment of validation test schemes for the situations where the split-sample
test is not sufficient. The only rigorous and comprehensive methodology reported
in the literature is that of Klemeš (1986b). It may correctly be argued that the
procedures outlined for the proxy-basin and the differential split-sample tests,
where tests have to be carried out using data from similar catchments, from a
purely theoretical point of view are weaker than the usual split-sample test, where
data from the specific catchment are available. However, no obviously better
testing schemes exist. Hence, this will have to be reflected in the performance
criteria in terms of larger expected uncertainties in the predictions.

One of the important practical fields of application for distributed models is
prediction of the effects of land use changes (Ewen and Parkin, 1996; Parkin et
al., 1996). Many such studies have been reported; however, most of them can be
characterised as hypothetical predictions, because the models have not been
subject to adequate validation tests (Lørup et al., 1998). In this case it would
be necessary to apply a differential split sample test but the data requirements are
considerable and will be difficult to meet in practical applications without
detailed information on patterns of hydrological response under different land
use and climatic conditions.

It must be realised that the validation tests proposed in this chapter are so
demanding that many applications today would fail to meet them. This does not
imply that these modelling studies are not useful, only that their output should be
realised to be somewhat more uncertain than is often stated and that they should
not make use of the term 'validated model'.

Success criteria need to be clearly articulated for the model calibration and
validation that focus on each model output for which it is intended to make
predictions. Hence, multisite calibration/validation is needed if predictions of
spatial patterns are required, and multi-variable checks are required if predictions of the behaviour of individual sub-systems within the catchments are
needed. Thus, as shown also in the case study, a model should only be assumed
valid with respect to outputs that have been explicitly validated. This means,
for instance, that a model which is validated against catchment runoff cannot
automatically be assumed valid also for simulation of erosion on a hillslope
within the catchment, because smaller scale processes may dominate here; it will
need validation against hillslope soil erosion data. Furthermore, it should be
emphasised that with the present generation of distributed model codes, which
do not contain adequate up- or down-scaling methodologies, separate calibra-

tion and validation tests have to be carried out every time the element size is changed.

As discussed above, the validation methodologies presently used, even in research projects, are generally not rigorous and far from satisfactory. At the same time models are being used in practice and daily claims are being made on validity of models and on the basis of, at the best, not very strict and rigorous test schemes. An important question then, is how can the situation be improved in the future? As emphasised by Forkel (1996), improvements cannot be achieved by the research community alone, but requires an interaction between the three main "players", namely water resources managers, code developers and model users.

The key responsibilities of the water resources manager are to specify the objectives and define the acceptance limits of accuracy performance criteria for the model application. Furthermore, it is the manager's responsibility to define requirements for code verification and model validation. In many consultancy jobs, accuracy criteria and validation requirements are not specified at all, with the result that the model user implicitly defines them in accordance with the achieved model results. In this respect it is important in the terms of reference for a given model application to ensure consistency between the objectives, the specified accuracy criteria, the data availability and the financial resources. In order for the manager to make such evaluations, some knowledge of the modelling process is required.

The model user has the responsibility for selection of a suitable code as well as for construction, calibration and validation of the site-specific model. In particular, the model user is responsible for preparing validation documents in such a way that the domain of applicability and the range of accuracy of the model are explicitly specified. Furthermore, the documentation of the modelling process should ideally be done in enough detail that it can be repeated several years later, if required. The model user has to interact with the water resources manager on assessments of realistic model accuracies. Furthermore, the model user must be aware of the capabilities and limitations of the selected code and interact with the code developer with regard to reporting of user experience such as shortcomings in documentation, errors in code, market demands for extensions, etc.

The key responsibilities of the developer of the model code are to develop and verify the code. In this connection it is important that the capabilities and limitations of the code appear from the documentation. As code development is a continuous process, code maintenance and regular updating with new versions, improved as a response to user reactions, become important. Although a model code should be comprehensively documented, doubts will, in practice, always occur once in a while on its functioning, even for experienced users. Hence, active support to and dialogue with model users are crucial for ensuring operational model applications at a high professional level.

Although the different players have different roles and functions, a special responsibility lies with the research community. Unless we take a lead in improving the situation within our own community, the overall credibility of hydrological modelling is at risk. Thus a major challenge for the coming decade is to further develop suitable rigorous validation schemes and impose them to all hydrological modelling projects. Part of this challenge lies in the collection and use of spatial patterns of key model inputs, parameters and outputs so that the calibration and validation exercises can fully quantify the model capabilities.

14

Summary of Pattern Comparison and Concluding Remarks

Rodger Grayson and Günter Blöschl

14.1 INTRODUCTION

It also seems obvious that search for new measurement methods that would
yield areal distributions, or at least reliable areal totals or averages, of hydrologic variables such as precipitation, evapotranspiration, and soil moisture
would be a much better investment for hydrology than the continuous pursuit
of a perfect massage that would squeeze the nonexistent information out of the
few poor anaemic point measurements... Klemeš (1986a, p. 187S)

...the collection of data without the benefit of a unifying conception (embo-
died in a model or theory) may submerge us in an ever deepening sea of
seemingly unrelated facts. Hillel (1986, p. 38)

These two wonderful quotes from 1986 encapsulate the motivation for the work
presented in this book. Just what was so special about 1986 is difficult to say,
perhaps it was to do with the combined developments in computer technology
and the desire for a sounder scientific base for hydrology. In any case, while these
calls were not new, they were restated in a powerful way. Progress was being
stymied by a lack of appropriate data and the often weak links between those
who undertook the modelling and measurement. Careful observation and measurement is of course the foundation on which science is built and hydrology is
not short of striking examples. Pioneering observations of runoff processes in the
1960s and early 70s by Emmett (1970), Betson (1964), Dunne and Black
(1970a,b) and others, expanded the view of how runoff was produced (although
many ideas were established much earlier, e.g. Hursh and Brater, 1941). But as
noted by Betson and Ardis (1978) these new concepts took a long time to be
explicitly incorporated into hydrological models, although their bulk effects
could be represented implicitly via calibration of parameters that the modeller
probably did not associate with the process. Interest lay primarily in getting the
catchment runoff right and the simpler a model the better, when this is the aim

(Dawdy, 1969). But when used for investigative purposes, the need for models to mimic the real processes and be "right for the right reasons" (Klemeš, 1986a) becomes paramount. However, mimicking real processes adds complexity, which in turn expands the amount and type of data needed for testing.

In the early chapters of this book we argued that in catchment hydrology, the measurement of *spatial patterns* is necessary to further our understanding of hydrological processes and to properly test and develop spatially explicit hydrological models. The case studies represent some of the few attempts to test this assertion by combining detailed spatial observations and modelling in a catchment hydrology context. The studies also serve to test whether the response of funding agencies and organisations to those powerful calls, that continued through to the 90s, can be vindicated. The case studies cover an extraordinary range of dominant processes, catchment sizes, data types and modelling approaches. Environments range from the semi-arid, convective storm dominated region of Arizona, through the tropical forests of the Amazon, to catchments in Australia, France, Belgium, Norway, Denmark and Canada; from the steep mountains of Austria to the rolling country of Idaho. Catchment sizes range from less than 1 hectare to more than 10,000. Data types include simple nested stream gauging data, numerous point samples of soil moisture using a number of methods, piezometric level, snow water equivalent, runoff detectors, soil chemical and vegetative indicators of recharge and discharge, and a range of remote sensing techniques (satellite SAR, airborne passive microwave, multispectral data, RADAR precipitation, and aerial photography). Most measurements have been quantitative but some have been descriptive or binary. The models have also covered a wide range of dynamic modelling approaches with different distributed structures, as well as stochastic and distribution function approaches. The one thing all of these studies can claim in common is the rare honour of comparing *observed* to *simulated* patterns.

14.2   WHAT HAVE WE LEARNED FROM THE COMPARISONS OF OBSERVED AND SIMULATED PATTERNS?

All case studies of this book have been concerned with comparing simulated and observed patterns of hydrologic variables to inform modelling. A summary of the most important conclusions reached on the basis of these comparisons in each of the chapters is given in Table 14.1.

In the following section, we attempt to compile a bigger picture from the outcomes of these studies in terms of the more general contributions to hydrological science. These fall into three main categories related to *processes*, *data*, and *modelling*.

Processes

A number of the studies have shown that often, a single process dominates hydrological response in a particular catchment. This dominant process depends on the climate and other environmental factors. In the arid/semi-arid climate of

Table 14.1. Summary of findings from the patterns comparisons

Chapter	Pattern comparison	What was learnt from the comparison
4. Patterns and organisation in precipitation	RADAR precipitation versus downscaled (statistically disaggregated) simulations based on large scale patterns from both RADAR data and atmospheric model output. (Figs 4.7, 4.9)	Spatio-temporal rainfall patterns exhibit dynamic scaling behaviour. With appropriate normalisation and dynamic scaling methods, both space and space-time variability can be simulated using stochastic simulation. Storm scale atmospheric models were found to simulate patterns that showed less variability in space and time than observed, indicating that the models *may* need some modification.
5. Patterns and organisation in evaporation	Point flux measurements of evaporation versus simulations from models based on remotely sensed surface temperature and surface cover. (Fig. 5.2)	Reasonable agreement with surface flux stations obtained. Uncertainty in effect of heterogeneity on flux measurements and lack of directly measured patterns of evaporative flux prevented detailed analysis of predictive capability or possible model errors. State of data and knowledge of spatial interactions insufficient at present for confident spatial evaporation estimation.
6. Runoff, precipitation and soil moisture at Walnut Gulch	Rainfall interpolated from dense rain-gauge network versus estimated from remotely sensed surface brightness temperature (ESTAR). (Fig. 6.12) Soil moisture measured by passive microwave versus simulated using various sources of variability and different amounts of data assimilation in a distributed model. (Figs 6.13, 6.15)	Remotely sensed brightness temperature looks to be a promising estimator of precipitation patterns in semi-arid environments. Rainfall patterns are the dominant control on soil moisture patterns and must be represented to model hydrological response in this environment. Spatial variability at the sub-hectare scale was influential on runoff processes. Estimating soil properties from soil type *deteriorated* simulations of soil moisture by a distributed model as compared to using uniform soil properties. Newtonian nudging was best able to assimilate PBMR soil moisture estimates with model estimates to correct spatial simulations. The representation of channel losses is critical to predicting runoff accurately in this environment.
7. Spatial snow cover processes at Kühtai and Reynolds Creek	Aerial photographs of snow cover versus simulations from a distributed model. (Figs 7.3, 7.4, 7.5)	Inclusion of topographically varied energy inputs and wind drift in a distributed model enabled the spatial variability of basic cover patterns to be reproduced. Refined representations of avalanching, wind drift and reflected and emitted radiation from adjacent areas are needed to further improve simulated patterns.

(continued)

357

Table 14.1 (*continued*)

Chapter	Pattern Comparison	What was learnt from the comparison
	Observed snow water equivalent from intensive point measurements versus simulations from a distributed model. (Fig. 7.12)	Patterns of SWE could be reproduced only when the process of wind drift was simulated – this process dominates spatial patterns in a rangeland environment. Other factors such as topographic variations in energy inputs were insignificant compared to drift.
8. Variable source areas, soil moisture and active microwave observations at Zwalmbeek and Coët-Dan	A field survey of saturated area versus estimated pattern from the standard deviation, and from PCA, of multi-temporal SAR images. (Figs 8.5, 8.11) Mapped soils that are characteristic of wet areas versus estimated patterns from the standard deviation, and from PCA, of multi-temporal SAR images. (Figs 8.6, 8.8)	Soil moisture and wet areas cannot be retrieved from single images on vegetated surfaces. The hypothesis that wet areas should be identifiable as areas of low variance in multi-temporal SAR images was confirmed. In areas where terrain variability is high, it dominates the SAR response. PCA applied to multitemporal images can be used to isolate the component of the backscatter coefficient that is dominated by variations in soil moisture, providing qualitative patterns of areas likely to be wet.
9. Soil moisture and runoff processes at Tarrawarra	Soil moisture in the top 30 cm from intensively sampled point measurements versus simulated soil moisture from a distributed model. (Figs 9.6, 9.9, 9.10, 9.11, 9.12)	A distributed model that represented the effect of spatial variability in topography on lateral surface and subsurface flow, and radiation exposure on evaporation, was able to represent the spatial and temporal variation in soil moisture. Introduction of variability in soil properties via mapping of soil type did not improve simulations. Preferential flow through cracks in the soil in Autumn needs to be represented to improve observed spatial patterns during this period. An additional soil layer needs to be represented, and better ET procedures included to further improve soil moisture estimates in the Spring.
10. Storm runoff generation at La Cuenca	Runoff occurrence from intensive network of runoff detectors versus simulated runoff occurrence from a distributed model (Figs 10.15, 10.16)	Simulations using deterministic patterns of soil hydraulic properties could not reproduce observed patterns of runoff occurrence. Distributed models in this tropical environment could represent the functional behaviour of spatial runoff response only by combining deterministic patterns and random realisations of soil hydraulic properties.

Table 14.1 (*continued*)

Chapter	Pattern comparison	What was learnt from the comparison
11. Shallow groundwater response at Minifelt	Water levels from a dense network of piezometers versus simulations from a quasi distributed model. (Figs 11.4, 11.12)	Topography and soil depth alone could not reproduce observed patterns. Measured patterns of piezometric level could be simulated only through the use of spatially variable soil porosity and hydraulic conductivity. These soil parameters needed to be "back-calculated" from spatial water table response and their patterns could not be interpreted physically. Model structure uncertainty complicated the assessment of the value of different data types for constraining the uncertainty in simulations. In the dry parts of the catchment, patterns were more useful than times series at a point for constraining the uncertainty in simulations but the reverse was true in the wet parts of the catchment.
12. Groundwater–vadose zone interactions at Trochu	Observations of long term recharge and discharge locations based on piezometric levels and vegetative, biological and soil chemical indicators, versus simulations from an equilibrium vadose zone model linked to a distributed groundwater model. (Figs 12.6, 12.7, 12.8, 12.9, 12.10)	Coupling of the vadose zone and groundwater behaviour was able to reproduce observed patterns of recharge and discharge. Recharge/discharge patterns are more dependent on local topography when strong coupling exists and on regional topography when weak coupling exists. Low bedrock transmissivities increase the area of a catchment over which coupling is important
13. Towards a formal approach to calibration and validation of models using spatial data	Multiple point observations of ground-water level and stream flow versus simulations from a large scale distributed model (Figs 13.5, 13.6, 13.7)	Predictive performance for stream gauges and wells that were not used for calibration was poor due to problems with the simulated groundwater divide. Use of even limited spatial data provides a severe test for distributed models and should become part of modelling practice of credibility of results is to be assured.

Walnut Gulch (Chapter 6) rainfall space-time variability was vastly more important than other controls, based on an analysis of rainfall data at the sub-hectare
scale and a sensitivity analysis using the KINEROS model. For the very different
climate of Reynolds Creek (Chapter 7) it was demonstrated that wind drift is by
far the most important process affecting space-time patterns of snow water
equivalent, by comparing observed snow water equivalent patterns with two
simulated scenarios, with and without a representation of snow drift. In the
humid climate of the Tarrawarra catchment (Chapter 9) saturated source area
runoff was the dominant runoff mechanism as concluded from simple initial
analyses of the observed TDR soil moisture patterns and later confirmed by
Thales simulations. The simulations also confirmed that subsurface water movement changes abruptly in spring and autumn. During summer (dry), vertical
water movement was dominant while in winter (wet), lateral water movement
was dominant. Finally, a comparison of simulated and observed recharge/dis-
charge patterns in the Prairie climate of Trochu (Chapter 12) indicated that
recharge and discharge patterns were controlled by the coupling of the regional
aquifer with the surface through the unsaturated zone, which dominated the local
water budget. This finding was corroborated by a sensitivity study comparing
scenarios with and without coupling to observed recharge/discharge patterns.
The scenario without coupling could be made to match observed patterns only
when unrealistically high values of recharge were assumed.

Comparisons of simulated and observed patterns have also shed light on the
nature of space-time variability of hydrologic variables that probably would not
have been possible by simple visualisation of the data alone. For example, analyses of RADAR rainfall data in Chapter 4 suggested that the space-time variability of rainfall is characterised by dynamic scaling, i.e. the rainfall fluctuations
in space and time can be represented by a power law when plotted against scale
after appropriate renormalisation. This property was used for generating rainfall
patterns by means of stochastic simulations (downscaling) that when compared
to RADAR rainfall patterns, looked realistic. Analyses of remotely sensed
(ESTAR) soil moisture patterns in Walnut Gulch (Chapter 6) indicated that,
following a rainstorm, these patterns were organised but this organisation
faded away after the storm, and the pattern became random. The authors suggested that this change-over is a reflection of the changing control on soil moisture of rainfall versus patterns of surface soil characteristics. A similar change-
over in the variability of soil moisture, however this time on a seasonal basis, was
identified at Tarrawarra (Chapter 9). Spatially organised patterns that were
related to terrain occurred in winter while spatially random patterns occurred
in summer. This change-over was identified by visual inspection of the TDR
measurements and further explained by comparison with model results. In La
Cuenca (Chapter 10) where infiltration excess and pipe flow were important
runoff mechanisms, the type of spatial variability in soil hydraulic conductivity
was inferred from a comparison of observed patterns of frequency of runoff
occurrence with a number of scenarios with different types of soil variability
represented. It was found that a combined deterministic (by soil type) and

stochastic pattern of conductivity produced patterns of runoff occurrence that were most similar to the observed patterns.

Clearly, the detail of these insights into catchment behaviour was made possible by the availability of measured patterns.

Data

While it is clear that the patterns were useful in their own right, several studies showed that they are even more useful if used in combination with time series data. Patterns and time series are therefore complementary and the case studies have shown that these different types of data (space variability and time variability, respectively) can be used to identify different properties of the catchment behaviour. In Chapter 7 conventional runoff hydrographs were used to identify the snow melt runoff *volume* from the catchment. In a similar fashion, runoff was used to close the water balance of the Tarrawarra catchment (Chapter 9) by enabling an estimation of deep drainage into bedrock. In La Cuenca (Chapter 10) runoff hydrographs were used to complement the spatial patterns, but this time at the event scale, to calibrate the Manning roughness parameter. In each of these cases, the information available in the time series was used for things that could not be well identified from the spatial data. Use of the two types of information together was the key to realistic simulation of space-time patterns of processes in each study. Similarly, in Minifelt (Chapter 11) soil porosity (related to the dynamics) was calibrated from mainly time series data (borehole data of the shallow groundwater table), while hydraulic conductivity was calibrated from mainly snapshots of spatial patterns of the groundwater table. It was the complementary nature of spatial pattern and time series data that enabled successful modelling.

Perhaps surprisingly, patterns of binary data were used in about half of the case studies; i.e. in the trade off between spatial resolution and information from a particular point (discussed in Chapter 2), the scales were tipped towards spatial resolution. All of these studies showed that a wealth of information can be revealed from a binary pattern. In Kühtai (Chapter 7) snow cover patterns (snow/no snow) were used; in Zwalmbeek and Coët-Dan (Chapter 8) patterns of saturated source areas (saturated/not saturated) were used; in La Cuenca (Chapter 10) patterns of runoff occurrence (for a single event, runoff occurred/ did not occur) were used; and in Trochu (Chapter 12) patterns of recharge/discharge (either recharge or discharge) were used. The data used at Trochu are particularly interesting as they have been derived from qualitative observation including chemical/vegetation indicators. These indicators integrate over time so are representative of the long-term mean of recharge/discharge conditions (Tóth, 1966). Although water tables for a given point in time (snap shots) would have been easier to measure, the binary recharge/discharge data were much more appropriate to test the equilibrium vadose zone model used in Chapter 12.

The spatial variability of physical soil properties is particularly critical in catchment hydrology, yet we have relatively poor ways of estimating them at the catchment scale. It is therefore not surprising that a number of case studies in this book have scrutinised the reliability of soils data and their effect on the representation of catchment response. For Walnut Gulch (Chapter 6), where the infiltration excess runoff mechanism dominates, TOPLATS was used to simulate scenarios of soil moisture patterns. One of the scenarios was based on uniform soil hydraulic properties, while the other scenario used pedotransfer functions from the literature to estimate the soil hydraulic properties from mapped soil type. A comparison of the soil moisture patterns from the two scenarios with observed soil moisture patterns from airborne PBMR, indicated that the one based on soil type was too patchy and the scenario using uniform soil properties was more consistent with the observations. At Tarrawarra (Chapter 9) one scenario used soil type to spatially distribute hydraulic conductivity measurements, assuming uniform conductivity within each soil type zone. This scenario produced artificially high soil moisture values at the interface of the soil types that could be identified by comparisons with observed soil moisture patterns. A similar comparison at La Cuenca (Chapter 10) indicated that the assumption of uniform conductivity in each of their three land types was not appropriate and a random component had to be added to the deterministic pattern imposed by land type. Clearly, the variability of soil physical properties within soil types can be as large or larger than the variability between soil types. This suggests that the widespread practice in distributed modelling of allocating soil hydraulic properties on the basis of soils type (using either pedo-transfer functions or typical measurements from each soil) is likely to result in poor simulations of patterns in soil moisture and runoff.

With respect to data issues, the case studies have highlighted the value of *complementary* data (spatial patterns and time series), the utility of *binary* patterns (which are often simple to collect compared to quantitative patterns) and some particular problems in representing soil properties in models. We next address the utility of these data for informing model development.

Modelling

An important reason for comparing simulated and observed patterns was to assess the credibility of the distributed catchment models, i.e. how well can they represent individual processes that operate in the catchment, and which processes are perhaps not represented very well? This assessment resulted in suggestions for changes in model structure or model parameters (or perhaps inputs) that are needed to refine the model simulations.

Most of the chapters in this book concluded that the models worked quite well, albeit after calibration, and that the main processes were very well represented. However, they also concluded that it is possible to use subtle differences between simulated and observed patterns to inform us about how the models could be improved. At Kühtai (Chapter 7), for example, the comparison of snow cover patterns suggested that the model underestimated snow water equivalent in cirques. This was traced back to not representing emitted radiation from surrounding terrain. Similarly, a tendency to overestimate (and underestimate) snow cover on south facing (and north facing) slopes, was interpreted as evidence that the model should account for the dependence of snow albedo on energy input. At Reynolds Creek (Chapter 7) the comparison of simulated and observed patterns of snow water equivalent suggested that the simulated drift is more sharply defined than the observed drift. This was traced back to differences in the wind conditions between the year used for calibrating the model and the year where the model was tested. A suggested remedy was to use a more sophisticated deterministic wind drift model that takes into account differences in wind conditions from year to year, although more data would probably be needed to properly test this idea. At Tarrawarra (Chapter 9) subtle differences between simulated and observed rates of the temporal change in soil moisture patterns in autumn suggested that the lateral soil hydraulic conductivity may in fact change with time. This was indicated by faster subsurface redistribution in early autumn than in late autumn. It was suggested that this was due to temporal changes in conductivity caused by the closing of cracks that had formed over summer, thereby reducing lateral conductivity. However, testing of this would need additional data. At Minifelt (Chapter 11) where shallow water table patterns were used to calibrate the spatial patterns of soil physical properties in TOPMODEL, it was difficult to physically interpret the calibrated patterns. This was suggested to be evidence that there may be structural problems with the TOPMODEL approach for Minifelt, and relaxing the TOPMODEL assumptions may improve spatial predictions. Also, the uncertainty analysis based on spatially uniform soils parameters (which is a more common TOPMODEL application) gave different predictions and different uncertainty bounds depending on whether time series of borehole data or patterns of piezometer data were used to constrain the model parameters. This also suggested that there may be substantial structural uncertainty with TOPMODEL as applied in the Minifelt example.

In about half of the case studies (Reynolds Creek, Tarrawarra, La Cuenca, Minifelt) comparisons of simulated and observed patterns were used not only to assess the reliability of the model, but also to calibrate some of the model parameters as mentioned above. Both assessment and calibration were done by a visual pattern comparison. Some of the case studies, however, used more objective and sophisticated methods for model testing and parameter identification. At Walnut Gulch (Chapter 6) four-dimensional data assimilation (4DDA) methods were used to update the model state variables of the TOPLATS model by using remotely sensed (PBMR) and in situ point measurements of soil moisture. It was concluded that 4DDA (already being in operational use in atmospheric modelling) holds substantial promise for operational use in spatially distributed hydrological modelling. There is an obvious parallel with operational runoff forecasting, where updating model state variables (albeit in the time domain) is common practice today. A formal parameter uncertainty analysis was performed in Chapter 11 based on the GLUE procedure which gave a very useful assessment of the reliability of model parameters and helped define the value of various data types in constraining the model parameter uncertainty. Although the method is computationally demanding it can handle nonlinear models and it can make use of observed spatial patterns. In Chapter 11, parameter uncertainty was plotted against the topographic wetness index, which allowed differences in uncertainty between the gully and ridge areas of the catchment to be examined. The validation tests were taken even further in Chapter 13 for the Karup catchment. This was a significantly larger application than the case study chapters in the book and focussed on use of models in a more practical context. The MIKE-SHE model was calibrated and than validated based on a formal procedure presented by the author, using data from a number of internal stream gauges and boreholes. In this respect it was more typical of what may be possible in practical applications of spatially distributed models outside small research catchments. The author concluded from the comparisons of observed and simulated hydrologic variables at a number of locations, that when a formal framework of validation tests is set a priori, it may be difficult to meet the validation criteria in practice. He concluded that formal protocols are needed for model validation and that implementation of these for practical applications will require more dialogue between model developers, users and the managers who use simulations in their decision making, so that capabilities and limitations are clearly articulated.

In this section, we have summarised in some detail, the conclusions of the case studies, highlighting where the use of measured patterns, often in combination with more traditional measurements, were useful in explaining processes and developing models within relatively small research catchments. As discussed in Chapter 13, the larger scale, more practically oriented problems to which distributed models are applied can also benefit from the use of pattern data, but that such data are much less common in the "real world". We predict that in the coming years, more effort will be placed on collecting and using patterns at the larger scale so that the benefits discussed in the case studies can be realised in more practical applications.

14.3   OUTLOOK

The use of spatial patterns in catchment hydrology is in its infancy, but initial results are encouraging and provide sound reasons to believe that there are great improvements to be made in our understanding of catchment hydrological processes; and in quantifying the way they affect, and are affected by, spatial variability across a range of scales. More specifically, the work presented in this book illustrates that to realise these improvements, we need *appropriate* data. *Appropriate* meaning that it tells us about system behaviour, tests critical assumptions in our understanding and in our models of that understanding, and provides enough information to resolve the problems of non-uniqueness and parameter identifiability inherent in complex models. Spatial patterns of hydrological response are an *appropriate* data source in this respect. So while collecting and collating large spatial data sets will be important to the development of spatial models and improved process understanding, just where are the specific areas where significant progress can be made? In the following few paragraphs we provide a brief assessment of key areas.

14.3.1 Improvements to Model Inputs

Numerous hydrological studies spanning over many decades have shown the importance of precipitation on hydrological response. Now with RADAR estimates of spatial patterns in precipitation and new methods such as those described in Chapter 4 for characterising space-time variability, we are on the threshold of a major advance in the use of spatial precipitation information in distributed models. To fully realise the potential of this information, we may well need changes to model structure, and will certainly need changes in the attitudes of hydrological modellers who have been firmly wedded to the use of raingauge data for calibration and testing. Several of the case studies showed that simple binary patterns can be powerful tests of distributed models and provide useful information on threshold phenomena such as saturated source areas. There would appear to be further scope for use of this type of data, but again some changes in attitude towards "non-quantitative" data, and possibly changes to model structure, may be needed. At least in the immediate future, remotely sensed (RS) data can be thought of in this context and might be best used to assist in reducing the degrees of freedom in distributed hydrological models by providing *patterns,* rather than absolute values, of important inputs and parameters. Model structures are being improved to better exploit pattern data via improved software engineering (such as integration with GIS platforms) and this should serve as encouragement for the development of hydrological algorithms that are specifically intended for the scale and nature of RS data. Chapter 5 clearly illustrated that we have a long way to go in fully understanding and being able to represent spatial patterns of evaporation. Given that evaporation can be over 90 % of the water budget in some environments, it is obvious that studies into dealing with spatial measurement and the role of land surface heterogeneity will continue to be critical to improvements in, particularly, large-scale models.

Spatially distributed modelling requires the use of interpolation for a number of purposes, including the matching of model and measurement scales of information used for input and testing. There is a need for improved interpolation methods that better enable us to incorporate our understanding of physical phenomena. While there are a range of methods already available, these need to be improved in their ability to represent organisation in spatial patterns of hydrological importance.

14.3.2 Improvements to Model Testing

It is envisaged that comparisons between observed and simulated patterns will eventually become part of standard procedures for model testing. As well as needing the observed patterns, we also need improved quantitative techniques for comparing the similarities and differences between patterns. Some simple methods have been presented at the end of Chapter 3, but few of these have so far been used in practice. With rich areas of research on topics such as pattern recognition, we expect that the sophistication of pattern comparisons will greatly increase as hydrologists come to realise the value of patterns for model development and testing. Comprehensive uncertainty analysis also needs to be further developed for spatially distributed models. The potential of these methods for assessing the separate sources of uncertainty (input information, parameter values, model structure and data used in testing) is large, but at present they are not computationally tractable for most distributed models. Methodological developments as well as improvements in computer power are likely to lead to wider use of such techniques.

14.3.3 Challenges for Model Conceptualisation

The primary challenge of hydrologists has been, and remains, the prediction of hydrological response in "ungauged" areas – i.e. areas for which we have no hydrological response information. There is still a need to improve methods for generalising results from small catchments such as those described in this book, to other catchments; from small catchments to large catchments; and for being able to predict hydrological response under changed land use and climatic conditions in catchments of all sizes. All of these needs can be met only with better understanding and representation of fundamental processes, and their spatial variability across a range of scales. Distributed modelling generally has moved beyond just trying to scale up small catchment models to large scales because of problems with identifiability and scale dependence. As has been suggested for some time by many authors, we need models for a range of scales that are parsimonious, but that reflect the manifestation of important processes at those different scales. In moving beyond the notion of "trying to model everything" we should be developing methods to identify dominant processes that control hydrological response in different environments (landscapes and climates) and at different scales, and then develop models to focus on these dominant processes (a notion we might call the "Dominant Processes Concept" (DPC)). This would provide a framework for the development and application of techniques specially designed to deal with those controls and help to avoid some of the overparameterisation problems that occur when processes that are not important are represented in models. Developments along the lines of the DPC may help with the generalisation problems that have haunted hydrologists since the science began.

14.4 FINAL REMARKS

As mentioned in the introduction, there have been many calls for data collection and analysis to go hand in hand, for improved understanding of processes, and for the scientific endeavour of measurement to be recognised. There is a range of evidence that these calls have elicited a response. For example, Water Resources Research has had "data notes" for some time (Hornberger, 1994) and the number being published is increasing. There is an increasing awareness that the development of a spatial model is not of itself useful, unless it can be properly tested so that it can provide more credible predictions, or more insight into process understanding. Large field campaigns are continuing and the valuable role of smaller catchment, process-focussed, studies is being recognised as the researchers have integrated their work with theoretical and modelling developments to ensure the results contribute to a wider understanding of patterns of hydrological variability. We hope that the case studies presented in this book, and the broader conclusions from this extraordinary range of studies, have unequivocally illustrated the value of this investment, and act as encouragement for more, and more innovative, studies into spatial patterns in catchment hydrology.

References

Abbott, M. B., Bathurst, J. C., Cunge, J. A., O'Connell, P. E. and Rasmussen, J. (1986). An introduction to the European Hydrologic System – Système Hydrologique Européen, "SHE", 1, History and philosophy of a physically-based, distributed modelling system. *J. Hydrology*, 87: 45–9.

Abbott, M. B. and Refsgaard, J. C. (1996). *Distributed Hydrological Modelling*. Kluwer, Dordrecht, The Netherlands, 321pp.

Abrahams, A. D., Parsons, A. J. and Luk, S. H. (1989). Distribution of depth of overland flow on desert hillslopes and its implications for modelling soil erosion. *J. Hydrology*, 106: 177–84.

Ahmed, S. and de Marsily, G. (1987). Comparison of geostatistical methods for estimating transmissivity using data on transmissivity and specific capacity. *Water Resour. Res.*, 23: 1717–37.

Air Weather Service (1979). *The Use of the Skew T, Log P Diagram in Analysis and Forecasting*. Air Weather Service Manual, Illinois.

Allard, D. (1993). On the connectivity of two random set models: the truncated Gaussian and the Boolean. In: Soares, A. (Ed.), Geostatistics Tróia '92. *Quantitative Geology and Geostatistics*. Kluwer, Dordrecht, The Netherlands, pp.467–78.

Allard, D. (1994). Simulating a geological lithofacies with respect to connectivity information using the truncated Gaussian model. In: Armstrong, M. and Down, P. A. (Eds), *Geostatistical Simulations*. Proc. Geostatistical Simulation Workshop, Fontainebleau, France, 27–28 May 1993. *Quantitative Geology and Geostatistics*. Kluwer, Dordrecht, The Netherlands, pp.197–211.

Altese, E., Bolognani, O., Mancini, M. and Troch, P. A. (1996). Retrieving soil moisture over bare soil from ERS-1 synthetic aperture radar data: Sensitivity analysis based on a theoretical surface scattering model and field data. *Water Resour. Res.*, 89(3): 653–61.

Ambroise, B., Beven, K. and Freer, J. (1996). Toward a generalization of the TOPMODEL concepts: topographic indices of hydrological similarity. *Water Resour. Res.*, 32(7): 2135–45.

Anderson, E. A. (1973). National Weather Service River Forecast System–Snow Accumulation and Ablation Model, NOAA Technical Memorandum NWS HYDRO-17, U.S. Dept. of Commerce.

Anderson, E. A. (1976). A Point Energy and Mass Balance Model of a Snow Cover, NOAA Technical report NWS 19, U.S. Dept. of Commerce.

368

Anderson, M. C., Norman, J. M., Diak, G. R., Kustas, W. P. and Mecikalski, J. R. (1997). A two source time-integrated model for estimating surface fluxes from thermal infrared satellite observations. *Remote Sens. Environ.*, 60: 195–216.

Anderson, M. G. and Burt, T. P. (1978a). Experimental investigations concerning the topographic control of soil water movement on hillslopes. *Z. Geomorph. N. F.,* Suppl. Bd. 29: 52–63.

Anderson, M. G. and Burt, T. P. (1978b). The role of topography in controlling throughflow generation. *Earth Surface Processes*, 3: 331–44.

Anderson, M. G. and Burt, T. P. (1978c). Toward more detailed field monitoring of variable source areas. *Water Resour. Res.*, 14(6): 1123–31.

Anderson, M. G. and Kneale, P. E. (1980). Topography and hillslope soil water relationships in a catchment of low relief. *J. Hydrology*, 47: 115–28.

Anderson, M. G. and Kneale, P. E. (1982). The influence of low-angled topography on hillslope soil-water convergence and stream discharge. *J. Hydrology*, 57: 65–80.

Anderson, M. G. and Rogers, C. C. M. (1987). Catchment scale distributed hydrological models – A discussion of research directions. *Progress in Physical Geography*, 11: 28–51.

Anderson, M. P. (1997). Characterization of geological heterogeneity. In: Dagan, G. and Neuman, S. P. (Eds), *Subsurface Flow and Transport*. Cambridge University Press, pp. 23–43.

Anderson, M. P. and Woessner, W. W. (1992). The role of postaudit in model validation. *Adv. Water Resour.*, 15: 167–73.

Armstrong, B. and Williams, K. (1986). *The Avalanche Book*, Fulcrum Inc, Golden, Colorado, 256pp.

Armstrong, M. (1998). *Basic Linear Geostatistics*. Springer, Berlin.

Attema, E. (1991). The active microwave instrument on-board the ERS-1 satellite. *Proc. IEEE*, 79: 791–9.

Austin, P. M. and Houze, R. A. (1972). Analysis of the structure of precipitation patterns in New England. *J. Appl. Meteor.*, 11: 926–35.

Avissar, R. (1992). Conceptual aspects of a statistical-dynamical approach to represent landscape subgrid-scale heterogeneities in atmospheric models. *J. Geophys. Res.*, 97: 2729–42.

Avissar, R. (1995). Scaling of land–atmosphere interactions: An atmospheric modelling perspective. *Hydrological Processes*, 9: 679–95.

Avissar, R. (1998). Which type of soil-vegetation-atmosphere transfer scheme is needed for general circulation models: a proposal for a higher-order scheme. *J. Hydrology*, 212–13: 136–54.

Avissar, R. and Liu, Y.-Q. (1996). Three dimensional numerical study of shallow convective clouds and precipitation induced by land surface forcing. *J. Geophys. Res.*, 101(D3): 7499–518.

Bair, E. S. (1994). Model (in)validation – a view from the courtroom. *Ground Water*, 32(4): 530–1.

Baldocchi, D. D. (1993). Scaling water vapor and carbon dioxide exchange from leaves to a canopy: rules and tools. In: Ehleringer, J. R. and Field, C. B. (Eds), *Scaling Physiological Processes Leaf to Globe*. Academic Press, New York, pp.77–114.

Baldocchi, D., Hicks, B. B. and Meyers, T. P. (1988). Measuring biosphere–atmosphere exchanges of biologically related gases with micrometeorological methods. *Ecology*, 69(5): 1331–40.

Band, L. E. and Moore, I. D. (1995). Scale: Landscape attributes and Geographical Information Systems. *Hydrological Processes*, 9: 401–22.

Band, L. E., Patterson P., Nemani, R. and Running, S. (1993). Forest ecosystem processes at the watershed scale: incorporating hillslope hydrology. *Agric. For. Meteorol.*, 63: 93–126.

Bárdossy, A. and Lehmann, W. (1998). Spatial distribution of soil moisture in a small catchment. Part 1: geostatistical analysis. *J. Hydrology*, 206: 1–15.

Barling, R. D., Moore, I. D. and Grayson, R. B. (1994). A quasi-dynamic wetness index for characterising the spatial distribution of zones of surface saturation and soil water content. *Water Resour. Res.*, 30: 1029–44.

Barros, A. P. and Lettenmaier, D. P. (1993). Dynamic modelling of the spatial distribution of precipitation in remote mountainous areas. *Monthly Weather Review*, 121(April): 1195–214.

Barros, A. P. and Lettenmaier, D. P. (1994). Dynamic modelling of orographically induced precipitation. *Rev. Geophys.*, 32(3): 265–84.

Bastiaanssen, W. G. M., Feddes, R. A. and Holtslag, A. A. M. (1998). A remote sensing surface energy balance algorithm for land (SEBAL) Part 1: Formulation. *J. Hydrology*, 212–213(1–4): 198–212.

Bathurst, J. C. and O'Connell, P. E. (1992). Future of distributed modelling: The Système Hydrologique Europeen. *Hydrological Processes*, 6: 265–77.

Bathurst, J. C., Wicks, J. M. and O'Connell, P. E. (1995). The SHE/SHESED basin scale water flow and sediment transport modelling system. In: Singh, V. P. (Ed.), *Computer Models of Watershed Hydrology*. Water Resources Publications, Highlands Ranch, Colorado, pp. 563–94.

Beck, M. B. (1987). Water quality modeling: a review of the analysis of uncertainty. *Water Resour. Res.*, 23: 1393–442.

Becker, A., Blöschl, G. and Hall, A. (1999). Land surface heterogeneity and scaling in hydrology. *J. Hydrology* (special issue), 217(3–4).

Bergström, S. (1991). Principles and confidence in hydrological modelling. *Nordic Hydrology*, 22: 123–36.

Betson, R. P. (1964). What is watershed runoff? *J. Geophys. Res.*, 69(8): 1541–52.

Betson, R. P. and Ardis, C. V. (1978). Implications for modelling surface-water hydrology. Chapter 8 in: Kirkby, M. J. (Ed.), *Hillslope Hydrology*. Wiley, Chichester, pp.295–323.

Beven, K. J. (1986). Runoff production and flood frequency in catchments of order *n*: an alternative approach. In: Gupta, V. K., Rodriguez-Iturbe, I. and Wood, E. F. (Eds), *Scale Problems in Hydrology*. Reidel, Dordrecht, pp.107–31.

Beven, K. J. (1987). Towards a new paradigm in hydrology. In: *Water for the future*: *Hydrology in perspective*. IAHS Publ. No. 164: 393–403.

Beven, K. (1989). Changing ideas in hydrology – the case of physically-based models. *J. Hydrology*, 105: 157–72.

Beven, K. (1995). Linking parameters across scales: Subgrid parameterizations and scale dependent hydrological models. *Hydrological Processes*, 9: 507–25.

Beven, K. (1996). A discussion of distributed hydrological modelling. In: Abbott, M. B. and Refsgaard, J. C. (Eds), *Distributed Hydrological Modelling*, Kluwer, Dordrecht, The Netherlands, pp. 255–78.

Beven, K. (1997). TOPMODEL: A critique. *Hydrological Processes*, 11: 1069–85.

Beven, K. J. and Binley, A. M. (1992). The future of distributed models: model calibration and uncertainty prediction. *Hydrological Processes*, 6: 279–98.

Beven, K. J. and Fisher, J. (1996). Remote sensing and scaling in hydrology. In: Stewart, J. B., Engman, E. T., Feddes, R. A. and Kerr, Y. (Eds), *Scaling up in Hydrology using Remote Sensing*. Wiley, London, pp.1–18.

Beven, K. J. and Kirkby, M. J. (1979). A physically-based, variable contributing area model of basin hydrology. *Hydrol. Sci. Bull.*, 24: 43–69.

Beven, K. J., Lamb, R., Quinn, P., Romanowicz, R. and Freer, J. (1995). TOPMODEL. In: Singh, V. P. (Ed.), *Computer Models of Watershed Hydrology*. Water Resources Publications, Highlands Ranch, Colorado, pp.627–68.

Beven, K. J. and Moore, I. D. (Eds) (1992). *Terrain Analysis and Distributed Modelling in Hydrology*. Wiley, New York, 249pp.

Binley, A. M. and Beven, K. J. (1991). Physically-based modelling of catchment hydrology: a likelihood approach to reducing predictive uncertainty. In: Farmer, D. G. and Rycroft, M. J. (Eds), *Computer Modelling in the Environmental Sciences*, pp.75–88.

Binley, A., Beven, K. and Elgy, J. (1989). A physically based model of heterogeneous hill-slopes. 2. Effective hydraulic conductivities. *Water Resour. Res.*, 25: 1227–33.

Blackman, R. B. and Tukey, J. W. (1958). *The Measurement of Power Spectra*. Dover Publ., New York, 190pp.

Blöschl, G. (1996). *Scale and Scaling in Hydrology*. Institut für Hydraulik, Gewässerkunde und Wasserwirtschaft. Wiener Mitteilungen, Wasser-Abwasser-Gewässer, Vol. 132, Technical University of Vienna, 346pp.

Blöschl, G. (1999). Scaling issues in snow hydrology. *Hydrological Processes*, 13, 2149–75.

Blöschl, G., Grayson, R. B. and Sivapalan, M. (1995). On the representative elementary area (REA) concept and its utility for distributed rainfall-runoff modelling. *Hydrological Processes*, 9: 313–30.

Blöschl, G., Gutknecht, D., Grayson, R. B., Sivapalan, M. and Moore, I. D. (1993). Organisation and randomness in catchments and the verification of distributed hydrologic models. *Eos, Transactions of the American Geophysical Union*, 74(43): 317.

Blöschl, G., Gutknecht, D. and Kirnbauer, R. (1991a). Distributed snowmelt simulations in an alpine catchment. 2. Parameter study and model predictions. *Water Resour. Res.*, 27(12): 3181–8.

Blöschl, G. and Kirnbauer, R. (1991). Point snowmelt models with different degrees of complexity – Internal processes. *J. Hydrology*, 129: 127–47.

Blöschl, G. and Kirnbauer, R. (1992). An analysis of snow cover patterns in a small alpine catchment. *Hydrological Processes*, 6: 99–109.

Blöschl, G., Kirnbauer, R. and Gutknecht, D. (1991b). Distributed snowmelt simulations in an Alpine catchment. 1. Model evaluation on the basis of snow cover patterns. *Water Resour. Res.*, 27: 3171–9.

Blöschl, G. and Sivapalan, M. (1995). Scale issues in hydrological modelling – a review. *Hydrological Processes*, 9: 251–90.

Blyth, E. M. and Harding, R. J. (1995). Application of aggregation models to surface heat flux from the Sahelian tiger brush. *Agric. For. Meteorol.*, 72: 213–35.

Bonell, M. and Gilmour, D. A. (1978). The development of overland flow in a tropical rain-forest catchment. *J. Hydrology*, 39: 365–82.

Bow, S.-T. (1992). *Pattern Recognition and Image Processing*. Dekker, New York, 558pp.

Bowser, W. E., Peters, T. W. and Newton, J. D. (1951). Soil survey of the Red Deer sheet. *Alberta Soil Survey*, 16: 86pp.

Bras, R. L. (1990). *Hydrology, An Introduction to Hydrologic Science*. Addison-Wesley, Reading, MA, 643pp.

Braun, L. N. (1985). Simulation of snowmelt-runoff in lowland and lower alpine regions of Switzerland. Zürcher Geographische Schriften 21, Geographisches Institut der Eidgenössischen Technischen Hochschule Zürich.

Bristow, K. L. and Campbell, G. S. (1984). On the relationship between incoming solar radiation and the daily maximum and minimum temperature. *Agric. For. Meteorol.*, 31: 159–66.

Bronstert, A. and Plate, E. (1997). Modelling of runoff generation and soil moisture dynamics for hillslopes and micro-catchments. *J. Hydrology*, 198: 177–95.

Brooks, R. H. and Corey, A. T. (1966). Properties of porous media affecting fluid flow. *J. Irrig. Drain. Div. Am. Soc. Civ. Eng.*, 92(IR2): 61–88.

Brubaker, K. L. and Entekhabi, D. (1996). Analysis of feedback mechanisms in land–atmosphere interaction. *Water Resour. Res.*, 32(5): 1343–57.

Brun, C., Bernard, R., Vidal-Madjar, D., Gascuel-Odoux, C., Mérot, P., Duchesne, J. and Nicolas, H. (1990). Mapping saturated areas with a helicopter-borne C band scatterometer. *Water Resour. Res.*, 26(5): 945–55.

Brun, E., David, P., Sudul, M. and Brunot, G. (1992). A numerical model to simulate snow-cover stratigraphy for operational avalanche forecasting. *J. Glaciology*, 28(128): 13–22.

Bruneau, P., Gascuel-Odoux, C., Robin, P., Mérot, P. and Beven, K. J. (1995). The sensitivity to space and time scales of a hydrological model using digital elevation data. *Hydrological Processes*, 9: 69–82.

Brutsaert, W. (1982). *Evaporation into the Atmosphere*. Kluwer, Dordrecht, Holland, 299pp.

Buol, S. W., Hole, F. D. and McCracken, R. J. (1989). *Soil Genesis and Classification*. Iowa State University Press: Ames.

Burt, T. P. and Butcher, D. P. (1985). Topographic controls of soil moisture distributions. *J. Soil Science*, 36: 469–86.

Burt, T. P. and Butcher, D. P. (1986). Development of topographic indices for use in semi-distributed hillslope runoff models. *Z. für Geomorph. N. F.,* Suppl. Bd. 58: 1–19.

Calver, A. and Cammeraat, L. H. (1993). Testing a physically-based runoff model against field observations on a Luxembourg hillslope. *Catena*, 20(3): 273–88.

Calver, A. and Wood, W. L. (1995). The Institute of Hydrology Distributed Model. In: Singh, V. P. (Ed.), *Computer Models of Watershed Hydrology*, Water Resources Publications, Highlands Ranch, Colorado, pp.595–626.

Carlson, T. N., Gillies, R. R. and Perry, E. M. (1994). A method to make use of thermal infrared temperature and NDVI measurements to infer soil water content and fractional vegetation cover. *Remote Sens. Rev.*, 52: 45–59.

Carlson, T. N., Perry, E. M. and Schmugge, T. J. (1990). Remote estimation of soil moisture availability and fractional vegetation cover for agricultural fields. *Agric. For. Meteor.*, 52: 45–69.

Carlson, V. A. (1969). Bedrock Topography of the Drumheller Map-Area, NTS 82P, Alberta, Research Council of Alberta.

Charney, J. G., Halem, M. and Jastrow, R. (1969). Use of incomplete historical data to infer the present state of the atmosphere. *J. Atmos. Sci.*, 26: 1160–3.

Chen, F. (1990a). Turbulent characteristics over a rough natural surface I: Turbulent structures. *Boundary-Layer Meteorol.*, 52: 151–75.

Chen, F. (1990b). Turbulent characteristics over a rough natural surface II: Responses of profiles to turbulence. *Boundary-Layer Meteorol.*, 52, 301–11.

Chow, V. T., Maidment, D. R. and Mays, L. W. (1988). *Applied Hydrology*. McGraw Hill.

Cognard, A.-L., Loumagne, C., Normand, M., Olivier, P., Ottlé, C., Vidal-Madjar, D., Louahala, S. and Vidal, A. (1995). Evaluation of the ERS 1/synthetic aperture radar capacity to estimate surface soil moisture: Two-year results over the Naizin watershed. *Water Resour. Res.*, 31(4): 975–82.

Colbeck, S. C. (1978). The physical aspects of water flow through snow. *Adv. Hydroscience*, 11: 165–206.

Colbeck, S. C. (1988). Snowmelt increase through albedo reduction. Workshop on Snow Hydrology, November 1988, Manali (India), Central Water Commission and Himachal Pradesh State Electricity Board, India.

Colbeck, S. C. (1991). The layered character of snow covers. *Rev. Geophys.*, 29(1): 81–96.

Coles, N. A., Sivapalan, M., Larsen, J. E., Linnet, P. E. and Fahrner, C. K. (1997). Modelling runoff generation on small agricultural catchments: can real world runoff responses be captured? *Hydrological Processes*, 11: 111–36.

Collier, C. G. and Knowles, J. M. (1986). Accuracy of rainfall estimates by radar III. Application for short-term flood forecasting. *J. Hydrology* 83: 237–49.

Cooley, K. R. (1988). Snowpack variability on Western Rangelands. In: Western Snow Conference Proceedings, Kalispell, Montana, April 18–20.

Cooley, R. L. (1997). Confidence intervals for ground-water models using linearization, likelihood and bootstrap methods. *Groundwater*, 35(5): 869–80.

Coppin, P. A., Raupach, M. R. and Legg, B. J. (1986). Experiments on scalar dispersion within a model plant canopy Part II: An elevated plane source. *Boundary-Layer Meteorol.*, 35: 167–91.

Copty, N., Rubin, Y. and Mavko, G. (1993). Geophysical–hydrological identification of field permeabilities through Bayesian updating. *Water Resour. Res.*, 29: 2813–25.

Costa-Cabral, M. C. and Burges, S. J. (1994). Digital elevation model networks (DEMON): A model of flow over hillslopes for computation of contributing and dispersal areas. *Water Resour. Res.*, 30: 1681–92.

Crawford, N. H. and Linsley, R. S. (1966). Digital simulation in hydrology: The Stanford Watershed model IV. Technical Report No. 39. Dept. Civil Engineering, Stanford University, Palo Alto, California.

Cressie, N. (1991). *Statistics for Spatial Data.* Wiley, New York, 900pp.

Cunning, J. B. (1986). The Oklahoma–Kansas Preliminary Regional Experiment for STORM-Central, *Bull. Am. Meteor. Soc.*, 67(12): 1478–86.

Daley, R. (1991). *Atmospheric Data Analysis.* Cambridge University Press, Cambridge, Massachusetts, 457pp.

Danskin, W. R. (1988). Preliminary evaluation of the hydrogeologic system in Owens Valley, California. U.S.G.S. Water-Resources Investigations Report 88-4003.

Davis, R. E., McKenzie, J. C. and Jordan, R. (1995). Distributed snow process modelling: an image processing approach. *Hydrological Processes*, 9: 865–75.

Dawdy, D. R. (1969). Considerations involved in evaluating mathematical modelling of urban water systems. U.S. Geol. Survey Water Supply Paper 159I-D.

Dawes, W. and Short, D. L. (1994). The significance of topology for modeling the surface hydrology of fluvial landscapes. *Water Resour. Res.*, 30(4): 1045–55.

Dawes, W. R., Zhang, L., Hatton, T. J., Reece, P. H., Beale, G. T. H. and Packer, I. (1997). Evaluation of a distributed parameter ecohydrological model (Topog_IRM) on a small cropping rotation catchment. *J. Hydrology*, 191: 64–86.

de Bruin, H. A. R., Bink, N. J. and Kroon, L. J. M. (1991). Fluxes in the surface layer under advective conditions. In: Schmugge, T. J. and André, J. C. (Eds), *Land Surface Evaporation Measurement and Parameterization.* Springer-Verlag, New York, pp.157–69.

de Marsily, G. (1986). *Quantitative Hydrogeology.* Academic Press, San Diego, 440pp.

de Marsily, G., Combes, P. and Goblet, P. (1992). Comments on 'Ground-water models cannot be validated', by L. F. Konikow and J. D. Bredehoeft. *Adv. Water Resour.*, 15: 367–9.

De Troch, F. (1977). Studie van de oppervlaktewaterhydrologie van een Oost-Vlaams stroombekken: de Zwalmbeek te Nederzwalm. PhD dissertation, University of Gent, Faculty of Applied Sciences (in Dutch).

De Troch, F. P., Troch, P. A., Su, Z. and Lin, D. S. (1996). Application of remote sensing for hydrological modelling. In: Abbott, M. B. and Refsgaard, J. C. (Eds), *Distributed Hydrological Modelling.* Kluwer, Dordrecht, The Netherlands, pp.165–91.

Dee, D. P. (1995). A pragmatic approach to model validation. In: Lynch, D. R. and Davies, A. M. (Eds), *Quantitative Skill Assessment of Coastal Ocean Models.* AGU, Washington, pp.1–13.

Deutsch, C. V. and Journel, A. G. (1997). *GSLIB, Geostatistical Software Library and User's Guide.* Oxford University Press, New York, 384pp.

Diak, G. R., Mecikalski, J. R., Anderson, M. C. and Bland, W. L. (1998). *Satellite Estimates of Terrestrial Radiation for Agricultural Applications.* 23rd Conference on Agricultural and Forest Meteorology, Albuquerque, NM. American Meteorological Society, Boston, MA, pp.1–2.

Dickinson, R. E., Henderson-Sellers, A. and Kennedy, P. J. (1993). Biosphere–Atmosphere Transfer Scheme (BATS) Version 1e as coupled to the NCAR Community Climate Model. NCAR/TN-387 + STR, National Center for Atmospheric Research.

Dingman, S. L. (1994). *Physical Hydrology*, Macmillan, 575pp.

Diskin, M. H. and Lane, L. J. (1972). A basinwide stochastic model for ephemeral stream runoff in South-eastern Arizona, *Bull. IASH,* 17(1): 61–76.

Diskin, M. H. (1970). Definition and uses of the linear regression model. *Water Resour. Res.,* 6(6): 1668–73.

Dolman, A. J. and Wallace, J. S. (1991). Lagrangian and *K*-theory approaches in modelling evaporation from sparse canopies. *Quart. J. R. Meteorol. Soc.,* 117: 1325–40.

Dozier, J. (1979). A solar radiation model for a snow surface in mountainous terrain. In: Colbeck, S. C. and Ray, M. (Eds), *Modeling Snow Cover Runoff.* Proc. Conf. Hanover. U.S. Army CRREL, pp.144–53.

Dozier, J. (1987). Recent research in snow hydrology. *Rev. Geophys.,* 25(2): 153–61.

Dozier, J. and Frew, J. (1990). Rapid calculation of terrain parameters for radiation modelling from digital elevation data. *IEEE Transactions on Geoscience and Remote Sensing* 28(5): 963–9.

Droegemeier, K. K. (1997). The numerical prediction of thunderstorms: challenges, potential benefits and results from realtime operational tests, *WMO Bull.,* 46: 324–36.

Droegemeier, K. K., Xue, M., Sathye, A., Brewster, K., Bassett, G., Zhang, J., Liu, Y., Zou, M., Crook, A., Wong, V. and Carpenter, R. (1996a). *Realtime Numerical Prediciton of Storm-Scale Weather during VORTEX '95, Part I: Goals and Methodology.* Preprints, 18th Conf. on Severe Local Storms, 15–20 Jan., Amer. Meteor. Soc., San Francisco, CA, pp. 6–10.

Droegemeier, K. K., Xue, M., Brewster, K., Liu, Y., Park, S. K., Carr, F., Mewes, J., Zong, J., Sathye, A., Bassett, G., Zou, M., Carpenter, R., McCarthy, D., Andra, D., Janish, P., Graham, R., Sanielvici, S., Brown, J., Loftis, B. and McLain, K. (1996b). *The 1996 CAPS Spring Operational Forecasting Period – Realtime Storm-Scale NWP, Part I: Goals and Methodology.* Preprints, 11th Conf. on Num. Wea. Pred., 19–23 August, Norfolk, VA, Amer. Meteor. Soc., San Francisco, CA, pp. 294–6.

Duan, Q., Sorooshian, S. and Gupta, V. (1992). Effective and efficient global optimization for conceptual rainfall-runoff models. *Water Resour. Res.,* 28(4): 1015–31.

Dubayah, R., Dozier, J. and Davis, F. W. (1990). Topographic distribution of clear-sky radiation over the Konza Prairie, Kansas. *Water Resour. Res.,* 26(4): 679–90.

Dunne, T. (1978). Field studies of hillslope flow processes. In: Kirkby, M. J. (Ed.), *Hillslope Hydrology.* Wiley, Chichester, pp.227–93.

Dunne, T. and Black, R. D. (1970a). An experimental investigation of runoff production in permeable soils. *Water Resour. Res.,* 6(2): 478–90.

Dunne, T. and Black, R. D. (1970b). Partial area contributions to storm runoff in a small New England watershed. *Water Resour. Res.,* 6(5): 1296–311.

Dunne, T., Moore, T. R. and Taylor, C. H. (1975). Recognition and prediction of runoff-producing zones in humid regions. *Hydrol. Sci. Bull.,* 20: 305–27.

Eagleson, P. S. (1978a). Climate soil and vegetation, 1, Introduction to water balance dynamics. *Water Resour. Res.,* 14(5): 705–12.

Eagleson, P. S. (1978b). Climate soil and vegetation, 2, The distribution of annual precipitation derived from observed storm sequences. *Water Resour. Res.,* 14(5): 713–21.

Eagleson, P. S. (1978c). Climate soil and vegetation, 3, A simplified model of soil moisture movement in the liquid phase. *Water Resour. Res.,* 14(5): 722–30.

Eagleson, P. S. (1978d). Climate soil and vegetation, 4, The expected value of annual evapotranspiration. *Water Resour. Res.,* 14(5): 731–40.

Eagleson, P. S. (1978e). Climate soil and vegetation, 5, A derived distribution of storm surface runoff. *Water Resour. Res.,* 14(5): 741–8.

Eagleson, P. S. (1978f). Climate soil and vegetation, 6, Dynamics of the annual water balance. *Water Resour. Res.,* 14(5): 749–64.

Eagleson, P. S., Fennessey, N. M., Qinling, W. and Rodriguez-Iturbe, I. (1987). Application of spatial Poisson models to airmass thunderstorm rainfall. *J. Geophys. Res.,* 92(D8): 9661–78.

Elder, K. (1995). Snow Distribution in Alpine Watersheds. Ph.D Thesis, Department of Geography, University of California.

Elder, K., Michaelsen, J. and Dozier, J. (1995). Small basin modelling of snow water equivalence using binary regression tree methods. In: Tonnessen, K. A. et al. (Eds), *Biogeochemistry of Seasonally Snow-Covered Catchments*, Proc. Boulder Symposium, July 3–14. IAHS Publ. No. 228, pp.129–39.

Elder, K., Rosenthal, W. and Davis, R. (1998). Estimating the spatial distribution of snow water equivalence in a montane watershed. *Hydrological Processes*, 12(10–11): 1793–808.

Elsenbeer, H., Cassel, K. and Castro, J. (1992). Spatial analysis of soil hydraulic conductivity in a tropical rainforest catchment. *Water Resour. Res.*, 28(12): 3201–14.

Elsenbeer, H., Cassel, K. and Zuniga, L. (1994). Throughfall in the terra firme forest of western Amazonia. *J. Hydrology (N.Z.)*, 32(2): 30–44.

Elsenbeer, H. and Lack, A. (1996). Hydrometric and hydrochemical evidence for fast flow-paths at La Cuenca, Western Amazonia. *J. Hydrology*, 180: 237–50.

Elsenbeer, H., Lack, A. and Cassel, K. (1995). Chemical fingerprints of hydrological compartments and flowpaths at La Cuenca, Western Amazonia. *Water Resour. Res.*, 31(12): 3051–8.

Elsenbeer, H. and Vertessy, R. A. (2000). Stormflow generation and flowpath characteristics in an Amazonian rainforest catchment. *Hydrological Processes*, 14, 2367–2381.

Emmett, W. W. (1970). The hydraulics of overland flow on hillslopes. U.S. Geol. Survey Prof. Paper 662-A, 47pp.

Engman, E. T. (1986). Roughness coefficients for routing surface runoff. *J. Irrigation and Drainage Div., Proc ASCE*, 112: 39–53.

Engman, E. T. (1991). Applications of microwave remote sensing of soil moisture for water resources and agriculture. *Remote Sens. Environ.*, 35: 213–26.

Engman, E. T. (1995). Recent advances in remote sensing in hydrology. *Reviews of Geophysics, Supplement*, 967–75.

Entekhabi, D. and Eagleson, P. S. (1989). Land surface hydrology parameterization for atmospheric general circulation models including subgrid scale spatial variability. *J. Climate*, 2: 816–31.

Erichsen, B. and Myrabø, S. (1990). Studies of the relationship between soil moisture and topography in a small catchment. In: Gambolati, G. (Ed.), *Proc. 8th International Conference on Computational Methods in Water Resources*, Venice.

Essery, R., Li, L. and Pomeroy, J. (1999). A distributed model of blowing snow over complex terrain. *Hydrological Processes*, 13: 2423–38.

Ewen, G. O. and Parkin, G. (1996). Validation of catchment models for predicting land-use and climate change impacts. 1. Method. *J. Hydrology*, 175: 583–94.

Famiglietti, J. S. and Wood, E. F. (1994). Multiscale modeling of spatially variable water and energy balance processes. *Water Resour. Res.*, 30: 3061–78.

Famiglietti, J. S. and Wood, E. F. (1995). Effects of spatial variability and scale on areally averaged evapotranspiration. *Water Resour. Res.*, 31: 699–712.

Fan, Y. and Bras, R. L. (1995). On the concept of a representative elementary area in catchment runoff. *Hydrological Processes*, 9: 821–32.

Farajalla, N. M. and Vieux, B. E. (1995). Capturing the essential spatial variability in distributed hydrological modelling: infiltration parameters. *Hydrological Processes*, 9: 55–68.

Faurès, J. M., Goodrich, D. C., Woolhiser, D. A. and Sorooshian, S. (1995). Impact of small-scale spatial rainfall variability on runoff modeling, *J. Hydrology*, 173: 309–26.

Federico, V. D. and Neuman, S. P. (1997). Scaling of random fields by means of truncated power variograms and associated spectra. *Water Resour. Res.*, 33(5): 1075–85.

Ferré, P. A., Knight, J. H., Rudolph, D. L. and Kachanoski, R. G. (1998). The sample areas of conventional and alternative time domain reflectometry probes. *Water Resour. Res.*, 34(11): 2971–9.

Finn, D., Lamb, B., Leclerc, M. Y. and Horst, T. W. (1996). Experimental evaluation of analytical and lagrangian surface-layer flux footprint models. *Boundary-Layer Meteorol.*, 80: 282–308.

Flavelle, P. (1992). A quantitative measure of model validation and its potential use for regulatory purposes. *Adv. Water Resour.*, 15: 5–13.

Flury, M., Flühler, H., Jury, W. A. and Leuenberger, J. (1994). Susceptibility of soils to preferential flow of water: A field study. *Water Resour. Res.*, 30: 1945–54.

Forkel, C. (1996). Das numerische Modell – ein schmaler Grat zwischen vertrauenswürdigem Werkzeug und gefährlichem Spielzeug. Presented at the 26th IWASA, RWTH Aachen, 4–5 January 1996.

Foufoula-Georgiou, E. and Georgakakos, K. P. (1991). Recent advances in space-time precipitation modeling and forecasting. Chapter 3 in Bowles, D. and O'Connell, E. (Eds), *Recent Advances in the Modeling of Hydrological Systems*. Reidel Publ. Co., pp. 47–65.

Foufoula-Georgiou, E. and Krajewski, W. (1995). Recent advances in rainfall modeling, estimation and forecasting. *Rev. Geophys. IUGG, U.S National Report, 1991–1994 – Contributions in Hydrology.*

Foufoula-Georgiou, E. and Tsonis, T. (Eds) (1996). Space-time variability and dynamics of rainfall. *J. Geophys. Res.* (Special issue), 101(D21): 26161–538.

Franks, S., Beven, K. J. and Lamb, R. (1997). Assessing the utility of distributed data in constraining the behaviour of hydrological models in the Wijlegemse subcatchment. In: De Troch, F. P., Troch, P. A. and Su, Z. (Eds), *Spatial and Temporal Soil Moisture Mapping from ERS-1 and JERS-1 SAR Data and Macroscale Hydrological Modelling*. Final Research Report, EC Environment Research Program EV5V-CT94-0446. Gent, Belgium, pp.334–57.

Franks, S. W., Gineste, Ph., Beven, K. J. and Mérot, Ph. (1998). On constraining the predictions of a distributed model: The incorporation of fuzzy estimates of saturated areas into the calibration process. *Water Resour. Res.*, 34(4): 787–97.

Freer, J., Beven, K. and Ambroise, B. (1996). Bayesian-estimation of uncertainty in runoff prediction and the value of data: An application of the GLUE approach. *Water Resour. Res.*, 32(7): 2161–73.

Freeze, R. A. (1971). Three-dimensional, transient, saturated-unsaturated flow in a groundwater basin. *Water Resour. Res.*, 7(2): 347–66.

Freeze, R. A. (1980). A stochastic-conceptual analysis of rainfall-runoff processes on a hillslope. *Water Resour. Res.*, 16(2): 391–408.

Freeze, R. A. and Harlan, R. L. (1969). Blueprint for a physically-based, digitally simulated hydrologic response model. *J. Hydrology*, 9: 237–58.

Freeze, R. A. and Witherspoon, P. A. (1966). Theoretical analysis of regional groundwater flow: 1. Analytical and numerical solutions to the mathematical model. *Water Resour. Res.*, 2(4): 641–56.

Frew, J. E. (1990). The Image Processing Workbench, PhD Thesis, University of California, Santa Barbara.

Friedl, M. A. (1997). Examining the effects of sensor resolution and sub-pixel heterogeneity on spectral vegetation indices: Implications for biophysical modeling. In: Quattrochi, D. A. and Goodchild, M. F. (Eds), *Scale in Remote Sensing and GIS*. CRC Lewis Pub., New York, pp.113–39.

Fritsch, J. M., Houze, R. A. Jr., Adler, R., Bluestein, H., Bosart, L., Brown, J., Carr, F., Davis, C., Johnson, R. H., Junker, N., Kuo, Y. H., Rutledge, S., Smith, J., Toth, Z., Wilson, J. W., Zipser, E. and Zrnic, D. (1998). Quantitative Precipitation Forecasting: Report of the Eight Prospectus Development Team, U.S. Weather Research Program, *Bull. Amer. Meteor. Soc.*

Gardner, W. R. (1958). Some steady state solutions of the unsaturated moisture flow equation with application to evaporation from a water table. *Soil Science*, 85(4): 228–32.

Garratt, J. R. (1978). Flux profile relations above tall vegetation. *Quart. J. R. Meteorol. Soc.*, 104: 199–211.

Garratt, J. R. (1980). Surface influences upon vertical profiles in the atmospheric near-surface layer. *Quart. J. R. Meteorol. Soc.*, 106: 803–19.

Garratt, M. J. and Spencer-Jones, D. (1981). *Yan Yean Geological Map*. Geological Survey of Victoria, Melbourne.

Gelhar, L. W. (1993). *Stochastic Subsurface Hydrology*. Prentice Hall, Englewood Cliffs, NJ, 390pp.

Gelman, A., Carlin, J. B., Stren, H. S. and Rubin, D. B. (1997). *Bayesian Data Analysis*. Chapman and Hall, p.526.

General Meteorological Package (GEMPAK) (1992). Version 5.1, Reference manual, Unidata Program Center, UCAR, Boulder.

Gessler, P. E, Moore, I. D., McKenzie, N. J. and Ryan, P. J. (1995). Soil landscape modelling and spatial prediction of soil attributes. *Int. J. Geogr. Inf. Sys.*, 9(4): 421–32.

Ghan, S. J., Liljegren, J. C., Shaw, W. J., Hubbe, J. H. and Doran, J. C. (1997). Influence of subgrid variability on surface hydrology. *J. Climate*, 10: 3157–66.

Gillies, R. R. and Carlson, T. N. (1995). Thermal remote sensing of surface soil water content with partial vegetation cover for incorporation into climate models. *J. Appl. Meteorol.*, 34: 745–56.

Gillies, R. R., Cui, J., Carlson, T. N., Kustas, W. P. and Humes, K. S. (1997). Verification of a method for obtaining surface soil water content and energy fluxes from remote measurements of NDVI and surface radiant temperature. *Int. J. Remote Sens.*, 18: 3145–66.

Gineste, P., Puech, C. and Mérot, P. (1998). Radar remote sensing of the source areas from the Coët-Dan catchment. *Hydrological Processes*, 12: 267–84.

Giorgi, F. and Avissar, R. (1997). Representation of heterogeneity effects in earth system modeling: Experience from land surface modeling. *Rev. Geophys.*, 35: 413–38.

Gomendy, V., Galhi, M., Western, A., Grayson, R. and White, R. (in preparation). Description and spatial variability of a yellow Duplex and Grey massive earth soil toposequence at the small catchment scale in Victoria.

Goodrich, D. C. (1990). Geometric simplification of a distributed rainfall-runoff model over a range of basin scales. PhD Thesis, The University of Arizona, 361pp.

Goodrich, D. C., Faurés, J.-M., Woolhiser, D. A., Lane, L. J. and Sorooshian, S. (1995). Measurement and analysis of small-scale convective storm rainfall variability. *J. Hydrology*, 173: 283–308.

Goodrich, D. C., Lane, L. J., Shillito, R. A., Miller, S. N., Syed, K. H. and Woolhiser, D. A. (1997). Linearity of basin response as a function of scale in a semi-arid watershed. *Water Resour. Res.* 33(12): 2951–65.

Goodrich, D. C., Schmugge, T. J., Jackson, T. J., Unkrich, C. L., Keefer, T. O., Parry, R., Bach, L. B. and Amer, S. A. (1994). Runoff simulation sensitivity to remotely sensed initial soil water content. *Water Resour. Res.*, 30(5): 1393–405.

Goodrich, D. C., Stone, J. J. and Van der Zweep, R. (1993). Validation Strategies Based on Model Application Objectives. Proc. Fed. Interagency Workshop on Hydrologic Modelling Demands for the 90's, U.S.G.S. Water Resources Invest. Report 93-4018, Ft. Collins, CO, June 7-9, pp.8-1–8-8.

Grayson, R. B., Blöschl, G., Barling, R. D. and Moore, I. D. (1993). Process, scale and constraints to hydrological modelling in GIS. In: Kovar, K. and Nachtnebel, H. P. (Eds), *Applications of Geographic Information Systems in Hydrology and Water Resources Management*. Proc. Vienna Symp., April 1993, IAHS Publ. No. 211, pp. 83–92.

Grayson, R. B., Blöschl, G. and Moore, I. D. (1995). Distributed parameter hydrologic modelling using vector elevation data: THALES and TAPES-C. Chapter 19 in: Singh, V. P. (Ed.), *Computer Models of Watershed Hydrology*. Water Resources Publications, Highlands Ranch, Colorado, pp.669–96.

Grayson, R. B., Moore, I. D. and McMahon, T. A. (1992a). Physically based hydrologic modeling, 1. A terrain-based model for investigative purposes. *Water Resour. Res.*, 28: 2639–58.

Grayson, R. B., Moore, I. D. and McMahon, T. A. (1992b). Physically based hydrologic modeling, 2. Is the concept realistic? *Water Resour. Res.*, 28: 2659–66.

Grayson, R. B. and Western, A. W. (1998). Towards areal estimation of soil water content from point measurements: time and space stability of mean response. *J. Hydrology*, 207: 68–82.

Grayson, R. B., Western, A. W., Chiew, F. H. S. and Blöschl, G. (1997). Preferred states in spatial soil moisture patterns: Local and nonlocal controls. *Water Resour. Res.*, 33(12): 2897–908.

Gupta, V. K. and Waymire, E. (1979). A stochastic kinematic study of subsynoptic space-time rainfall. *Water Resour. Res.*, 15(3): 637–44.

Gupta, V. K. and Waymire, E. (1990). Multiscaling properties of spatial rainfall in river flow distributions. *J. Geophys. Res.,* 95: 1999–2009.

Gupta, H. V., Sorooshian, S. and Yapo, P. O. (1998). Toward improved calibration of hydrologic models: Multiple and noncommensurable measures of information. *Water Resour. Res.*, 34(4): 751–63.

Gyasi-Agyei, Y., Willgoose, G. and De Troch, F. P. (1995). Effects of vertical resolution and map scale of digital elevation models on geomorphologic parameters used in hydrology. *Hydrological Processes*, 9: 363–82.

Hardy, J., Albert, M. and Marsh, P. (Eds) (1999). Snow hydrology, *Hydrological Processes* (special issue), 13: Nos 12–15.

Hatton, T. J., Dawes, W. R. and Vertessy, R. A. (1995). The importance of landscape position in scaling SVAT models to catchment scale hydroecological prediction. In: Feddes, R. A (Ed.), *Space and Time Scale Variability and Interdependencies in Hydrological Processes*. Cambridge University Press, New York, pp.43–53.

Henebry, G. M. (1997). Advantages of principal components analysis for land cover segmentation from SAR image series. In: *Proceedings of the 3rd ERS symposium*. Florence, Italy, pp.175–8.

Herschy, R. W. (Ed.) (1999). *Hydrometry: Principles and Practice*. Wiley, 384pp.

Hewlett, J. D. and Hibbert, A. R. (1967). Factors affecting the response of small watersheds to precipitation in humid areas. In: Sopper, W. E. and Lull, H. W. (Eds), *Forest Hydrology*. Pergamon Press, Oxford, pp.275–90.

Hillel, D. (1986). Modelling in soil physics: A critical review. In: *Future Developments in Soil Science Research*. A collection of Soil Sci. Soc. Am. Golden Anniversary contributions presented at Annual Meeting, New Orleans, pp.35–42.

Hipps, L. E., Humes, K. and Kustas, W. P. (1995). *Defining Scales of Inhomogeneity for Surface Properties: Implications towards Studies of Surface–Atmosphere Interactions*. Proceedings of Conference on Hydrology, Dallas, TX, American Meteorological Society, Boston, MA, pp.141–2.

Hipps, L. E., Or, D. and Neale, C. M. U. (1996). Spatial structure and scaling of surface fluxes in a Great Basin ecosystem. In: Stewart, J. B., Engman, E. T., Feddes, R. A. and Kerr, Y. (Eds), *Scaling up in Hydrology using Remote Sensing*. Wiley, London, pp.113–25.

Hipps, L. E., Swiatek, E. and Kustas, W. P. (1994). Interactions between regional surface fluxes and the atmospheric boundary layer over a heterogeneous watershed. *Water Resour. Res.*, 30(5): 1387–92.

Hochstöger, F. (1989). Ein Beitrag zur Anwendung und Visualisierung digitaler Geländemodelle (On the application and visualization of digital terrain models). Geowissenschaftliche Mitteilungen 34, Veröffentlichungen des Institutes für Photogrammetrie und Fernerkundung, Technische Universität Wien.

Hornberger, G. M. (1994). Data and analysis note: A new type of article for Water Resources Research. *Water Resour. Res.*, 30(12): 3241–2.

Horst, T. W. and Weil, J. C. (1992). Footprint estimation for scalar flux measurements in the atmospheric surface layer. *Boundary-Layer Meteorol.*, 59: 279–96.

Horton, R. E. (1933). The role of infiltration in the hydrologic cycle. *Trans. Am. Geophys. Union*, 14: 446–60.

Houser, P. R. (1996). Remote Sensing Soil Moisture using 4-Dimensional Data Assimilation. Ph.D dissertation, Department of Hydrology and Water Resources, The University of Arizona.

Houser, P. R., Gupta, H. V., Shuttleworth, W. J. and Famiglietti, J. S. (2000). Multi-objective calibration and sensitivity of a distributed land–surface water and energy balance model. *Water Resour. Res.* (in press)

Houser, P. R., Shuttleworth, W. J., Gupta, H. V., Famiglietti, J., Syed, K. H. and Goodrich, D. C. (1998). Integration of soil moisture remote sensing and hydrologic modelling using data assimilation. *Water Resour. Res.*, 34(12): 3405–20.

Huff, F. A. (1967). Time distribution of rainfall in heavy storms. *Water Resour. Res.*, 3(4): 1007–19.

Humes, K. S., Kustas, W. P. and Goodrich, D. C. (1997). Spatially distributed sensible heat flux over a semiarid watershed. Part I: use of radiometric surface temperatures and spatially uniform resistance. *J. Appl. Meteorol.*, 36: 281–92.

Huntingford, C., Allen, S. J. and Harding, R. J. (1995). An intercomparison of single and dual-source vegetation–atmosphere transfer model applied to transpiration from sahelian savannah. *Boundary-Layer Meteorol.*, 74: 397–418.

Hursh, C. R. and Brater, E. F. (1941). Separating storm-hydrographs from small drainage-areas into surface- and subsurface-flow. *Trans. American Geophysical Union*, 22: 863–70.

Hutchinson, M. F. (1989). A new method for gridding elevation and stream line data with automatic removal of spurious pits. *J. Hydrology*, 106: 211–32.

Hutchinson, M. F. (1991). The application of thin plate smoothing splines to continent-wide data assimilation. In: Jasper, J. D. (Ed.), *Data Assimilation Systems*, BMRC Research Report 27, Bureau of Meteorology, Melbourne, 104–13.

Hutchinson, M. F. (1993). On thin plate splines and kriging. In: Tarter, M. E. and Lock, M. D. (Eds), *Computing Science and Statistics*, Vol.25. Interface Foundation of North America, University of California, Berkeley, pp.55–62.

Hutchinson, M. F. and Gessler, P. E. (1994). Splines – more than just a smooth interpolator. *Geoderma*, 62: 45–67.

IAHR (1994). Publication of guidelines for validation documents and call for discussion. *International Association for Hydraulic Research Bulletin*, 11:41.

Iorgulescu, I. and Musy, A. (1997). Generalisation of TOPMODEL for a power law transmissivity profile. *Hydrological Processes*, 11: 1353–5.

Islam, S., Bras, R. L. and Rodriguez-Iturbe, I. (1988). Multidimensional modelling of cumulative rainfall: Parameter estimation and model adequacy through a continuum of scales. *Water Resour. Res.*, 24(7): 985–92.

Jackson, T. H. R. (1994). A Spatially Distributed Snowmelt-Driven Hydrologic Model applied to the Upper Sheep Creek Watershed. Ph.D Thesis, Utah State University.

Jackson, T. J. and Le Vine, D. E. (1996). Mapping surface soil moisture using an aircraft-based passive microwave instrument: algorithm and example. *J. Hydrology*, 184: 57–84.

Jackson, T. J., Le Vine, D. M., Griffis, A. J., Goodrich, D. C., Schmugge, T. J., Swift, C. T. and O'Neill, P. E. (1993). Soil moisture and rainfall estimation over a semiarid environment with the ESTAR microwave radiometer. *IEEE Trans. Geosci. Remote Sens.*, 31(4): 836–41.

Jacobs, B. L., Rodriguez-Iturbe, I. and Eagleson, P. S. (1988). Evaluation of a homogeneous point process description of Arizona thunderstorm rainfall. *Water Resour. Res.*, 24(7): 1174–86.

Jain, S. K., Storm, B., Bathurst, J. C., Refsgaard, J. C. and Singh, R. D. (1992). Application of the SHE to catchments in India. Part 2. Field experiments and simulation studies with the SHE on the Kolar subcatchment of the Narmada River. *J. Hydrology,* 140: 25–47.

Jakeman, A. and Hornberger, G. M. (1993). How much complexity is warranted in a rainfall–runoff model? *Water Resour. Res.,* 29(8): 2637–49.

James, L. D. and Burges, S. J. (1982). Selection, calibration and testing of hydrological models. In: Haan, C. T., Johnson, H. P. and Brakensiek, D. L. (Eds), *Hydrological Modelling of Small Watersheds.* Am. Soc. Agr. Eng. Monogr. Series Vol. 5, pp.435–72.

Jarvis, P. G. and McNaughton, K. G. (1986). Stomatal control of transpiration: Scaling up from leaf to region. *Adv. Ecological Research,* 15: 1–49.

Jayatilaka, C. J., Storm, B. and Mudgway, L. B. (1998). Simulation of water flow on irrigation bay scale with MIKE SHE, *J. Hydrology,* 208: 108–30.

Jensen, H. (1989). Räumliche Interpolation der Stundenwerte von Niederschlag, Temperatur und Schneehöhe. Zürcher Geographische Schriften 35. Geographisches Institut der Eidgenössischen Technischen Hochschule Zürich, 70pp.

Jensen, K. H. and Mantoglou, A. (1993). Future of distributed modelling. Chapter 12 in: Beven, K. J. and Moore, I. D. (Eds), *Terrain Analysis and Distributed Modelling in Hydrology.* Wiley, New York, pp.203–12.

Jones, E. B. (1983). Snowpack Ground-Truth Manual, Report No. CR 170584, National Aeronautics and Space Administration (NASA), Goddard Space Flight Center, Greenbelt, MD.

Jordan, J. P. (1994). Spatial and temporal variability of stormflow generation processes on a Swiss catchment. *J. Hydrology,* 153: 357–82.

Jordan, R. (1991). A one-dimensional temperature model for a snow cover. Technical documentation for SNTHERM.89, special technical report 91–16, U.S. Army CRREL.

Journel, A. G. and Deutsch, C. V. (1993). Entropy and spatial disorder. *Math. Geol.,* 25: 329–55.

Journel, A. G. and Huijbregts, C. J. (1978). *Mining Geostatistics,* Academic Press, London, 600pp.

Kabat, P., Hutjes, R. W. A. and Feddes, R. A. (1997). The scaling characteristics of soil parameters: From plot scale heterogeneity to subgrid parameterization. *J. Hydrology,* 190: 363–96.

Kader, B. A. and Yaglom, A. M. (1990). Mean fields and fluctuation moments in unstably stratified turbulent boundary layers. *J. Fluid Mech.,* 212: 637–62.

Kain, J. S. and Fritsch, J. M. (1992). The role of convective "trigger function" in numerical weather forecasts of mesoscale convective systems. *Meteor. Atm. Phys.,* 40: 93–106.

Karnieli, A. M., Diskin, M. H. and Lane, L. J. (1994). CELMOD5 – a semi-distributed cell model for conversion of rainfall into runoff in semi-arid watersheds. *J. Hydrology,* 157: 61–85.

Katul, G., Goltz, S. M., Hsieh, C., Cheng, Y., Mowry, F. and Sigmon, J. (1995). Estimation of surface heat and momentum fluxes using the flux-variance method above uniform and non-uniform terrain. *Boundary-Layer Meteorol.,* 74: 237–60.

Kavvas, M. L. and Delleur, J. W. (1981). A stochastic cluster model for daily rainfall sequences. *Water Resour. Res.,* 17(4): 1151–60.

Keller, R. (Ed.) (1978). *Hydrologischer Atlas der Bundesrepublik Deutschland, Deutsche Forschungsgemeinschaft.* Boldt Verlag, Boppard, Germany, 68pp.

Kendall, C. and McDonnell, J. J. (Eds) (1998). *Isotope Tracers in Catchment Hydrology.* Elsevier, 839pp.

Khanna, S. and Brasseur, J. G. (1997). Analysis of Monin–Obukhov similarity from large-eddy simulation. *J. Fluid Mech.,* 345: 251–86.

Kiefer, C. J. and Chu, H. H. (1957). Synthetic storm pattern for drainage design. *J. Hydr. Div., ASCE,* 83(HY4): 1–25.

Kim, C. P., Salvucci, G. D. and Entekhabi, D. (1999). Groundwater–surface water interaction and the climatic spatial patterns of hillslope hydrological response. *Hydrology and Earth Systems Science*, 3(3): 375–84.

Kirkby, M. (1975). Hydrograph modelling strategies. In: Peel, R., Chisholm, M. and Haggett, P. (Eds), *Processes in Physical and Human Geography*, pp.69–90.

Kirkby, M. J. (1988). Hillslope runoff processes and models. *J. Hydrology*, 100: 315–40.

Kirkby, M. J. (1997). TOPMODEL: a personal view. *Hydrological Processes*, 11(9): 1087–97.

Kirkby, M. J., Callan, J., Weyman, D. and Wood, J. (1976). Measurement and modeling of dynamic contributing areas in a very small catchment. University of Leeds, School of Geography, Working Paper 167, Leeds, UK.

Kirnbauer, R. and Blöschl, G. (1990). A lysimetric snow pillow station at Kühtai/Tyrol. In: Lang, H. and Musy, A. (Eds), *Hydrology in Mountainous Regions. I – Hydrological Measurements; the Water Cycle*. Proc. Lausanne Symp., August 1990, IAHS Publ. no. 193, pp.173–80.

Kirnbauer, R. and Blöschl, G. (1994). Wie ähnlich sind Ausaperungsmuster von Jahr zu Jahr? (How similar are snow cover patterns from year to year?) *Deutsche Gewässerkundliche Mitteilungen*, 37(5/6): 113–21.

Kirnbauer, R., Blöschl, G. and Gutknecht, D. (1994). Entering the era of distributed snow models. *Nordic Hydrology*, 25: 1–24.

Klemeš, V. (1983). Conceptualisation and scale in hydrology. *J. Hydrology*, 65: 1–23.

Klemeš, V. (1986a). Dilettantism in hydrology: transition or destiny? *Water Resour. Res.*, 22: 177S–88S.

Klemeš, V. (1986b). Operational testing of hydrological simulation models. *Hydrol. Sci. J.*, 31(1–3): 13–24.

Kneizys, F. X., Shettles, E. P., Abreu, L. W., Chetwynd, J. H., Anderson, G. P., Anderson, W. O., Selby, J. E. A. and Clough, S. A. (1988). Users Guide to LOWTRAN 7, AFGL-TR-0177, Air Force Geophysical Laboratory.

Knisel, W. G. (Ed.) (1980). *CREAMS: A field-scale model for chemical runoff and erosion from agricultural management system*. USDA-SEA Conservation Research Report No. 26, 643pp.

Koltermann, C. E. and Gorelick, S. M. (1996). Heterogeneity in sedimentary deposits: A review of structure-imitating, process-imitating, and descriptive approaches. *Water Resour. Res.*, 32: 2617–58.

König, M. and Sturm, M. (1998). Mapping snow distribution in the Alaskan Arctic using aerial photography and topographic relationships. *Water Resour. Res.*, 34(12): 3471–83.

Konikow, L. F. (1978). Calibration of groundwater models. In: *Verification of Mathematical and Physical Models in Hydraulic Engineering*. American Society of Civil Engineering, New York, pp.87–93.

Konikow, L. F. and Bredehoeft, J. D. (1992). Ground-water models cannot be validated. *Adv. Water Resour.*, 15: 75–83.

Koterba, M. T. (1986). Differential influences of storm and watershed characteristics on runoff from ephemeral streams in southeastern Arizona. Ph.D dissertation, Department of Hydrology and Water Resources, University of Arizona, Tucson, Arizona.

Kottegoda, N. T. and Rosso, R. (1996). *Introductory Statistical, Probability and Reliability Methods for Civil and Environmental Engineers*. McGraw-Hill, New York.

Kouwen, N. and Garland, G. (1989). Resolution considerations in using radar rainfall data for flood forecasting. *Can. J. Civ. Eng.*, 16: 279–89.

Krajewski, W. F., Venkataraman, L., Georgakakos, K. P. and Jain, S. C. (1991). A Monte Carlo study of rainfall sampling effect on a distributed catchment model. *Water Resour. Res.*, 27(1): 119–28.

Kraus, K. and Blöschl, G. (1998). Fernerkundung zur Sicherung der Wiener Wasservorräte (Remote sensing for safeguarding the Viennese water supply). In: Ricica, K. and Haslinger, U. (Eds). *Wasserspuren – Nachhaltige Zukunftspfade*, MA22 Umweltschutz, Wien, pp.26–9.

Krige, D. G. (1951). A statistical approach to some basic mine evaluation problems on the Witwatersrand. *J. Chem. and Metall. Soc. of South Africa*, 52: 119–39.

Kuczera, G. and Parent, E. (1998). Monte Carlo assessment of parameter uncertainty in conceptual catchment models: The Metropolis Algorithm. *J. Hydrology*, 211(1–4): 69–85.

Kumar, P. and Foufoula-Georgiou, E. (1993a). A multicomponent decomposition of spatial rainfall fields: 1. Segregation of large and small-scale features using wavelet transforms. *Water Resour. Res.*, 29(8): 2515–32.

Kumar, P. and Foufoula-Georgiou, E. (1993b). A multicomponent decomposition of spatial rainfall fields: 2. Self-similarity in fluctuations. *Water Resour. Res.*, 29(8): 2533–44.

Kupfersberger, H. and Blöschl, G. (1995). Estimating aquifer transmissivities – on the value of auxiliary data. *J. Hydrology*, 165: 85–99.

Kustas, W. P. and Daughtry, C. S. T. (1990). Estimation of the soil heat flux/net radiation ratio from multispectral data. *Agric. For. Meteorol.*, 49: 205–23.

Kustas, W. P., Daughtry, C. S. T. and van Oevelen, P. J. (1993). Analytical treatment of the relationships between soil heat flux/net radiation ratio and vegetation indices. *Remote Sens. Environ.*, 46: 319–30.

Kustas, W. P. and Goodrich, D. C. (1994). Preface to special section on Monsoon '90. *Water Resour. Res.*, 30(5): 1211–25.

Kustas, W. P., Hipps, L. E. and Humes, K. S. (1995). Calculation of basin-scale surface fluxes by combining remotely sensed data and atmospheric properties in a semiarid landscape. *Boundary-Layer Meteorol.*, 73: 105–24.

Kustas, W. P. and Humes, K. S. (1996). Variations in the surface energy balance for a semi-arid rangeland using remotely sensed data at different spatial resolutions. In: Stewart, J. B., Engman, E. T., Feddes, R. A. and Kerr, Y. (Eds), *Scaling up in Hydrology using Remote Sensing.*, Wiley, London, pp.127–45.

Kustas, W. P. and Norman, J. M. (1996). Use of remote sensing for evapotranspiration monitoring over land surfaces. *Hydrol. Sci. J.*, 41: 495–516.

Kustas, W. P., Prueger, J. H., Hipps, L. E., Ramalingam, K., Hatfield, J. L., Schmugge, T. J., Rango, A., Ritchie, J. C. and Havstad, K. M. (1998). *Application of Monin–Obukhov similarity over a mesquite dune site in the Jornada Experimental Range.* Proceedings of 23rd Conference on Agric. & Forest Meteorol., American Meteorological Society, Boston, MA, pp.216–19.

Kustas, W. P., Rango, A. and Uijlenhoet, R. (1994). A simple energy budget algorithm for the snowmelt runoff model. *Water Resour. Res.*, 30(5): 1515–27.

Kutchment, L. S., Demidov, V. N., Naden, P. S., Cooper, D. M. and Broadhurst, P. (1996). Rainfall-runoff modelling of the Oure basin, North Yorkshire: an application of a physically based distributed model. *J. Hydrology*, 181: 323–42.

Lamb, R., Beven, K. and Myrabø, S. (1997). Discharge and water table predictions using a generalized TOPMODEL formulation. *Hydrological Processes*, 11(9): 1145–67.

Lamb, R., Beven, K. and Myrabø, S. (1998a). A generalised topographic-soils hydrological index. In: Lane, S. N., Richards, K. S. and Chandler, J. H. (Eds), *Landform Monitoring, Modelling and Analysis.* Wiley, Chichester, pp.263–78.

Lamb, R., Beven, K. and Myrabø, S. (1998b). Use of spatially distributed water table observations to constrain uncertainty in a rainfall-runoff model. *Adv. Water Resour.*, 22(4): 305–17.

Lane, L. J. (1982). Distributed model for small semiarid watersheds. *J. Hydraulics Div., ASCE,* 108(HY10): 1114–31.

Lane, L. J. (1983a). Transmission losses. Chapter 19 in: *SCS National Engr. Handbook.* US Government Printing Office, Washington, DC, pp.19-1–19-21.

Lane, L. J. (1983b). *SPUR Hydrology Component: Water routing and sedimentation,* USDA-ARS Misc Pub. No. 1431, pp.62–7.

Lane, L. J. and Renard, K. G. (1972). Evaluation of a basin wide stochastic model for ephemeral runoff from semiarid watersheds. *Trans. ASAE,* 15(1): 280–3.

Laur, H., Bally, P., Meadows, P., Sanchez, J., Schaettler, B. and Lopinto, E. (1997). ERS SAR calibration: derivation of backscattering coefficient σ^0 in ESA ERS SAR PRI products (Tech. Rep.). Frascati, Italy: European Space Agency.

Leavesley, G. H. and Stannard, L. G. (1995). The precipitation-runoff modelling system – PRMS. Chapter 9 in: Singh, V. P. (Ed.), *Computer Models of Watershed Hydrology*. Water Resources Publications, Highlands Ranch, Colorado, pp.281–310.

Leclerc, M. Y. and Thurtell, G. W. (1990). Footprint prediction of scalar fluxes using a Markovian analysis. *Boundary-Layer Meteorol.*, 52: 247–58.

Lee, J. and Hoppel, K. (1992). Principal components transformation of multifrequency polarimetric SAR imagery. *IEEE Trans. Geosc. Rem. Sens.*, 80(4): 686–96.

Levine, J. B. and Salvucci, G. D. (1999a). Equilibrium analysis of groundwater–vadose zone interactions and the resulting spatial distribution of hydrologic fluxes across a Canadian prairie. *Water Resour. Res.*, 35(5): 1369–83.

Levine, J. B. and Salvucci, G. D. (1999b). Characteristic rate scale and time scale of supply-limited transpiration under a Richards–Cowan framework. *Water Resour. Res.*, 35(12): 3947–54.

Lhomme, J. P., Chehbouni, A. and Monteney, B. (1994). Effective parameters of surface energy balance in heterogeneous landscape. *Boundary Layer Meteorol.*, 71: 297–309.

Lhomme, J. P., Monteny, B. and Bessemoulin, P. (1997). Inferring regional surface fluxes from convective boundary layer characteristics in a Sahelian environment. *Water Resour. Res.*, 33: 2563–9.

Linsley, Jr., R. K., Kohler, M. A. and Paulhus, J. L. H. (1982). *Hydrology for Engineers*. McGraw-Hill.

Liston, G. E. and Sturm, M. (1998). A snow-transport model for complex terrain. *J. Glaciology* 44: 498–516.

Loague, K. and Kyriakidis, P. C. (1997). Spatial and temporal variability in the R-5 infiltration data set: Déjà vu and rainfall-runoff simulations. *Water Resour. Res.*, 33(12): 2883–95.

Lopez, A., Laur, H. and Nezry, E. (1990). Statistical distribution and texture in multilook and complex SAR images. In: *Proc. International Geoscience and Remote Sensing Symposium (IGARSS)*. Washington DC., USA, pp.2427–30.

Lopes, V. L., Nearing, M. A., Foster, G. R., Finkner, S. C. and Gilley, J. E. (1989). The Water Erosion Prediction Project: Model Overview. Proc. ASCE National Water Conf., Irrigation and Drain. and Water Resources Planning and Mgmt. Divs., Newark, DE, pp.503–10.

Lørup, J. K., Refsgaard, J. C. and Mazvimavi, D. (1998). Assessing the effect of land use change on catchment runoff by combined use of statistical tests and hydrological modelling: Case studies from Zimbabwe. *J. Hydrology*, 205: 147–63.

Los, H. and Gerritsen, H. (1995). Validation of water quality and ecological models. Presented at the 26th IAHR Conference, London, 11–15 Sept. 1995, Delft Hydraulics, 8pp.

Luce, C. H., Tarboton, D. G. and Cooley, K. R. (1997). Spatially distributed snowmelt inputs to a semi-arid mountain watershed. Proc. Western Snow Conference, Banff, Canada, May 5–8.

Luce, C. H., Tarboton, D. G. and Cooley, K. R. (1998). The influence of the spatial distribution of snow on basin-averaged snowmelt. *Hydrological Processes*, 12(10–11): 1671–83.

Luce, C. H., Tarboton, D. G. and Cooley, K. R. (1999). Sub-grid parameterization of snow distribution for an energy and mass balance snow cover model. *Hydrological Processes*, 13: 1921–33.

McClung, D. and Schaerer, P. (1993). *The Avalanche Handbook*, 2nd Edition. Mountaineers Books, 271pp.

McDonald, M. G. and Harbaugh, A. W. (1996). User's documentation for MODFLOW-96, an update to the U.S. Geological Survey Modular Finite-Difference Ground-Water Flow Model, U.S. Geological Survey Open-File Report 96-485, 56pp.

McDonnell, J. J., Bonell, M., Stewart, M. K. and Pearce, A. J. (1990). Deuterium variations in storm rainfall: implications for stream hydrograph separation. *Water Resour. Res.*, 26(3): 455–8.

McKay, G. A. and Gray, D. M. (1981). The distribution of snowcover. Chapter 5 in: Gray, D. M. and Male, D. H. (Eds), *Handbook of Snow, Principles, Processes, Management and Use*. Pergamon Press, Willowdale, Canada.

McLaughlin, D. and Townley, L. R. (1996). A reassessment of the groundwater inverse problem. *Water Resour. Res.*, 32(4): 1131–61.

McNaughton, K. G. and Jarvis, P. G. (1983). Predicting effects of vegetation changes on transpiration and evaporation. In: Kozlowski, T. T. (Ed.), *Water Deficit and Plant Growth*, Vol. VII. Academic Press, New York, pp.2–47.

McNaughton, K. G. and Spriggs, T. W. (1986). A mixed-layer model for regional evaporation. *Boundary Layer Meteorol.*, 34: 243–62.

McNaughton, K. G. and Spriggs, T. W. (1989). An evaluation of the Priestley and Taylor equation and the complementary relationship using results from a mixed-layer model of the convective boundary layer. In: Black, T. A., Spittlehouse, D. L., Novak, M. D. and Price, D. T. (Eds), *Estimation of Areal Evapotranspiration*. IAHS Publ. no 177, Wallingford, pp.89–104.

McNaughton, K. G. and Van den Hurk, B. J. J. M. (1995). A 'Lagrangian' revision of the resistors in the two-layer model for calculating the energy budget of a plant canopy. *Boundary-Layer Meteorol.*, 74: 262–88.

Mahrt, L. (1998). Flux sampling errors for aircraft and towers. *J. Atmospheric and Oceanic Technology*, 15: 416–29.

Male, D. H. and Gray, D. M. (1981). Snowcover ablation and runoff. Chapter 9 in: Gray D. M. and Male, D. H. (Eds), *Handbook of Snow, Principles, Processes, Management and Use*. Pergamon Press, Willowdale, Canada, pp.360–436.

Mallat, S. (1989). A theory for multiresolution signal decomposition: the wavelet representation. *IEEE Trans. on Pattern Anal. and Mach. Intel.*, 11(7): 674–93.

Mantoglou, A. and Wilson, J. L. (1981). Simulation of random fields with the turning bands method. Report No. 264, Department of Civil Engineering, Massachusetts Institute of Technology, Cambridge, Massachusetts, 199pp.

Marks, D. and Dozier, J. (1992). Climate and energy exchange at the snow surface in the alpine region of the Sierra Nevada, 2. Snow cover energy balance. *Water Resour. Res.*, 28(11): 3043–54.

Marsan, D., Schertzer, D. and Lovejoy, S. (1996). Causal space-time multifractal processes: Predictability and forecasting of rain fields. *J. Geophys. Res.*, 101(D21): 26333–46.

Marshall, S. E. and Warren, S. G. (1987). Parameterization of snow albedo for climate models. In: Goodison, R. E., Barry, R. G. and Dozier, J. (Eds) *Large Scale Effects of Seasonal Snowcover*, Proc. Vancouver Symposium, August 1987, IAHS Publ. No. 166, pp. 43–50.

Matheron, G. (1965). *Les variables régionalisées et leur estimation*. Masson, Paris. Cited in de Marsily (1986).

Matheron, G. (1973). The intrinsic random functions and their applications. *Adv. in Appl. Prob.*, 5: 438–68.

Mattikalli, N. M., Engman, E. T., Jackson, T. J. and Ahuja, L. R. (1998). Microwave remote sensing of temporal variations of brightness temperature and near-surface soil water content during a watershed-scale field experiment, and its application to the estimation of soil physical properties. *Water Resour. Res.*, 34(9): 2289–99.

Mecikalski, J. R., Diak, G. R., Anderson, M. C. and Norman, J. M. (1999). Estimating fluxes on continental scales using remotely-sensed data in an atmospheric–land exchange model. *J. Appl. Meteorol.*, 38(9): 1352–69.

Meijerink, A. M. J., Brouwer, H. A. M., Mannaerts, C. M. and Valenzuela, C. R. (1994). Introduction to the use of geographic information systems for practical hydrology. Report No. 23, International Institute for Aerospace Survey and Earth Sciences, Enschede, 243pp.

Mérot, P., Crave, A., Cascuel-Odoux, C. and Louhala, S. (1994). Effect of saturated areas on backscattering coefficient of the ERS 1 synthetic aperture radar: First results. *Water Resour. Res.*, 30(2): 175–9.

Merz, B. and Bárdossy, A. (1998). Effects of spatial variability on the rainfall runoff process in a small loess catchment. *J. Hydrology*, 212–213: 303–17.

Merz, B. and Plate, E. J. (1997). An analysis of the effects of spatial variability of soil and soil moisture on runoff. *Water Resour. Res.*, 33(12): 2909–22.

Michaud, J. D. (1992). Distributed rainfall-runoff modelling of thunderstorm generated floods: A case study in a mid-sized, semiarid watershed in Arizona. Ph.D dissertation, Department of Hydrology and Water Resources, University of Arizona, Tucson, Arizona.

Michaud, J. D. and Shuttleworth, W. J. (1997). Aggregate descriptions of land atmosphere interactions. *J. Hydrology* (special issue), 190(3–4).

Michaud, J. D. and Sorooshian, S. (1994a). Effect of rainfall-sampling errors on simulations of desert flash floods. *Water Resour. Res.*, 30(10): 2765–75.

Michaud, J. D. and Sorooshian, S. (1994b). Comparison of simple versus complex distributed runoff models on a midsized semiarid watershed. *Water Resour. Res.*, 30(3): 593–605.

Miljøstyrelsen (1983). Karup river investigation. An investigation of the hydrological impacts of irrigation. Miljøprojekter 51, National Agency of Environmental Protection and the counties of Viborg, Ringkjøbing and Århus (in Danish).

Milly, P. C. D. and Eagleson, P. S. (1987). Effects of spatial variability on annual average water balance. *Water Resour. Res.*, 23(11): 2135–43.

Milly, P. C. D. and Eagleson, P. S. (1988). Effects of storm scale on surface runoff volume. *Water Resour. Res.*, 24(4): 620–24.

Molnar, L. (1992). *Principles for a new edition of the digital elevation modeling system, SCOP.* International Archives for Photogrammetry and Remote Sensing, XXIX/B4, Washington, pp.962–8.

Monteith, J. L. (1995a). A reinterpretation of stomatal responses to humidity. *Plant, Cell and Environment*, 18: 357–64.

Monteith, J. L. (1995b). Accommodation between transpiring vegetation and the convective boundary layer. *J. Hydrology*, 166: 251–63.

Monteith, J. L. and Unsworth, M. H. (1990). *Principles of Environmental Physics*, 2nd edition. Edward Arnold, New York, 290pp.

Moore, I. D. and Burch, G. J. (1986). Sediment transport capacity of sheet and rill flow: Application of unit stream power theory. *Water Resour. Res.*, 22: 1350–60.

Moore, I. D., Burch, G. J. and Mackenzie, D. H. (1988a). Topographic effects on the distribution of surface soil water and the location of ephemeral gullies. *Trans. American Society of Agricultural Engineers*, 31(4): 1098–107.

Moore, I. D. and Foster, G. R. (1990). Hydraulics and overland flow. Chapter 7 in: Anderson, M. G. and Burt, T. P. (Eds), *Process Studies in Hillslope Hydrology*. Wiley, pp.215–54.

Moore, I. D. and Grayson, R. B. (1991). Terrain-based catchment partitioning and runoff prediction using vector elevation data. *Water Resour. Res.*, 27: 1177–91.

Moore, I. D., Grayson, R. B. and Ladson, A. R. (1991). Digital terrain modelling: A review of hydrological, geomorphological, and biological applications. *Hydrological Processes*, 5: 3–30.

Moore, I. D., O'Loughlin, E. M. and Burch, G. J. (1988b). A contour based topographic model for hydrological and ecological applications. *Earth Surface Processes*, 13: 305–20.

Moore, R. D. and Thompson, J. C. (1996). Are water-table variations in a shallow forest soil consistent with the TOPMODEL concept? *Water Resour. Res.*, 32(3): 663–9.

Moore, R. J., Harding, R. J., Austin, R. M., Bell, V. A. and Lewis, D. R. (1996). Development of improved methods for snowmelt forecasting. R&D Note 402, Research Report, Institute of Hydrology, Wallingford.

Morton, A. (1993). Mathematical models: Questions of trustworthiness. *Brit. J. Phil. Soc.*, 44: 659–74.

Mott, K. A. and Parkhurst, D. F. (1992). Stomatal responses to humidity in air and helox. *Plant, Cell and Environment*, 14: 509–15.

Mroczkowski, M., Raper, G. P. and Kuczera, G. (1997). The quest for more powerful validation of conceptual catchment models. *Water Resour. Res.*, 33(10): 2325–35.

Munley, W. G. and Hipps, L. E. (1991). Estimation of regional evaporation for a tallgrass prairie from measurements of properties of the atmospheric boundary layer. *Water Resour. Res.*, 27: 225–30.

Myers, D. E. (1994). Spatial interpolation: An overview. *Geoderma*, 62: 17–28.

Myrabø, S. (1986). Runoff studies in a small catchment. *Nordic Hydrology*, 17: 335–46.

Myrabø, S. (1988). Automation in Hillslope Hydrology. NHP-rapport 22, pp.36–45.

Myrabø, S. (1997). Temporal and spatial scale of response area and groundwater variation in till. *Hydrological Processes*, 11: 1861–80.

Nash, J. E. and Sutcliffe, J. V. (1970). River flow forecasting through conceptual models, I. A discussion of principles. *J. Hydrology*, 10: 228–90.

Negri, A. J., Anagnostou, E. N. and Adler, R. F. (2000). A 10-year climatology of Amazonian rainfall derived from passive microwave satellite observations. *J. of Applied Meteorology*, 39: 42–56.

Neuman, S. P., Fogg, G. E. and Jacobson, E. A. (1980). A statistical approach to the inverse problem of aquifer hydrology: 2. Case Study. *Water Resour. Res.*, 16(1): 32–58.

Neuman, S. P. and Yakowitz, S. (1979). A statistical approach to the inverse problem of aquifer hydrology: 1. Theory. *Water Resour. Res.*, 15(4): 845–60.

Neuwirth, F. (1982). Beziehungen zwischen den kurzwelligen Strahlungskomponenten auf die horizontale Fläche und der Bewölkung an ausgewählten Stationen in Österreich (Relationships between the components of shortwave radiation at the horizontal surface and cloudiness at selected stations in Austria). *Archives for Meteorology, Geophysics, and Bioclimatology, Ser. B*, 30: 29–43.

New, M., Hulme, M. and Jones, P. (1999). Representing twentieth-century space-time climate variability, Part I: Development of a 1961–90 mean monthly terrestrial climatology. *J. Climate*, 12(3): 829–56.

Nezry, E., Lopez, A. and Touzy, R. (1991). Detection of structural and textural features for SAR images filtering. In: *Proc. International Geoscience and Remote Sensing Symposium (IGARSS)*. Espoo, Finland, pp.2169–72.

Norman, J. M. (1993). Scaling processes between leaf and canopy levels. In: Ehleringer, J. P. and Field, C. B. (Eds), *Scaling Physiological Processes. Leaf to Globe*. Academic Press, New York, pp.41–76.

Norman, J. M. and Becker, F. (1995). Terminology in thermal infrared remote sensing of natural surfaces. *Agric. For. Meteorol.* 77: 153–66.

Norman, J. M., Kustas, W. P. and Humes, K. S. (1995). A two-source approach for estimating soil and vegetation energy fluxes from observations of directional radiometric surface temperature. *Agric. For. Meteorol.*, 77: 263–93.

Northcote, K. H. (1979). *A Factual Key for the Recognition of Australian Soils*. Rellim Technical Publications, Coffs Harbour, NSW, Australia, 123pp.

NRCS (1998). http://www.wcc.nrcs.usda.gov/. Snow Survey and Water Supply Forecasting Program Manager, Natural Resources Conservation Service, 101 SW Main Street, Suite 1600, Portland, Oregon 97204-3224, U.S.A.

Obled, C. (1990). Hydrological modeling in regions of rugged relief. In: Lang, H. and Musy, A. (Eds), *Hydrology in Mountainous Regions. I – Hydrological Measurements; the Water Cycle*, Proc. Lausanne Symp., Aug. 1990, IAHS Publ. 193, 599–613.

Obled, C. and Harder, H. (1979). A review of snowmelt in the mountain environment. In: Colbeck, S. C. and Ray, M. (Eds), *Modeling Snow Cover Runoff*. Proc. Conf. Hanover. U.S. Army CRREL, pp.179–204.

Obled, C., Wendling, J. and Beven, K. (1994). The sensitivity of hydrologic models to spatial rainfall patterns: An evaluation using observed data. *J. Hydrology* 159: 305–33.

O'Callaghan, J. F. and Mark, D. M. (1984). The extraction of drainage networks from digital elevation data. *Computer Vision, Graphics and Image Processing*, 28: 323–44.

Ogden, F. L. and Julien, P. Y. (1993). Runoff sensitivity to temporal and spatial rainfall variability at runoff plane and small basin scale. *Water Resour. Res.*, 29(8): 2589–97.

Ogden, F. L. and Julien, P. Y. (1994). Runoff model sensitivity to radar rainfall resolution. *J. Hydrology.*, 158: 1–18.

Olyphant, G. A. (1986). Longwave radiation in mountainous areas and its influence on the energy balance of alpine snowfields. *Water Resour. Res.*, 22(1): 62–6.

Ophori, D. U. and Tóth, J. (1989). Characterization of ground-water flow by field mapping and numerical simulation, Ross Creek Basin, Alberta, Canada. *Ground Water*, 27(2): 193–201.

Oreskes, N., Shrader-Frechette, K. and Belitz, K. (1994). Verification, validation and confirmation of numerical models in the earth sciences. *Science*, 264: 641–6.

Osborn, H. B. (1964). Effect of storm duration on runoff from rangeland watersheds in the semiarid southwestern United States. Bull. IASH IX(4): 40–7.

Osborn, H. B. (1977). Point to area convective rainfall simulation. *Proc. Weather–Climate Modelling for Real-Time Applications in Agriculture and Forest Meteorology*, 13th Agriculture and Forest Meteorology Conference, American Meteorological Society, pp. 51–2.

Osborn, H. B. and Lane, L. J. (1972). Depth–area relationships for thunderstorm rainfall in Southeastern Arizona. *Trans. ASAE*, 15(4): 670–3, 680.

Osborn, H. B. and Lane, L. J. (1981). Point–area–frequency conversions for summer rainfall in Southeastern Arizona, Hydrology and water resources in Arizona and the Southwest. Office of Arid Land Studies, Univ. of Arizona, Tucson, 11: 39–42.

Osborn, H. B., Shirley, E. D., Davis, D. R. and Koehler, R. B. (1980). Model of time and space distribution of rainfall in Arizona and New Mexico. USDA-SEA Agricultural Reviews and Manuals, ARM-W-14, 27pp.

Over, T. M. and Gupta, V. K. (1994). Statistical analysis of mesoscale rainfall: Dependence of a random cascade generator on large-scale forcing. *J. Appl. Meteorol.*, 33(12): 1526–42.

Over, T. M. and Gupta, V. K. (1996). A space-time theory of mesoscale rainfall using random cascades. *J. Geophys. Res.*, 101(D21): 26319–31.

Palacios-Vélez, O. L., Gandoy-Bernasconi, W. and Cuevas-Renaud, B. (1998). Geometric analysis of surface runoff and the computation order of unit elements in distributed hydrological models. *J. Hydrology*, 211(1–4): 266–74.

Paniconi, C. and Wood, E. F. (1993). A detailed model for simulation of catchment scale subsurface hydrologic processes. *Water Resour. Res.*, 29(6): 1601–20.

Parkin, G., O'Donnell, G. O., Ewen, J., Bathurst, J. C., O'Connell, P. E. and Lavabre, J. (1996). Validation of catchment models for predicting land-use and climate change impacts. 2. Case study for a Mediterranean catchment. *J. Hydrology*, 175: 595–613.

Parsons, A. J., Abrahams, A. D. and Luk, S.-H. (1990). Hydraulics of interrill overland flow on a semi-arid hillslope, southern Arizona. *J. Hydrology*, 117: 255–73.

Peck, A., Gorelick, S., de Marsily, G., Foster, S. and Kovalevsky, V. (1988). *Consequences of Spatial Variability in Aquifer Properties and Data Limitations for Groundwater Modelling Practice*. IAHS Publ. No. 175, IAHS Press, Wallingford.

Peck, E. L. and Schaake, J. C. (1990). Network design for water supply forecasting in the West. *Water Resour. Bull.*, 26(1): 87–99.

Perica, S. (1995). A model for multiscale disaggregation of spatial rainfall based on coupling meteorological and scaling descriptions. PhD Thesis, University of Minnesota, Minneapolis.

Perica, S. and Foufoula-Georgiou, E. (1996a). Linkage of scaling and thermodynamic parameters of rainfall: Results from midlatitude mesoscale convective systems. *J. Geophys. Res.*, 101(D3): 7431–48.

Perica, S. and Foufoula-Georgiou, E. (1996b). A model for multiscale disaggregation of spatial rainfall based on coupling meteorological and scaling descriptions. *J. Geophys. Res.*, 101(D21): 26347–61.

Pessoa, M. L., Bras, R. L. and Williams, E. R. (1993). Use of weather radar for flood forecasting in the Sieve River Basin: A sensitivity analysis. *J. Appl. Meteorol.*, 32(3): 462–75.

Philip, J. R. (1957). The theory of infiltration. The infiltration equation and its solution. *Soil Science*, 83: 345–57.

Philip, J. R. (1975). Some remarks on science in catchment prediction. In: Chapman, T. G. and Dunin, F. X. (Eds), *Prediction in Catchment Hydrology: A National Symposium on Hydrology*. Australian Academy of Science, pp.23–30.

Pilgrim, D. H. and Cordery, I. (1975). Rainfall temporal patterns for design floods. *J. Hydr. Div., ASCE*, 101(HY1): 81–95.

Pitman, A. J., Henderson-Sellers, A. and Yang, Z. L. (1990). Sensitivity of regional climates to localized precipitation in global models. *Nature*, 346: 734–7.

Pomeroy, J. W. and Gray, D. M. (1995). Snowcover: accumulation, relocation and management. National Hydrology Research Institute Science Report No. 7, Saskatoon, Saskatchewan, Canada, 144pp.

Popper, K. (1977). *The Logic of Scientific Discovery*. Routledge, 480pp.

Porcello, L., Massey, N., Innes, R. and Marks, J. (1976). Speckle reduction in synthetic aperture radars. *J. Opt. Soc. Am.*, 66(11): 1305–11.

Press, W. H., Teukolsky, S. A., Vetterling, W. T. and Flannery, B. P. (1986). *Numerical Recipes in FORTRAN: The Art of Scientific Computing*, 2nd Edition. Cambridge University Press.

Price, A. G. and Dunne, T. (1976). Energy balance computations of snowmelt in a subarctic area. *Water Resour. Res.*, 12(4): 686–94.

Price, J. C. (1990). Using spatial context in satellite data to infer regional scale evapotranspiration. *IEEE Trans. Geosci. Remote Sens.*, GE-28: 940–8.

Priestley, C. H. B. and Taylor, R. J. (1972). On the assessment of the surface heat flux and evaporation using large-scale parameters. *Mon. Weather Rev.*, 100: 81–92.

Puckett, W. E., Dane, J. H. and Hajek, B. F. (1985). Physical and mineralogical data to determine soil hydraulic properties. *Soil Sci. Soc. Am. J.*, 49: 831–6.

Quinn, P. and Anthony, S. (in review). Digital terrain analysis and hydrological modelling. In: *Spatial Analysis and Hydrological Modelling*. Wiley.

Quinn, P. F. and Beven, K. J. (1993). Spatial and temporal predictions of soil moisture dynamics, runoff, variable source areas and evapotranspiration for Plynlimon, Mid-Wales. *Hydrological Processes*, 7: 425–48.

Quinn, P. F., Beven, K., Chevallier, P. and Planchon, O. (1991). The prediction of hillslope flow paths for distributed hydrological modelling using digital terrain models. *Hydrological Processes*, 5: 59–79.

Quinn, P. F., Beven, K. J. and Lamb, R. (1995). The $\ln(a/\tan\beta)$ index: how to calculate it and how to use it within the TOPMODEL framework. *Hydrological Processes*, 9(2): 161–82.

Rawls, W. J., Brakensiek, D. L. and Miller, N. (1983). Green-Ampt infiltration parameters from soils data. *J. Hydraul. Eng.*, 109: 62–70.

Refsgaard, J. C. (1996). Terminology, modelling protocol and classification of hydrological model codes. In: Refsgaard, J. C. and Abbott, M. B. (Eds), *Distributed Hydrological Modelling*, Kluwer, Dordrecht, The Netherlands, pp.17–40.

Refsgaard, J. C. (1997). Parameterisation, calibration and validation of distributed hydrological models. *J. Hydrology*, 198(1–4): 69–97.

Refsgaard, J. C. and Knudsen, J. (1996). Operational validation and intercomparison of different types of hydrological models. *Water Resour. Res.*, 32(7): 2189–202.

Refsgaard, J. C. and Storm, B. (1995). MIKE SHE. In: Singh, V. P. (Ed.), *Computer Models of Watershed Hydrology*, Water Resources Publications, Highlands Ranch, Colorado, pp.809–46.

Refsgaard, J. C. and Storm, B. (1996). Construction, calibration and validation of hydrological models. In: Abbott, M. B. and Refsgaard, J. C. (Eds), *Distributed Hydrological Modelling*. Kluwer, Dordrecht, The Netherlands, pp.41–54.

Refsgaard, J. C., Storm, B. and Abbott, M. B. (1996). Comment on: "A discussion of distributed hydrological modelling" by K. Beven. In: Refsgaard, J. C. and Abbott, M. B. (Eds), *Distributed Hydrological Modelling*. Kluwer, Dordrecht, The Netherlands, pp.279–87.

Renard, K. G., Lane, L. J., Simanton, J. R., Emmerich, W. E., Stone, J. J., Weltz, M. A., Goodrich, D. C. and Yakowitz, D. S. (1993). Agricultural impacts in an arid environment: Walnut Gulch case study. *Hydrol. Sci. Tech.*, 9(1–4): 145–90.

Rhea, J. O. (1978). Orographic precipitation model for hydrometeorological use. Ph.D Thesis, Colorado State University, Fort Collins, Colorado.

Richards, J. A. (1986). *Remote Sensing Digital Image Analysis*. Springer-Verlag.

Rieger, W. (1998). A phenomenon-based approach to upslope contributing area and depressions in DEMs. *Hydrological Processes*, 12: 857–72.

Rignot, E. J. M. and van Zyl, J. J. (1993). Change detection techniques for ERS-1 SAR data. *IEEE Trans. Geosc. Rem. Sens.*, 31(4): 896–906.

Riley, S. J., Crozier, P. and Blong, R. J. (1981). An inexpensive and easily installed runoff plot. *J. Soil Conservation Service, New South Wales*, 37: 144–7.

Robins, J. S., Kelly, L. L. and Hamon, W. R. (1965). Reynolds Creek in Southwest Idaho: An outdoor hydrologic laboratory. *Water Resour. Res.*, 1(3): 407–13.

Rodhe, A. (1981). Spring flood meltwater or ground-water? *Nordic Hydrology*, 12(1): 21–30.

Rohrer, M. B. (1992). Die Schneedecke im Schweizer Alpenraum und ihre Modellierung. (Snow cover modelling in the Swiss Alps.) Zürcher Geographische Schriften, Heft 49. Geographisches Institut der Eidgenössischen Technischen Hochschule Zürich, 178pp.

Romano, N. and Santini, A. (1997). Effectiveness of using pedo-transfer functions to quantify the spatial variability of soil water retention characteristics. *J. Hydrology*, 202: 137–57.

Romanowicz, R., Beven, K. J. and Tawn, J. A. (1994). Evaluation of predictive uncertainty in nonlinear hydrological models using a Bayesian approach. In: Barnett, V. and Feridum Turkman, K. (Eds), *Statistics for the Environment 2: Water Related Issues*, pp.297–317.

Rosenzweig, C. and Abramopoulos, F. (1997). Land-surface model development for the GISS GCM. *J. Climate*, 10(8): 2040–54.

Rosso, R. (1994). An introduction to spatially distributed modelling of basin response. In: Rosso, R., Peano, A., Becchi, I. and Bemporad, G. A. (Eds), *Advances in Distributed Hydrology*. Water Resources Publications, Highlands Ranch, Colorado, pp.3–30.

Roth, M. and Oke, T. R. (1995). Relative efficiencies of turbulent transfer of heat, mass, and momentum over a patchy urban surface. *J. Atmos. Sci.*, 52: 1863–74.

Salahshour Dehchali, J. (1993). Etude hydrologique du bassin versant du Coët-dan à Naizin; cartographie et modélisation des zones humides. Mémoire présenté en vue de l'obtention du Diplôme d'Agronomie Approfondie, INRA-ENSAR, 43pp.

Salvucci, G. D. (1998). Limiting relations between soil moisture and texture with implications for measured, modelled and remotely sensed estimates. *Geophys. Res. Letters*, 25(10): 1757–60.

Salvucci, G. D. and Entekhabi, D. (1994a). Equivalent steady moisture profile and the time compression approximation in water balance modelling. *Water Resour. Res.*, 30(10): 2737–49.

Salvucci, G. D. and Entekhabi, D. (1994b). Comparison of the Eagleson statistical-dynamical water balance model with numerical simulations. *Water Resour. Res.*, 30(10): 2751–7.

Salvucci, G. D. and Entekhabi, D. (1995). Hillslope and climatic controls on hydrologic fluxes. *Water Resour. Res.*, 31(7): 1725–39.

Salvucci, G. D. and Entekhabi, D. (1997). Corrections to "Hillslope and climatic controls on hydrologic fluxes". *Water Resour. Res.*, 33(1): 277.

Satterlund, D. R. (1979). An improved equation for estimating long-wave radiation from the atmosphere. *Water Resour. Res.*, 15: 1643–50.

Saulnier, G.-M., Beven, K. and Obled, C. (1997c). Digital elevation analysis for distributed hydrological modelling: Reducing scale dependence in effective hydraulic conductivity values. *Water Resour. Res.*, 33(9): 2097–101.

Saulnier, G.-M., Beven, K. and Obled, C. (1997a). Including spatially variable effective soil depths in TOPMODEL. *J. Hydrology*, 202(1–4): 158–72.

Saulnier, G. M., Obled, C. and Beven, K. (1997b). Analytical compensation between DTM grid resolution and effective values of saturated hydraulic conductivity within the TOPMODEL framework. *Hydrological Processes*, 11(9): 1331–46.

Schertzer, D. and Lovejoy, S. (1987). Physical modeling and analysis of rain and clouds by anisotropic scaling multiplicative processes, *J. Geophys. Res.*, 92(D8): 9693–714.

Schlesinger, S., Crosbie, R. E., Gagné, R. E., Innis, G. S., Lalwani, C. S., Loch, J., Sylvester, J., Wright, R. D., Kheir, N. and Bartos, D. (1979). Terminology for model credibility. SCS Technical Committee on Model Credibility. *Simulation*, 32(3): 103–4.

Schmid, H. P. (1994). Source areas for scalars and scalar fluxes. *Boundary-Layer Meteorol.*, 67: 293–318.

Schmugge, T., Jackson, T. J., Kustas, W. P., Roberts, R., Parry, R., Goodrich, D. C., Amer, S. A. and Weltz, M. A. (1994). Push broom microwave observations of surface soil moisture in Monsoon '90. *Water Resour. Res.*, 30(5): 1321–7.

Schmugge, T. J., Kustas, W. P. and Humes, K. S. (1998). Monitoring land surface fluxes using ASTER observations. *IEEE Trans. Geosci. Remote Sens.*, 36(5): 1–10.

Schuepp, P. H., Leclerc, M. Y., Macpherson, J. I. and Desjardins, R. L. (1990). Footprint prediction of scalar fluxes from analytical solutions of the diffusion equation. *Boundary-Layer Meteorol.*, 50: 355–73.

Schulz, G. A. (1988). Remote sensing in hydrology. *J. Hydrology*, 100: 239–65.

Seaman, N. L. (1990). Newtonian Nudging: A Four-Dimensional Approach to Data Assimilation. Mesoscale Data Assimilation, 1990 Summer Colloquium, National Center for Atmospheric Research, Boulder, Colorado, 6 June–3 July.

Seed, A. W., Srikanthan, R. and Menabde, M. (1999): A space and time model for design storm rainfall. *J. Geophys. Res.* 104 (D24): 31623–30.

Seibert, J., Bishop, K. H. and Nyberg, L. (1997). A test of TOPMODEL's ability to predict spatially distributed groundwater levels. *Hydrological Processes*, 11(9): 1131–44.

Sellers, P. J., Heiser, M. D., Hall, F. G., Goetz, S. J., Strebel, D. E., Verma, S. B., Desjardins, R. L., Schuepp, P. H. and MacPherson, J. I. (1995). Effects of spatial variability in topography, vegetation cover and soil moisture on area-averaged surface fluxes: A case study using FIFE 1989 data. *J. Geophys. Res.*, 100(D12): 25607–29.

Seyfried, M. S. and Wilcox, B. P. (1995). Scale and the nature of spatial variability: Field examples having implications for hydrologic modelling. *Water Resour. Res.*, 31(1): 173–84.

Shaw, E. M. and Lynn, P. P. (1972). Areal rainfall evaluation using two surface fitting techniques. *Bull. Int. Assoc. Hydrol. Sci.*, XVII(4) 12: 419–33.

Shen, H. W. and Julien, P. Y. (1992). Erosion and sediment transport. Chapter 12 in: Maidment, D. (Ed.), *Handbook of Hydrology*. McGraw-Hill, New York, pp.12.1–12.61.

Shi, Z. and Fung, K. B. (1994). A comparison of digital speckle filters. In: *Proc. International Geoscience and Remote Sensing Symposium (IGARSS)*. Pasadena, USA, pp.2129–33.

Shuttleworth, W. J. (1991). Insight from large-scale observational studies of land/atmosphere interactions. *Surv. Geophys.*, 29: 585–606.

Shuttleworth, W. J. and Wallace, J. S. (1985). Evaporation from sparse crops – An energy combination theory. *Quart. J. R. Meteorol. Soc.*, 111: 839–55.

Singh, A. (1989). Digital change detection techniques using remotely-sensed data. *Int. J. Rem. Sens.*, 10(6): 989–1003.

Singh, V. P. (Ed.) (1995). *Computer Models of Watershed Hydrology*. Water Resources Publications, Highlands Ranch, Colorado, 1130pp.

Sivapalan, M., Beven, K. J. and Wood, E. F. (1987). On hydrologic similarity, 2, A scaled model of storm runoff production. *Water Resour. Res.*, 23(2): 2266–78.

Sivapalan, M. and Blöschl, G. (1998). Transformation of point rainfall to areal rainfall: intensity–duration–frequency curves. *J. Hydrology*, 204: 150–67.

Sivapalan, M. and Woods, R. A. (1995). Evaluation of the effects of general circulation model's subgrid variability and patchiness of rainfall and soil moisture on land surface water balance fluxes. In: Kalma, J. D. and Sivapalan, M. (Eds), *Scale issues in Hydrological Modelling*. John Wiley, New York, pp.453–73.

Smith, M., Allen, R., Monteith, J. L., Perrier, A., Santos Pereira, L. and Segeren, A. (1992). Expert consultation on revision of FAO methodologies for crop water requirements. Land and Water Development Division, Food and Agriculture Organization of the United Nations, Rome.

Smith, R. E., Goodrich, D. C. and Woolhiser, D. A. (1990). Areal effective infiltration dynamics for runoff on small catchments. *Trans. 14th Inter. Congress of Soil Sci.*, Volume I: Commission I, Kyoto, Japan, Aug. 1990, pp.22–7.

Smith, R. E., Goodrich, D. R., Woolhiser, D. A. and Simanton, J. R. (1994). Comment on "Physically based hydrological modelling 2: Is the concept realistic?" by Grayson, Moore and McMahon. *Water Resour. Res.*, 30: 851–4.

Smith, R. E., Goodrich, D. R., Woolhiser, D. A. and Unkrich, C. L. (1995). KINEROS – A KINematic Runoff and EROSion Model. Chapter 20 in: Singh, V. P. (Ed.), *Computer Models of Watershed Hydrology*. Water Resources Publications, Highlands Ranch, Colorado, pp.697–732.

Smith, R. E. and Hebbert, R. H. B. (1979). A Monte Carlo analysis of the hydrologic effects of spatial variability of infiltration. *Water Resour. Res.*, 15(2): 419–29.

Smith, R. E. and Parlange, J.-Y. (1978). A parameter-efficient hydrologic infiltration model. *Water Resour. Res.*, 14(33): 533–8.

Sorooshian, S. and Dracup, J. (1980). Stochastic parameter estimation procedures for hydrological rainfall–runoff models: correlated and heteroscedastic error cases. *Water Resour. Res.*, 16(2): 430–42.

Sorooshian, S. and Gupta, V. K. (1995). Model calibration. Chapter 2 in: Singh, V. P. (Ed.), *Computer Models of Watershed Hydrology*. Water Resources Publications, Highlands Ranch, Colorado, pp.23–63.

Sorooshian, S., Gupta, H. V. and Rodda, J. C. (1997). *Land Surface Processes in Hydrology – Trials and Tribulations of Modeling and Measuring*. Springer, Berlin, 497pp.

Stannard, D. I. (1993). Comparison of Penman–Monteith, Shuttleworth–Wallace, and modified Priestley–Taylor evapotranspiration models for wildland vegetation in semiarid rangeland. *Water Resour. Res.*, 29: 1379–92.

Stannard, D. I., Blanford, J. H., Kustas, W. P., Nichols, W. D., Amer, S. A., Schmugge, T. J. and Weltz, M. A. (1994). Interpretation of surface flux measurements in heterogeneous terrain during Monsoon '90. *Water Resour. Res.*, 30(5): 1227–39.

Stauffer, D. R. and Seaman, N. L. (1990). Use of four-dimensional data assimilation in a limited-area mesoscale model. Part I: Experiments with synoptic-scale data. *Mon. Weather Rev.*, 118: 1250–77.

Stendal, M. M. (1978). Hydrological data – Norden, Karup representative basin, Denmark. Data volume 1965–77. Danish National Committee for the IHD, Copenhagen.

Stephenson, G. R. and Freeze, R. A. (1974). Mathematical simulation of subsurface flow contributions to snowmelt and runoff, Reynolds Ck. Watershed, Idaho. *Water Resour. Res.*, 10: 284–94.

Stewart, J. B. and Verma, S. B. (1992). Comparison of surface fluxes and conductances at two contrasting sites within the FIFE area. *J. Geophys. Res.*, 97: 18623–8.

Stoertz, M. W. and Bradbury, K. R. (1989). Mapping recharge areas using a ground-water flow model – A case study. *Ground Water*, 27(2): 220–8.

Stone, J. J., Lane, L. J., Shirley, E. D. and Renard, K. G. (1986). A runoff-sediment yield model for semiarid regions. *Proc. 4th Fed. Interagency Sedimentation Conf.*, Las Vegas, NV, pp. 6-75–6-84.

Stull, R. (1988). *An Introduction to Boundary Layer Meteorology*. Kluwer, Boston, MA.

Styczen, M. (1995). Validation of pesticide leaching models. In: *Leaching models and EU registration*. Final report of the work of the Regulatory Modelling Work Group of FOCUS (FOrum for the Co-ordination of pesticide fate models and their USe), DOC.4592/VI/95, European Commission.

Styczen, M. and Storm, B. (1993). Modelling of N-movements on catchment scale – a tool for analysis and decision making. 1. Model description & 2. A case study. *Fertilizer Research*, 36: 1–17.

Sugita, M. and Brutsaert, W. (1991). Daily evaporation over a region from lower boundary layer profiles. *Water Resour. Res.*, 27: 747–52.

Swiatek, E. (1992). Estimating regional surface fluxes from measured properties of the atmospheric boundary layer in a semiarid ecosystem. M.S. Thesis, Utah State University, Logan, UT.

Syed, K. H. (1994). Spatial Storm Characteristics and Basin Response. M.S. Thesis, Department of Hydrology and Water Resources, The University of Arizona, Tucson, Arizona, 261pp.

Tabios, G. Q. and Salas, J. D. (1985). A comparative analysis of techniques for spatial interpolation of precipitation. *Water Resour. Bull.*, 21: 365–80.

Talsma, T. (1987). Re-evaluation of the well permeameter as a method for measuring hydraulic conductivity. *Australian J. Soil Research*, 25: 361–8.

Talsma, T. and Hallam, P. M. (1980). Hydraulic conductivity measurements of forest catchments. *Australian J. Soil Research*, 18: 139–48.

Tanner, M. H. (1992). *Tools for statistical inference: Observed data and data augmentation methods*. Lecture notes in statistics, 67, Springer-Verlag.

Tarboton, D. G. (1997). A new method for the determination of flow directions and upslope areas in grid digital elevation models. *Water Resour. Res.*, 33(2): 309–19.

Tarboton, D. G., Chowdhury, T. G. and Jackson, T. H. (1995). A spatially distributed energy balance snowmelt model. In: Tonnessen, K. A. et al. (Eds), *Biogeochemistry of Seasonally Snow-Covered Catchments*, Proc. Boulder Symposium, July 3–14, IAHS Publ. no. 228, pp.141–55.

Tarboton, D. G. and Luce, C. H. (1996). Utah Energy Balance Snow Accumulation and Melt Model (UEB), Computer model technical description and user's guide, Utah Water Research Laboratory and USDA Forest Service Intermountain Research Station (http://www.engineering.usu.edu/dtarb/).

Tessier, Y., Lovejoy, S. and Schertzer, D. (1993). Universal multifractals: Theory and observations for rain and clouds. *J. Appl. Meteor.*, 32(2): 223–50.

Thiessen, A. H. (1911). Precipitation for large areas. *Monthly Weather Review*, 39: 1082–4.

Tóth, J. (1962). A theory of groundwater motion in small drainage basins in central Alberta, Canada. *J. Geophys. Res.*, 67(11): 4375–87.

Tóth, J. (1963). A theoretical analysis of groundwater flow in small drainage basins. *J. Geophys. Res.*, 68(16): 4795–812.

Tóth, J. (1966). Mapping and interpretation of field phenomena for groundwater reconnaissance in a prairie environment, Alberta, Canada. *Bull. IASH*, 11(2): 20–68.

Tsang, C.-F. (1991). The modeling process and model validation. *Ground Water*, 29: 825–31.

Turner, J. V. and Macpherson, D. K. (1987). Mechanisms affecting streamflow and stream water quality: an approach via stable isotope, hydrogeochemical, and time series analysis. *Water Resour. Res.*, 26(12): 3005–19.

Tyndale-Biscoe, J. P., Moore, G. A. and Western, A. W. (1998). A system for collecting spatially variable terrain data. *Computers and Electronics in Agriculture*, 19(2): 113–28.

U.S. Army Corps of Engineers (1956). Snow Hydrology, Summary report of the Snow Investigations, U.S. Army Corps of Engineers, North Pacific Division, Portland, Oregon.

U.S. Senate Committee (1959). Report on findings on facility needs for soil and water research. 175pp.

Ulaby, F. T., Dubois, P. C. and van Zyl, J. (1996). Radar mapping of surface soil moisture. *J. Hydrology*, 184: 57–84.

Ulaby, F. T., Moore, R. K. and Fung, A. K. (1982). *Microwave Remote Sensing, Active and Passive, Vol. II : Radar Remote Sensing and Surface Scattering and Emission Theory*, Artech House, Inc.

van Straten, G. and Keesman, K. J. (1991). Uncertainty propagation and speculation in projective forecasts of environmental change: A lake-eutrophication example. *J. Forecasting*, 10: 163–90.

Valdes, J. B., Rodriguez-Iturbe, I. and Gupta, V. K. (1985). Approximations of temporal rainfall from a multidimensional model. *Water Resour. Res.*, 21(8): 1259–70.

Vanmarcke, E. (1983). *Random Fields: Analysis and Synthesis*. MIT Press, Cambridge, Massachusetts, 382pp.

Venugopal, V. (1999). Spatio-temporal organization and space-time downscaling of precipitation fields. PhD Thesis, University of Minnesota, Minneapolis.

Venugopal, V., Foufoula-Georgiou, E. and Sapozhnikov, V. (1999a). Evidence of dynamic scaling in space-time rainfall. *J. Geophys. Res.*, 104(D24): 31599–610.

Venugopal, V., Foufoula-Georgiou, E. and Sapozhnikov, V. (1999b). A space-time downscaling model for rainfall. *J. Geophys. Res.*, 104(D16): 705–21.

Verhoef, A., de Bruin, H. A. R. and Van Den Hurk, B. J. J. M. (1997). Some practical notes on the parameter kB-1 for sparse vegetation. *J. Appl. Meteorol.*, 36: 560–72.

Verhoest, N. E. C., Troch, P. A., Paniconi, C. and De Troch, F. P. (1998). Mapping basin scale variable source areas from multitemporal remotely sensed observations of soil moisture behavior. *Water Resour. Res.*, 34(12): 3235–44.

Verstraete, M. M. (1988). Radiation transfer in plant canopies: Scattering of solar radiation and canopy reflectance. *J. Geophys. Res.*, 93(D8): 9483–94.

Verstraete, M. M. (1987). Radiation transfer in plant canopies: Transmission of direct solar radiation and the role of leaf orientation. *J. Geophys. Res.*, 92(D9): 10985–95.

Verstraete, M. M., Pinty, B. and Dickinson, R. E. (1990). A physical model of the bidirectional reflectance of vegetation canopies, 1. Theory. *J. Geophys. Res.*, 95(D8): 11755–65.

Vertessy, R. A. and Elsenbeer, H. (1999). Distributed modeling of storm flow generation in an Amazonian rain forest catchment: Effects of model parameterization. *Water Resour. Res.*, 35(7): 2173–87.

Vertessy, R. A., Hatton, T. J., Benyon, R. J. and Dawes, W. R. (1996). Long term growth and water balance predictions for a mountain ash (*E. regnans*) forest subject to clearfelling and regeneration. *Tree Physiol.*, 16: 221–32.

Vertessy, R. A., Hatton, T. J., O'Shaughnessy, P. J. and Jayasuriya, M. D. A. (1993). Predicting water yield from a mountain ash forest catchment using a terrain analysis-based catchment model. *J. Hydrology*, 150: 665–700.

Vickers, D. and Mahrt, L. (1997). Quality control and flux sampling problems for tower and aircraft data. *J. Atmos. Oceanic Tech.*, 14: 512–26.

Viessman, Jr., W. and Lewis, G. L. (1996). *Introduction to Hydrology*. Harper Collins, New York.

Wagner, W. (1998). Soil moisture retrieval from ERS scatterometer data. Research Report EUR 18670, European Commission Joint Research Centre, Ispra, Italy, 101pp.

Wahba, G. and Wendelberger, J. (1980). Some new mathematical methods for variational objective analysis using splines and cross validation. *Monthly Weather Review*, 108: 36–57.

Wang, Q. J., McConachy, F. L. N., Chiew, F. H. S., James, R., de Hoedt , G. C. and Wright, W. J. (1998). *Climatic Atlas of Australia: Maps of Evapotranspiration*. To be published by the Bureau of Meteorology, Australia.

Wang, W. and Seaman, N. L. (1997). A comparison study of convective parameterization schemes in a mesoscale model. *Mon. Weath. Rev.*, 125: 252–78.

Warner, T. T. and Hsu, H.-M. (2000). Nested model simulation of moist convection: the impact of coarse-grid parameterized convection on fine-grid resolved convection. *Mon. Weath. Rev.*, 128(7): 2211–31.

Warrick, A. W., Zhang, R., Moody, M. M. and Myers, D. E. (1990). Kriging versus alternative interpolators: errors and sensitivity to model inputs. In: Roth, K. et al. (Eds), *Field-scale Water and Solute Flux in Soils*. Birkhäuser Verlag, Basel, pp.157–64.

Watson, F. G. R., Grayson, R. B., Vertessy R. A. and McMahon, T. A. (1998). Large scale distribution modelling and the utility of detailed data. *Hydrological Processes*, 12(6): 873–88.

Waymire, E., Gupta, V. K. and Rodriguez-Iturbe, I. (1984). A spectral theory of rainfall intensity at the meso-β scale. *Water Resour. Res.*, 20: 1453–65.

Weisman, M. L. and Klemp, J. B. (1982). The dependence of numerically simulated convective storms on vertical wind shear and buoyancy. *Mon. Wea. Rev.*, 110: 504–20.

Wen, X.-H. and Gómez-Hernández, J. J. (1996). Upscaling hydraulic conductivities in heterogeneous media: An overview. *J. Hydrology*, 183: ix–xxxii.

Western, A. W. and Blöschl, G. (1999). On the spatial scaling of soil moisture. *J. Hydrology*, 217: 203–24.

Western, A. W., Blöschl, G. and Grayson, R. B. (1998a). Geostatistical characterisation of soil moisture patterns in the Tarrawarra Catchment. *J. Hydrology*, 205: 20–37.

Western, A. W., Blöschl, G. and Grayson, R. B. (1998b). How well do indicator variograms capture the spatial connectivity of soil moisture? *Hydrological Processes*, 12: 1851–68.

Western, A. W., Blöschl, G. and Grayson, R. B. (2000). Towards capturing hydrologically significant connectivity in spatial patterns. *Water Resour. Res.*, in press.

Western, A. W. and Grayson, R. B. (1998). The Tarrawarra data set: Soil moisture patterns, soil characteristics and hydrological flux measurements. *Water Resour. Res.*, 34(10): 2765–8.

Western, A. W., Grayson, R. B., Blöschl, G., Willgoose, G. R. and McMahon, T. A. (1999a). Observed spatial organisation of soil moisture and its relation to terrain indices. *Water Resour. Res.*, 35(3): 797–810.

Western, A. W., Grayson, R. B. and Green, T. R. (1999b). The Tarrawarra Project: High resolution spatial measurement, modelling and analysis of soil moisture and hydrological response. *Hydrological Processes*, 13(5):689–700.

Whelan, M. J. and Anderson, J. M. (1996). Modelling spatial patterns of throughfall and interception loss in a Norway spruce (*Picea abies*) plantation at the plot scale. *J. Hydrology*, 186: 335–54.

Wiener, N. (1949). *Extrapolation, Interpolation and Smoothing of Stationary Time Series*. MIT Press, Cambridge, Massachusetts.

Wigmosta, M. S. and Lettenmaier, D. P. (1999). A comparison of simplified methods for routing topographically driven subsurface flow. *Water Resour. Res.*, 35(1): 255–64.

Wigmosta, M. S., Vail, L. W. and Lettenmaier, D. P. (1994). A distributed hydrology–vegetation model for complex terrain. *Wat. Resour. Res.*, 30(6): 1665–79.

Wilks, D. S. (1995). *Statistical Methods in the Atmospheric Sciences: An Introduction*. Academic Press, 464 pp.

Willgoose, G. and Kuczera, G. (1995). Estimation of sub-grid scale kinematic wave parameters for hillslopes. *Hydrological Processes*, 9: 469–82.

Williams, A. G. and Hacker, J. M. (1993). Interactions between coherent eddies in the lower convective boundary layer. *Boundary-Layer Meteorol.*, 64: 55–74.

Williams, K. S. and Tarboton, D. G. (1999). The ABC's of snowmelt: A topographically factorized energy component snowmelt model. *Hydrological Processes*, 13: 1905–20.

Williams, R. E. (1988). Comment on "Statistical theory of groundwater flow and transport: pore to laboratory, laboratory to formation, and formation to regional scale" by Gedeon Dagan. *Water Resour. Res.*, 24: 1197–200.

Winchell, M., Gupta, H. V. and Sorooshian, S. (1998). On the simulation of infiltration- and saturation-excess runoff using radar-based rainfall estimates: effects of algorithm uncertainty and pixel aggregation. *Water Resour. Res.*, 34(10): 2655–70.

Winstral, A., Elder, K. and Davis, R. (1999). Implementation of digital terrain analysis to capture the effects of wind redistribution in spatial snow modelling. In: 67th Western Snow Conference, South Lake Tahoe, CA.

Wiscombe, W. J. and Warren, S. G. (1981). A model of the spectral albedo of snow. I: Pure snow. *J. Atmos. Sci.*, 37: 2712–33.

WMO (1975). Intercomparison of conceptual models used in operational hydrological forecasting. WMO operational hydrology report no 7, WMO no 429. World Meteorological Organisation, Geneva.

WMO (1986). Intercomparison of models for snowmelt runoff. WMO operational hydrology report no 23, WMO no 646. World Meteorological Organisation, Geneva.

WMO (1992). Simulated real-time intercomparison of hydrological models. WMO operational hydrology report no 38, WMO no 779. World Meteorological Organisation, Geneva.

Wolock, D. M. and Price, C. V. (1994). Effects of digital elevation model map scale and data resolution on a topography-based watershed model. *Water Resour. Res.*, 30: 3041–52.

Woo, M., Heron, R., Marsh, P. and Steer, P. (1983). Comparison of weather station snowfall with winter snow accumulation in high arctic basins. *Atmos. Ocean*, 21(3): 312–25.

Wood, E. F., Lettenmaier, D. P. and Zartarian, V. G. (1992). A land-surface hydrology parameterization with subgrid variability for general circulation models. *J. Geophys. Res.*, 97(D3): 2717–28.

Wood, E. F., Sivapalan, M., Beven, K. and Band, L. (1988). Effects of spatial variability and scale with implications to hydrologic modeling. *J. Hydrology*, 102: 29–47.

Woods, R., Sivapalan, M. and Duncan, M. (1995). Investigating the Representative Elementary Area concept: An approach based on field data. *Hydrological Processes*, 9: 291–312.

Woods, R. A., Sivapalan, M. and Robinson, J. (1997). Modelling the spatial variability of subsurface runoff using a topographic index. *Water Resour. Res.*, 33(5): 1061–73.

Woolhiser, D. A. and Goodrich, D. C. (1988). Effect of storm rainfall intensity patterns on surface runoff. *J. Hydrology*, 102: 335–54.

Xue, M., Droegemeier, K. K., Wong, V., Shapiro, A. and Brewster, K. (1995). ARPS Version 4.0 User's Guide, 380pp. Available from the Center for Analysis and Prediciton of Storms, 100 East Boyd Street, Norman, OK 73019.

Xue, M., Droegemeier, K. K. and Wong, V. (2000a). The Advanced Regional Prediction System (ARPS) – A multiscale nonhydrostatic atmospheric simulaiton and prediction tool. Part I: Model dynamics and verification. *Meteor. and Atmos. Physics*, in press.

Xue, M., Droegemeier, K. K., Wong, V., Shapiro, A., Brewster, K., Carr, F., Weber, D., Liu, Y. and Wang, D.-H. (2000b). The Advanced Regional Prediction System (ARPS) – A multiscale nonhydrostatic atmospheric simulation and prediction tool. Part II: Model physics and applications. *Meteor. and Atmos. Physics*, in press.

Yang, D. and Woo, M. (1999). Representativeness of local snow data for large scale hydrologic investigations. *Hydrological Processes*, 13(12–13): 1977–88.

Zak, S. K., Beven, K. J. and Reynolds, B. (1997). Uncertainty in the estimation of critical loads: a practical methodology. *Water, Air and Soil Pollution*, 98: 297–316.

Zepeda-Arce, J., Foufoula-Georgiou, E., Droegemeier, K. (2000). Space-time rainfall organization and its role in validating quantitative precipitation forecast. *J. Geophys. Res.* 105(D8), 10129–46.

Zermeño-Gonzalez, A. and Hipps, L. E. (1997). Downwind evolution of surface fluxes over a vegetated surface during local advection of heat and saturation deficit. *J. Hydrology*, 192: 189–210.

Zhang, S. and Foufoula-Georgiou, E. (1997). Subgrid scale rainfall variability and effects on atmospheric and surface variable prediction. *J. Geophys. Res.*, 102(D16): 19559–73.

Zhang, W. and Montgomery, D. R. (1994). Digital elevation model grid size, landscape representation, and hydrologic simulations. *Water Resour. Res.*, 30: 1019–28.

Zhu, H. and Journel, A. G. (1993). Formatting and integrating soft data: stochastic imaging via the Markov–Bayes algorithm. In: Soares, A. (Ed.), *Geostatistics Tróia '92*. Kluwer, Dordrecht, The Netherlands, pp.1–12.

Index

Page numbers referring to tables are in *italics*; those referring to figures are in ***bold italics***.

DAT

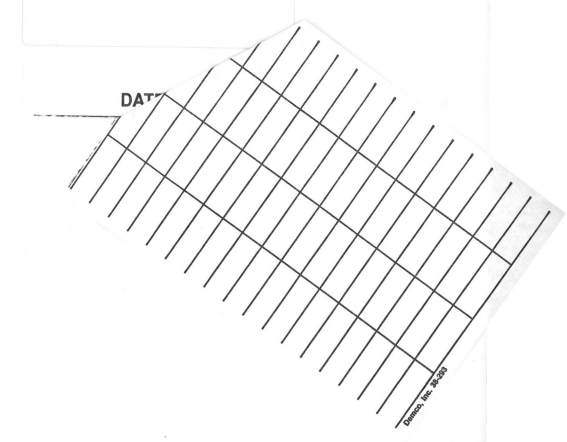

Demco, Inc. 38-293